Diagnosis *of* PLANT VIRUS DISEASES

Edited by

R.E.F. Matthews

Emeritus Professor of Microbiology
School of Biological Sciences
University of Auckland
Auckland, New Zealand

CRC Press
Boca Raton   Ann Arbor   London   Tokyo

Library of Congress Cataloging-in-Publication Data

Diagnosis of plant virus diseases / editor, R.E.F. Matthews.
    p.   cm.
   Includes bibliographical references and index.
   ISBN 0-8493-4284-8
   1. Virus diseases of plants--Diagnosis.   I. Matthews, R. E. F.
(Richard Ellis Ford), 1921– .
   SB736.D53   1993
632′ .8--dc20   92-38242
                                                      CIP

International Standard Book Number 0-8493-4284-8

Library of Congress Card Number 92-38242

Printed in the United States of America 1 2 3 4 5 6 7 8 9 0

Printed on acid-free paper

PREFACE

Plant diseases caused by viruses bring about substantial losses in agricultural and horticultural crops. These losses occur worldwide, but are particularly important in the tropics and subtropics. The first requirement for the control of such diseases is the identification of the causative virus or viruses. Over recent years such identification has been made easier by two developments: first, a comprehensive and internationally agreed classification system for viruses is now available; and second, a variety of technical improvements have been made in the methods that can be used for virus identification. Chapter 1 gives an overview of the present classification of plant viruses, and of the current diagnostic techniques.

In Chapter 2 J. Horváth discusses the use of host plants in diagnosis, with special reference to disease symptoms and host range. A. T. Jones in Chapter 3 describes various methods of experimental transmission used in diagnosis, while in Chapter 4 he deals with the sometimes difficult problem of virus transmission through soil. Some viruses and groups of viruses produce characteristic intracellular inclusion bodies, which may be diagnostic for particular viruses or groups. The use of these in diagnosis is described by J. R. Edwardson and his colleagues in Chapter 5.

The methods that can be used to purify a virus from diseased tissue sometimes give useful clues as to the kind of virus involved. These are discussed by R. Stace-Smith and R. R. Martin in Chapter 6. Serological procedures have developed and diversified rapidly in recent years, and have become one of the most important tools in virus diagnosis. These techniques, with practical details are given by M. H. V. Van Regenmortel and M-C. Dubs in Chapter 7.

In Chapter 8 R. G. Milne describes the use of electron microscopy to characterise virus particles, with particular reference to negative staining, immunosorbent and decoration procedures, and gold labeling. The theory and practice of nucleic acid hybridization in diagnosis is given by R. Hull in Chapter 9. Many plants showing symptoms of virus or virus-like disease contain various dsRNA species not found in healthy plants. Analysis of such dsRNA species by gel electrophoresis may provide diagnostic evidence of a nonspecific kind, as discussed by J. A. Dodds in Chapter 10.

The last three chapters deal with aspects of diagnosis that may present particular difficulties. Viroids are very small circular ssRNA molecules that are not thought to be related to viruses. However they frequently cause virus-like disease symptoms. For this reason, in Chapter 11, D. Hanold describes practical procedures for the diagnostic methods applicable to these agents. Some important diseases with virus-like symptoms have, as yet, no known etiology. The difficult problems posed by such diseases are described by J. W. Randles in Chapter 12. Finally A. F. Murant, in Chapter 13 is concerned with the recognition of diseases that may be caused by a complex of a transmission-dependent and a helper virus.

I wish to thank all the contributors to this volume for their ready cooperation in the project.

THE EDITOR

R. E. F. Matthews, Ph.D, is Emeritus Professor of Microbiology in the School of Biological Sciences, University of Auckland, New Zealand.

Dr. Matthews obtained his M.Sc. degree in Botany from Auckland University College in 1942. Postwar, he graduated Ph.D. in the field of plant virology from the University of Cambridge, England, in 1948. In 1948–1949 he was a postdoctoral Fellow in the Department of Plant Pathology, University of Wisconsin at Madison. From 1949 to 1961 he was employed as a plant virologist in the Plant Diseases Division of the Department of Scientific and Industrial Research in Auckland. From 1962 to 1986 he was Professor of Microbiology in the Department of Cell Biology, University of Auckland. During this period, while on sabbatical leave, he carried out research in Madison, (McArdle Laboratory for Cancer Research, 1965 to 1966); Tübingen, Germany (Max Planck Institut für Biologie, 1975 to 1976); and Strasbourg, France (Institut de Biologie Moleculaire et Cellulaire du C.N.R.S., 1983 to 1984).

Dr. Matthews is a member of the Biochemical Society, London, the American Association for the Advancement of Science, the American Phytopathological Society, the New York Academy of Sciences, the New Zealand Institute of Chemistry, and a Life Member of the Cancer Society of New Zealand. He was President of the International Committee for Taxonomy of Viruses from 1975 to 1981. He is a Life Member of that organization. He was awarded the Doctor of Science degree, University of Cambridge in 1964. He is a Fellow of the Royal Society of London, a Fellow of the Royal Society of New Zealand, and a Fellow of the New Zealand Institute of Chemists.

Dr. Matthews has published more than 140 scientific papers and several books, mainly in the fields of plant virology and general virus taxonomy.

CONTRIBUTORS

R. G. Christie
Department of Agronomy
University of Florida
Gainesville, Florida

J. A. Dodds
Department of Plant Pathology
University of California
Riverside, California

M. C. Dubs
Postdoctoral Fellow
Department of Immunochemistry
Institut de Biologie Moleculaire et
 Cellulaire, CNRS
Strasbourg, France

J. R. Edwardson
Department of Agronomy
University of Florida
Gainesville, Florida

D. Hanold
Senior Research Fellow
Department of Crop Protection
Waite Agricultural Research
 Institute
University of Adelaide
Glen Osmond, South Australia

J. Horváth
Department of Plant Pathology
Institute for Plant Protection
Pannonian University of
 Agricultural Science
Keszthely, Hungary

Roger Hull
Deputy Head
Virus Research Department
Virology Department
John Innes Institute
Norwich, England

A. T. Jones
Senior Principal Scientist
Scottish Crop Research Institute
Invergowrie, Dundee
Scotland

Robert R. Martin
Research Scientist
Agriculture Canada
Vancouver, British Columbia
Canada

R. E. F. Matthews
School of Biological Sciences
University of Auckland
Auckland, New Zealand

Robert G. Milne
Istituto di Fitovirologia Applicata
CNR (National Research Council)
Torino, Italy

A. F. Murant
Senior Principal Scientific Officer
Scottish Crop Research Institute
Invergowrie, Dundee
Scotland

M. Petersen
Department of Pathology
University of Florida
Gainesville, Florida

D. E. Purcifull
Department of Plant Pathology
University of Florida
Gainesville, Florida

J. W. Randles
Associate Professor
Department of Crop Protection
Waite Agricultural Research
 Institute
University of Adelaide
Glen Osmond, South Australia

Richard Stace-Smith
Research Scientist
Agriculture Canada
Vancouver, British Columbia
Canada

M. H. V. Van Regenmortel
Director
Immunochemistry Department
Institut de Biologie Moleculaire et
 Cellulaire, CNRS
Strasbourg, France

TABLE OF CONTENTS

Chapter 1

OVERVIEW

R. E. F. Matthews

TABLE OF CONTENTS

0-8493-4284-8/93/$0.00 + $.50

I. INTRODUCTION

Diseases due to plant viruses cause losses estimated at about $15 billion per year worldwide in both agricultural and horticultural crops. For various reasons, such losses may be particularly important in tropical and subtropical regions. In attempts to control these diseases, the first essential step is to establish the identity of the virus causing the disease and to determine its properties. Over recent years, this task has been made substantially less difficult by two kinds of development.

First, the physical, genetic, and biological properties of many viruses have been sufficiently well described for the International Committee on Taxonomy of Viruses (ICTV) to be able to arrange them in families and groups of viruses with similar properties, thus providing an overall framework for virus identification.

The second major factor facilitating diagnosis has been the development of improved technologies for virus identification. Refined serological and electron microscopic procedures, and the development of widely applicable methods using nucleic acid hybridization have become particularly important.

II. CLASSIFICATION AS A FRAMEWORK FOR DIAGNOSIS

The current internationally approved classification and nomenclature of viruses is given by Francki et al.[1] This publication forms a useful companion to this volume for those involved in diagnosis. Here, I will give a brief overview of the current classification. Figure 1 illustrates in outline drawings of particle morphology the 35 groups and families of plant viruses that have been delineated and approved by the ICTV.

This classification provides an excellent framework for diagnosis. There is now a reasonably high probability that a virus causing a disease found in a new crop, or from a new region, will belong in one of these 35 families or groups; and that it is likely to be a strain of a virus that has already been extensively studied. There have been 394 viruses allocated to these taxa, with a further 320 being given a provisional allocation (Table 1).

The Plant Virus Subcommittee of the ICTV is moving to adopt the family-genus-species classification for all plant viruses. Current proposals for families and genera already approved by the Executive Committee of the ICTV are summarized in Table 2. It is almost certain that these proposals will be adopted at the next meeting of the ICTV in 1993.

III. AN OVERVIEW OF TECHNIQUES

The main techniques currently available for the diagnosis of virus diseases in plants are dealt with in the 12 chapters which follow. Inoculation to the original host species or cultivar (Chapter 2) is necessary to demonstrate that

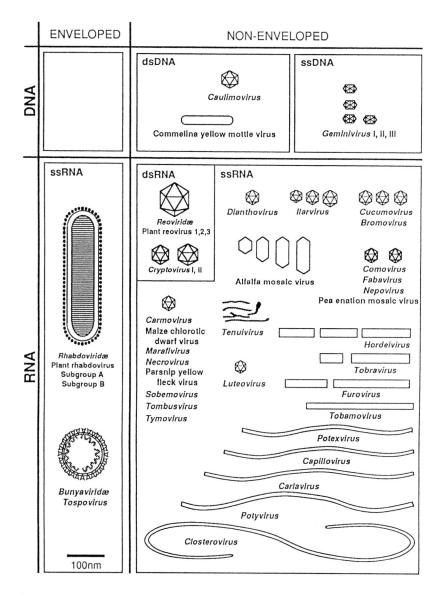

FIGURE 1. The families and groups of plant viruses. Outline diagrams drawn approximately to scale. (From Francki, R. I. B. et al., *Arch. Virol. Suppl.*, 2, 450, 1991. With permission.)

the virus identified by other means is in fact the cause of the disease in question (see Section V). Thus, positive diagnosis will almost always involve the application of two or more of the available techniques.

A knowledge of the size, shape, and overall properties of the virus particle are usually important for placing a virus in its correct family or group. Electron

Table 1. The 35 Families and Groups of Plant Viruses Approved by the ICTV[a]

Characterization	Family of group	No. of members	No. of probable or possible members	Total
dsDNA nonenveloped	*Caulimovirus*	11	6	17
	Commelina yellow mottle virus group	3	11	14
ssDNA nonenveloped	*Geminivirus*	35	13	48
dsRNA nonenveloped	*Cryptovirus*	20	10	30
	Reoviridae	6	2	8
ssRNA enveloped	Rhabdoviridae	14	76	90
	Bunyaviridae	1	0	1
ssRNA nonenveloped				
Monopartite genomes				
Isometric particles	*Carmovirus*	8	9	17
	Luteovirus	12	12	24
	Marafivirus	3	0	3
	MCDV group	1	2	3
	Necrovirus	2	2	4
	PYFV group	2	1	3
	Sobemovirus	10	6	16
	Tombusvirus	12	0	12
	Tymovirus	18	1	19

Rod-shaped particles	*Capillovirus*	2	2	4
	Closterovirus	10	12	22
	Carlavirus	27	29	56
	Potexvirus	18	21	39
	Potyvirus	74	84	158
	Tobamovirus	12	2	14
Bipartite genomes				
Isometric particles	*Comovirus*	14	0	14
	Dianthovirus	3	0	3
	Fabavirus	3	0	3
	Nepovirus	28	8	36
	PEMV	1	0	1
Rod-shaped particles	*Furovirus*	5	6	11
	Tobravirus	3	0	3
Tripartite genomes				
Isometric particles	*Bromovirus*	6	0	6
	Cucumovirus	3	1	4
	Ilarvirus	20	0	20
Isometric and bacilliform particles	AMV	1	0	1
Rod-shaped particles	*Hordeivirus*	4	0	4
Quadripartite genome				
Rod-shaped particles	*Tenuivirus*	3	4	7
		394	320	714

[a] Listed according to particle morphology and type of nucleic acid.

Data from Reference 1.

Table 2. Provisional Classification of Plant Viruses in Families and Genera

	Genera
Previously existing families	
Rhabdoviridae	Two genera corresponding to subgroups A and B, but as yet unnamed
Reoviridae	*Phytoreovirus, Fijivirus,* and *Oryzavirus*
Bunyaviridae	*Tospovirus*
Newly established families	
Cryptoviridae	*Alphacryptovirus* and *Betacryptovirus*
Geminiviridae	*Monogeminivirius, Intergeminivirus,* and *Bigeminivirus*
Tombusviridae	*Tombusvirus* and *Carmovirus*
Comoviridae	*Comovirus, Nepovirus,* and *Fabavirus*
Bromoviridae	*Bromovirus, Cucumovirus, Ilarvirus,* and *Alfamovirus*
Potyviridae	*Potyvirus, Bymovirus,* and *Rymovirus*
Ungrouped genera	
DNA viruses	*Caulimovirus* and *Badnavirus*
RNA viruses (isometric)	*Necrovirus, Dianthovirus, Sobemovirus, Luteovirus, Enamovirus, Tymovirus, Sequivirus, Marafivirus, Idaeovirus, Machlomovirus,* and a genus for Rice Tungro virus, but as yet unnamed
RNA viruses (rod-shaped)	*Tobamovirus, Tobravirus, Furovirus,* and *Hordeivirus*
RNA viruses (filamentous)	*Potexvirus, Carlavirus, Capillovirus, Closterovirus,* and *Tenuivirus*

microscopy is the most direct method now used for determining size and shape. This technique applied to virus preparations is detailed in Chapter 8. Viruses with large particles, particularly those with membranes, can be characterized in thin sections of infected tissue. I have not included a chapter on thin-sectioning techniques because many good accounts of these methods are available elsewhere.[4]

The value of particle morphology for placing a virus in its correct group or family varies significantly for different kinds of virus. Table 3 summarizes some properties of seven virus groups and families which may be quite readily distinguished by particle morphology.

Table 4 lists some properties of the rod-shaped plant viruses. Members of some of these groups, such as tobamoviruses, tobraviruses, and tenuiviruses can usually be placed tentatively in their correct group on particle morphology alone. For some others this would be much more difficult.

Table 5 details the 18 plant virus groups with isometric particles. With the exception of alfalfa mosaic virus, which has bacilliform particles as well as isometric, these groups may be difficult or impossible to distinguish on particle morphology alone. Most have particle diameters around 30 nm. Furthermore, different treatments may significantly affect the apparent particle diameter of an icosahedral virus.[5] Differences in surface morphology can be observed in negatively stained preparations between some of the virus groups listed in Table 5, provided great care is taken with the preparation procedure.[6]

Two kinds of techniques of wide applicability in virus diagnosis have now emerged. These are serological procedures, especially various forms of enzyme-linked immunosorbent assay (ELISA) (Chapter 7) and nucleic acid

Table 3. Seven Virus Groups and Families with Genomes Other Than ss Positive Sense RNA

Particle morphology	Particle dimensions (nm)	Genomic nucleic acid			Family, genus, or group
		Kind	No. of molecules	Mol Wt × 10^{-6}	
Enveloped					
Bacilliform	100–430 long, 45–100 wide	ssRNA −ve sense	1	4.2–4.6	Rhabdoviridae
Isometric or pleomorphic	≈85 diameter	ssRNA −ve and ambisense	3	2.7, 1.5, 0.9	Bunyaviridae (*Tospovirus*)
Nonenveloped					
Isometric	50	dsDNA	1	5.1	*Caulimovirus*
Bacilliform	130 × 30	dsDNA	1	4.8	Commelina yellow mottle virus group
Geminate	18 × 30	ssDNA	1 or 2	0.7–0.8	*Geminivirus*
Icosahedra	65–70	dsRNA	10 or 12	16–20	Reoviridae
Icosahedra	30–38	dsRNA	2	1.2–1.5, 0.97–1.38	*Cryptovirus*

From Francki, R. I. B., Fauquet, C. M., Knudson, D. L., and Brown, F., Eds., *Arch. Virol. Suppl.*, 2, 450, 1991. With permission.

Table 4. Ten Plant Virus Groups with ss Positive Sense RNA and Rod-Shaped Particles

No. of genomic RNAs	Rod lengths (nm)	Rod diameter (nm)	RNA Mol Wt × 10^-6	Coat protein Mol Wt × 10^-3	Group
			Rigid Rods		
1	300	18	2.0	≈17	*Tobamovirus*
2	180–215 and 46–114	≈22	2.4 and 0.6–1.4	22	*Tobravirus*
2	250–300, 92–160	≈20	1.8–2.4 and 1.2–1.8	≈20	*Furovirus*
3	110–150	≈20	1.2–1.05 1.01	≈22	*Hordeivirus*
			Flexuous Rods		
1	470 580	13	2.1	≈20	*Potexvirus*
1	640	12	2.5	≈27	*Capillovirus*
1	610–700	≈13	≈2.6	≈32	*Carlavirus*
1	680–900	11	≈3.3	≈34	*Potyvirus*
1	700–2000	12	2.5–6.5	23–43	*Closterovirus*
4	Varying lengths, sometimes branched	8	3.0, 1.6, 1.0, 0.9 (may be minus sense)	32	*Tenuivirus*

From Francki, R. I. B., Fauquet, C. M., Knudson, D. L., and Brown, F., Eds., *Arch. Virol. Suppl.*, 2, 450, 1991. With permission.

Table 5. Eighteen Plant Virus Groups with ss Positive Sense RNA and Isometric Particles

No. of genomic RNAs	Particle diameter (nm)	RNA Mol Wt $\times 10^{-6}$	Coat protein Mol Wt $\times 10^{-3}$	Group
1	33	1.3	38	*Carmovirus*
1	28	1.4	23–33	*Necrovirus*
1	30	1.4	30	*Sobemovirus*
1	30	1.5	43	*Tombusvirus*
1	25–30	2.0	24	*Luteovirus*
1	29	2.0	20	*Tymovirus*
1	31	2.2	22, 28	*Marafivirus*
1	30	3.2	22.5, 25, 34	Maize chlorotic dwarf virus group
1	30	3.5	23, 26, 31	Parsnip yellow fleck virus group
2	31–34	1.5, 0.5	40	*Dianthovirus*
2	28	1.7, 1.3	22	Pea enation mosaic virus group
2	28	1.9, 1.1	22, 42	*Comovirus*
2	30	2.1, 1.5	27, 43	*Fabavirus*
2	28	≈2.6, 1.3–2.4	55–60	*Nepovirus*
3	26	1.0, 0.90, 0.67	20	*Bromovirus*
3	29	1.1, 0.97, 0.70	26	*Cucumovirus*
3	26–35	1.1, 0.90, 0.70	25	*Ilarvirus*
3	18, with bacilliform particles, 56, 43, 35, and 30 nm long and isometric (18)	1.17, 0.83, 0.69	24	Alfalfa mosaic virus group

From Francki, R. I. B., Fauquet, C. M., Knudson, D. L., and Brown, F., Eds., *Arch. Virol. Suppl.*, 2, 450, 1991. With permission.

hybridization (Chapter 9). Table 6 summarizes the relative sensitivity of some of these procedures for the detection of plant viruses.

Besides the size and morphology of the virus particle as revealed by electron microscopy, various other physical properties of the virus particle have in the past been used in diagnosis. In the early years of plant virus research three so-called physical properties feature were commonly reported. These were the thermal death point, dilution end point, and longevity *in vitro*, all based on measurements of infectivity of the virus, usually in crude extracts. These properties have been shown to be quite unreliable as diagnostic criteria and are now rarely used.[9] Three more precisely determined physical properties have been reported for many viruses. These are (1) *density* of the virus measured in water, sucrose, or CsCl solutions, (2) sedimentation coefficient and diffusion coefficient, and (3) electrophoretic mobility of the virus. These properties can be somewhat laborious to determine if accurate numbers are required. Furthermore, while they may provide useful additional data for placing a new virus in its correct family or group, these properties do not often distinguish between individual viruses within a family or group. To describe adequately and identify a virus it is usually necessary to isolate the virus in reasonably purified form. The procedure that can be used for successful isolation will depend to a large extent on the properties of the virus

**Table 6. Sensitivity of Some Serological and
            Molecular Techniques for Plant Virus
            Detection and Identification**

	Detection range
Serological[a]	
Double immunodiffusion	2–20 μg/ml
Liquid precipitin tests	1–10 μg/ml
Radial immunodiffusion	0.5–10 μg/ml
Rocket electrophoresis	0.2–1.0 μg/ml
Complement fixation	50–100 ng/ml
Immuno-osmophoresis	50–100 ng/ml
Passive hemaglutination	20–50 ng/ml
Latex text	5–20 ng/ml
ELISA	1–10 ng/ml
Immunoelectron microscopy	1–10 ng/ml
Molecular[b]	
Western blotting	1–10 ng/ml
Molecular hybridization	<1 pg of RNA

[a] Data from Reference 7.
[b] Data from Reference 8.

particle and the concentration it reaches in the host plant. Thus, a method
that is successful for purification of a virus may sometimes itself provide a
useful taxonomic guideline (Chapter 6).

Besides information on the host range of a virus and the macroscopic
disease symptoms it causes (Chapter 2), host plants can be used to establish
other biological properties which may be useful in diagnosis. The ways in
which a virus is transmitted can give clues as to identity. These include:
experimental transmission, for example by mechanical inoculation, and transmission through seed and pollen (Chapter 3); transmission through soil and
soil-inhabiting organisms (Chapter 4); and transmission by arthropod vectors,
a topic not dealt with in this book. Many viruses can be placed in their correct
taxonomic group by using cytological techniques to reveal characteristic intracellular inclusions (Chapter 5).

Most plant viruses have single-stranded (ss) RNA genomes, which may
be in one or several pieces; and each genomic RNA is replicated on an RNA
strand of complementary base sequence. During the process of viral replication, double-stranded (ds) RNA species corresponding to each genome
segment may accumulate in infected tissue. In addition, dsRNAs corresponding to subgenomic messenger RNAs may sometimes be present. These various
dsRNA species are readily isolated, and may sometimes be useful as a taxonomic guide (Chapter 10).

Viroids are very small circular ssRNA molecules that are generally considered to constitute a class of disease-causing agents different from viruses.
This is because of their very small size, and because they code for no proteins,
and are therefore replicated using host enzymes. However, the diseases they
cause may be very much like those caused by viruses, and thus they may be
a confusing factor in diagnosis. For this reason diagnostic methods applicable
to viroids have been detailed (Chapter 11).

IV. DATABASES AS AIDS IN DIAGNOSIS

The series *Descriptions of Plant Viruses* produced by A. F. Murant and B. D. Harrison on behalf of the Association of Applied Virologists beginning in 1970, and now numbering 324, published in 20 sets, provides very useful descriptions for individual plant viruses (available from the Association of Applied Virologists, Subscriptions Department, National Vegetable Research Station, Wellsbourne, Warwick, UK CV35 9EF).

Beginning in the early 1980s A. J. Gibbs and his colleagues in Australia have been collecting data on plant viruses for storage and retrieval in a computer-based data bank — the Virus Identification Data Exchange (VIDE) system. The project uses the DELTA (*DE*scription *L*anguage for *TA*xonomy) database system to collect available diagnostic information for all the plant viruses that have been adequately described. Information on 570 possible characters has been sought from plant virologists worldwide.

The stored information is progressively being made available in hard copy. The first book was *Viruses of Plants in Australia,*[10] the second, *Viruses of Tropical Plants.*[11] The third, due in late 1992 will be called *Viruses of Plants.* It will give the complete information available for the 890 virus species, and 50 virus groups or genera at present in the database. It is also planned that the VIDE database will be available for distribution in 1993 to be used with personal computers having an IBM-compatible hard disk, using the INTKEY (*INT*eractive *KEY*) program. This program allows the investigator to use the database fully interactively as an identification aid or information source. It is planned that a second version of the VIDE-INTKEY system will have full color, high-resolution pictures of disease symptoms as an additional aid in identification. It is probable that the VIDE database will become the first component of the World Virus Data Base organized by the ICTV. Further information can be obtained from Dr. A. J. Gibbs, Research School of Biological Sciences, Australian National University, Canberra, ACT, Australia. Phone – (61) (6) 249-4211, Fax – (61) (6) 249-4437, EMail – gibbs @ rsbso.anu.edu.au.

V. KOCH'S POSTULATES IN RELATION TO DIAGNOSIS OF PLANT VIRUS DISEASES

For diagnosis of disease caused by a cellular pathogen, Koch's postulates are of classical importance. They have been formulated in many ways that differ in detailed wording. The following is an authoritative version for bacterial diseases.[12] "(1) The causal agent must be associated in every case with the disease as it occurs naturally, and conversely the disease must not appear without this agent. (2) The causal agent must be isolated in pure culture, and its specific characteristics determined. (3) When the host is inoculated under favorable conditions with suitable controls, the characteristic symptoms of the disease must develop. (4) The causal agent must be reisolated, usually by means of the technique employed for the first isolation, and identified as that first isolate.''

Since plant viruses cannot be cultured outside host cells, requirement (2) cannot be met. Nevertheless the postulates remain vital for diagnosis of plant virus diseases. They have been reworded by various authors to make them more appropriate to the experimental limitations of plant virology (for example see Chapter 2, Section V). Authors of various chapters in this book have used the ideas behind the postulates in slightly different ways, and I have not attempted to standardize the usage. For example, Randles, in the Introduction to Chapter 12, lists six steps that most virologists use to implicate a virus as the cause of a particular disease. In the rest of his chapter he outlines the problems for some diseases of unknown origin where these six steps cannot be applied.

VI. AGENTS CAUSING VIRUS-LIKE DISEASES

In the initial stages of planning an investigation of a new virus-like disease, it must always be borne in mind that a number of biological and physical agents, quite unrelated to viruses, may cause virus-like diseases. Besides the viroids, already mentioned, the most important are (1) mycoplasmas and similar organisms, (2) toxins produced by arthropods, (3) genetic abnormalities, (4) nutritional deficiencies, and (5) hormone damage. These are discussed by Matthews.[13]

VII. VIRUS DISEASES WITH MORE THAN ONE CAUSAL AGENT

A variety of examples are known among field infections in which more than one virus, or a virus and some other kind of agent acting together, are necessary to produce a particular disease. In such circumstances it is particularly important that Koch's postulates be fulfilled as closely as is possible.

A. TWO OR MORE STRAINS OF THE SAME VIRUS

Sometimes a disease may be caused by a mixture of two or more closely related strains of the same virus. For example the characteristic mosaic disease produced by turnip yellow mosaic *Tymovirus* in Chinese cabbage is caused by a variety of strains in the same infected plant, each producing different macroscopic leaf symptoms and cytological effects on the chloroplasts in various blocks of leaf tissue in the mosaic. Inoculation from these different blocks of tissue to fresh plants will produce different diseases.[13]

B. CHANCE INFECTIONS WITH TWO OR MORE UNRELATED VIRUSES

Many examples from field infections have been described where two unrelated viruses infecting a host species separately produce different kinds of disease, and when present as a double infection cause a third kind of disease. For example, this phenomenon was common in potatoes before virus-free stocks became available.

C. A VIRUS AND A SATELLITE RNA

The classic example here is a lethal necrotic disease of tomato, first observed in Alsace in 1972. Kaper and Waterworth[14] showed that this disease was brought about by a small satellite RNA in the presence of cucumber mosaic *Cucumovirus,* which, on its own, causes fern-leaf symptoms. Other examples are now known.

D. A VIRUS AND SOME NONINFECTIOUS AGENT

Instances have been described where the disease caused by a virus in a particular host plant is modified by some noninfectious agent, such as a chemical pollutant in soil or air, or nonoptimal mineral nutrition of the host plant.

E. TWO UNRELATED VIRUSES IN A FIXED RELATIONSHIP

In nature there are many examples where two unrelated viruses coexist in a fixed relationship in the host plant, because one has become dependent on the other for a vital function. This situation is best exemplified by the complexes of transmission-dependent and helper viruses discussed in Chapter 13. The groundnut rosette disease complex is perhaps the most interesting and instructive. This complex involves two viruses, groundnut rosette virus (GRV), and groundnut rosette assistor *Luteovirus* which is required by GRV for aphid transmission. The disease produced by these two viruses is modulated by a satellite RNA of GRV. All three agents must be inoculated to healthy groundnut plants to reproduce the original disease and the aphid transmissibility of the complex.

VIII. SUMMARY

Generally speaking, diagnosis of a plant virus disease is now an easier exercise than it was a few decades ago. This is because the current official classification of viruses provides a sound overall framework for diagnosis, and because new sensitive and specific diagnostic tools have been developed. The most important of these are various electron microscopic procedures, serological tests, particularly ELISA, and methods based on nucleic acid hybridization.

Nevertheless the diagnostician must always give appropriate attention to Koch's postulates in as far as they can be applied to plant virus diseases. The possibility must always be borne in mind that a new disease may be due to two or more agents acting together, for example a virus and a satellite RNA, or two viruses, one being in a dependent relationship.

REFERENCES

1. **Francki, R. I. B., Fauquet, C. M., Knudson, D. L., and Brown, F., Eds.,** Classification and Nomenclature of Viruses, Fifth Report of the International Committee for Taxonomy of Viruses, *Arch. Viol. Suppl.,* 2, 450, 1991.
2. **Barnett, O. W.,** Potyviridae, a proposed family of plant viruses, *Arch. Virol.,* 118, 139, 1991.
3. **Martelli, G. P.,** Classification and nomencalture of viruses: state of the art, *Arch. Virol.,* 1992.
4. **Hall, J. L. and Hawes, C., Eds.,** *Electron Microscopy of Plant Cells,* Academic Press, London, 1991.
5. **Roberts, I. M.,** Practical aspects of handling, preparing, and staining samples containing plant viruses for electron microscopy, in *Developments and Applications in Virus Testing,* Jones, R. A. C. and Torrance, L., Eds., Dev. Appl. Biol. Ser. 1, Association of Applied Biologists, Wellsbourne, 1986, 213.
6. **Hatt, T. and Francki, R. I. B.,** Differences in the morphology of isometric particles of some plant viruses stained with uranyl acetate as an aid to their identification, *J. Virol. Meth.,* 9, 237, 1984.
7. **van Regenmortel, M. H. V.,** *Serology and Immunochemistry of Plant Viruses,* Academic Press, New York, 1982, 302.
8. **Frenkel, M. J., Jilka, J. M., Shukla, D. D., and Ward, C. W.,** Differentiation of potyviruses and their strains by hybridization with the 3′ non-coding region of the viral genome, *J. Gen. Virol.,*
9. **Francki, R. I. B.,** Limited value of the thermal inactivation point, longevity *in vitro* and dilution end point as criteria for the characterization, identification and classification of plant viruses, *Intervirology,* 13, 91, 1980.
10. **Bücken-Osmond, C., Crabtree, K., Gibbs, A., and McLean, G.,** *Viruses of Plants in Australia,* Australian National University, Research School of Biological Sciences, Canberra, 1988, 590.
11. **Brunt, A., Crabtree, K., and Gibbs, A.,** *Viruses of Tropical Plants,* C.A.B. International, Wallingford, Oxon, U.K., 1990, 707.
12. **Riker, A. J.,** Inoculations with bacteria causing plant disease, in *Manual of Microbiolgoical Methods,* Society of American Bacteriologists, McGraw-Hill, New York, 1957, 286.
13. **Matthews, R. E. F.,** *Plant Virology,* 3rd ed, Academic Press, New York, 1991, 835.
14. **Kaper, J. M. and Waterworth, H. E.,** Cucumber mosaic virus associated RNA5. Causal agent for tomato necrosis, *Science,* 196, 429, 1977.

Chapter 2

HOST PLANTS IN DIAGNOSIS

J. Horváth

TABLE OF CONTENTS

0-8493-4284-8/93/$0.00 + $.50

I. INTRODUCTION

There are several ways in which host plants may assist in the diagnosis of diseases caused by viruses: (1) *disease symptoms,* either on plants in the field, or on specially selected experimental species known as indicator hosts; (2) the species of plants that can or cannot be infected by a virus, that is, the *host range,* may be more or less characteristic for a particular virus; (3) the phenomenon of *cross-protection* has sometimes been used as an aid in diagnosis; (4) an unknown virus is often extracted from either natural or experimental hosts, purified, and characterized *in vitro* by various means. It is then essential to carry out *back-inoculation* with the purified virus to the original host species in order to ensure that the virus studied was in fact the one causing the original disease. These various uses of host plants are discussed in this chapter. The discussion is confined to the 24 groups of plant viruses that are readily transmitted by mechanical means.

II. DISEASE SYMPTOMS

In the early years of plant virus research, virus detection and identification were based on symptom development. This usually turned out to be a very unreliable criterion. Symptoms on plants in the field may be unreliable because: (1) several viruses may cause similar symptoms in the same crop; (2) a single virus may cause highly variable symptoms, depending on virus strain; (3) a mixture of viruses, or virus strains, or the presence of a satellite RNA may greatly affect disease expression; (4) different crop cultivars may be affected differently by the same virus; and (5) different soil and weather conditions may alter disease expression.

For these reasons many searches have been made for experimental host plants, which under reasonably standardized conditions, usually in the glasshouse, will give consistent and distinguishing disease symptoms with particular viruses. Such plants are known as indicator species. However, even in a glasshouse, disease symptoms can be greatly influenced by various factors (e.g., temperature, day length, light intensity, age of plants, virus strains, plant cultivars, and accessions). Furthermore, it is known that the accessions or varieties of the same species may give different responses to the same virus.[1-7] Thus, symptomatology in itself is usually not sufficient for the exact identification of viruses.[8-18]

Latent infections may be a problem. Some viruses, for example alfalfa mosaic virus, cassava common mosaic *Potexvirus,* viola mottle *Potexvirus,* and several carlaviruses cause latent infections in most host species, and, in such species, cannot be studied symptomatologically.[19-21] In the course of testing the susceptibility of 456 plant species to 24 viruses, Horváth[22] discovered 1312 new susceptible host-virus combinations, 13% of which were latent infections.

The symptomatological identification of viruses is often made very difficult by the fact that more than one virus may occur in the same plant, and the symptoms may become either milder or more severe in such mixed infections. During the 10-year period of 1957 to 1967 in the U.S. 62% of the plant introductions indexed positive for one or more viruses.[23]

On the other hand, certain symptoms can be characteristic of particular host-virus combinations, for example:

1.   "Flower break" (*Chrysanthemum* spp. infected with cucumber mosaic *Cucumovirus; Matthiola incana, Petunia hybrida* infected with turnip mosaic *Potyvirus; Tulipa* spp. infected with tulip breaking *Potyvirus* syn.: tulip mosaic virus)
2.   "Fern-leaf" (*Lycopersicon esculentum* infected with a mixture of cucumber mosaic *Cucumovirus* and tomato mosaic *Tobamovirus*)
3.   "Enation" (*Pisum sativum* infected with pea enation mosaic virus)
4.   "Veinal necrosis" (*Nicotiana tabacum* infected with veinal necrosis strain of potato Y *Potyvirus*)

These provide useful information for virus diagnosis.[10,15-16,18,24-29]

III. HOST RANGE

A. DEFINITIONS

Matthews[10,24] points out that the reported host range of a virus can be rather meaningless because (1) in many of the published host range studies only positive results were recorded; (2) the lack of symptoms following inoculation of test plants was not always checked by back-inoculation to an indicator species, to test for masked infection; (3) the manner of inoculation may affect the results; (4) when studying large numbers of species it is usually practicable to make tests under only one set of conditions, but it is known that a given species may vary widely in susceptibility to a virus, depending on the conditions of growth; (5) even closely related strains of a virus may differ in the range of plants they infect; and (6) mesophyll protoplasts may be readily infected with a virus that causes little or no apparent infection when applied to intact leaves.

With respect to this last point, Matthews[10] defined an immune species (nonhost) as one in which the virus does not replicate in isolated protoplasts, nor in cells of the intact plant, even in the initially inoculated cells. Inoculum virus may be uncoated, but no progeny viral genomes are produced. He defined an infectible species (host) as one in which the virus can replicate in isolated protoplasts. If the virus which replicates in isolated protoplasts can replicate only in the initially infected cells of an inoculated leaf, no visible symptoms will be produced, and such a plant was defined by Matthews[10] as showing extreme hypersensitivity. This condition is thought to be due to the viral-coded movement protein being unable to function in the particular host species. No visible symptoms are seen with extreme hypersensitivity, and such a small proportion of cells in the inoculated leaf are infected that back inoculation to a test assay species would not usually reveal any infection.

There have been very few tests of host range using the infectibility of isolated protoplasts as the criterion. Thus, for almost all the nonhosts reported in the literature, it is not known whether they are immune or infectible, but showing extreme hypersensitivity, as defined above. Therefore, in this chapter, the terms nonhost or immune will include both the immune and extreme hypersensitive categories.

B. FACTORS AFFECTING EXPERIMENTAL HOST RANGE

Any factor that reduces or increases the susceptibility of a potential host species may affect the experimentally determined host range. These factors include growth conditions (light, temperature, day length, water supply, nutrition), cultivar or line of the host species, and method of inoculation.

Another important aspect may be the stage of development at which the plants are most suitable for inoculation. According to Pawley,[30] *Chenopodium amaranticolor* and *Nicotiana clevelandii* are in that state with 10 leaves, *N. glutinosa* with 7 leaves, and *N. tabacum* and *Gomphrena globosa* with 4 leaves. With cucumber (*Cucumis sativus*) and squash (*Cucurbita pepo*), cotyledons are usually inoculated; with bean (*Phaseolus vulgaris*) and cowpea (*Vigna sinensis*) primary leaves are often used, because older leaves of the plants are less susceptible, or not susceptible, to the viruses. Very young primary bean leaves and very young cucumber cotyledons are resistant to several viruses. These plants do not reach their optimum susceptibility until about 10 days from sowing of seed, and become resistant after about 15 days. Tobacco leaves remain susceptible much longer.[31] Further useful data are to be found in works by Noordam[17] and Hill[32] concerning the age (days from sowing at 20°C) and the best stage for use of test plants. Virus susceptibility of plants is also influenced by leaf position.[33-34]

In some genera, inhibitors are present in the leaves which may make mechanical transmission of particular viruses difficult, or prevent it altogether (e.g., *Capsicum, Chenopodium, Emex, Dahlia, Dianthus, Phytolacca*).[8,12,18,35] This question is further complicated by the fact that the reactions of inoculated test plants to a virus depend in a great measure on the plant in which the virus has been propagated. For example, van der Meer et al.[36] found that *Chenopodium quinoa* never showed symptoms when inoculated with sap from *Lonicera* species infected by *Lonicera* latent *Carlavirus*, even though the inocula induced a large number of lesions in *Nicotiana megalosiphon*. However, when *C. quinoa* plants were inoculated with sap from infected *Nicotiana megalosiphon* or *N. clevelandii,* a limited number of small lesions developed on the inoculated leaves. With further successive transmissions from *C. quinoa* to *C. quinoa* many more lesions were obtained and in these tests some isolates also caused systemic symptoms.

C. VARIATION IN THE EXTENT OF HOST RANGE

The extent of the natural host range of viruses varies greatly. There are viruses whose natural host range covers only a single plant; for example, Andean potato mottle *Comovirus,* potato M *Carlavirus* and potato S *Carlavirus,* all three being isolated exclusively from potato (*Solanum tuberosum*). On the other hand, the natural host range of some viruses includes many species; for example, cucumber mosaic *Cucumovirus* (476 species in 67 families), and alfalfa mosaic virus (46 species in 12 families).

With experimental host ranges, some plant families have species that can be infected by many viruses, and which are therefore useful in host range studies. These include the Aizoaceae, Amaranthaceae, Brassicaceae, Chenopodiaceae, Cucurbitaceae, Papilionaceae, Poaceae, and Solanaceae.

Host range studies for diagnostic purposes are most useful for those viruses with a relatively narrow host range.[10] These include members of the *Geminivirus* and *Sobemovirus* groups. In certain virus groups some viruses have wide experimental host ranges while for others it is narrow. For example, in the *Nepovirus* and *Tymovirus* groups, some of the viruses have a narrow host range (olive latent ringspot *Nepovirus,* cacao yellow mosaic *Tymovirus*), while others have a wide host range (e.g., arabis mosaic *Nepovirus,* cherry leaf roll *Nepovirus,* tobacco ringspot *Nepovirus,* belladonna mottle *Tymovirus*).

IV. CROSS-PROTECTION

It has been known for many years that when a plant is inoculated with a mild strain of a virus it may be protected from superinfection with a related severe strain of the virus. For a time, the phenomenon was thought to have some value for identification of related virus strains. However, in general, it is not a reliable test. There are many examples of incomplete or undetectable cross-protection between related virus strains, as well as examples of cross-protection between obviously distinct viruses.[37-43] According to Dodds,[44] different mechanisms are probably responsible for cross-protection between different groups of viruses.

V. KOCH'S POSTULATES

Bos[25] redefined Koch's postulates[45] to be applicable to plant viruses and virus diseases in the following way: (1) the virus must be concomitant with the disease; (2) it must be isolated from the diseased plant, separated from concomitants, multiplied in a propagation host, purified physicochemically, and the properties of the virus particle determined; (3) when inoculated into a healthy host plant it must reproduce the disease; and (4) the same virus must be demonstrated to occur in, and must be reisolated from the host plants inoculated with the purified virus. Appropriate host plants are essential for this process, especially when a disease in the field is caused by a mixture of viruses or virus strains, or a virus and a satellite RNA.

VI. SUGGESTED SPECIES FOR IDENTIFICATION OF VIRUS GROUPS

From a survey of the host plants of viruses, we can establish that there are some species susceptible to 40 to 50% or more of the viruses belonging to a given virus group. There are also plants susceptible to all the viruses of a given virus group. For example, *Chenopodium quinoa* is susceptible to all the 27 members of the *Nepovirus* group. On the other hand, some species are suitable for detecting only one or two viruses belonging to a given virus group. For example, among the 17 host plants of five viruses in the *Caulimovirus* group (blueberry red ringspot, carnation etched ring, cauliflower mosaic, Dahlia mosaic, and soybean chlorotic mottle), there is not a single species with which more than one of these caulimoviruses can be detected. With *Nicotiana clevelandii,* 1 of 45 host plants among thirteen sobemoviruses, only four can be reliably detected (see relevant descriptions[46]). Similarly, 2 of 41 *Tymovirus* hosts, *Chenopodium amaranticolor* and *Nicotiana clevelandii,* make possible the reliable detection of only four tymoviruses.

Thus, there are many uncertainties in using test plants to place an unknown virus in an established virus group. This problem has been discussed in an excellent work edited by Kurstak,[47] which deals with the diagnostic problems of some 25 plant virus groups. We should like to emphasize that the host plants and nonhost plants on their own do not provide a reliable basis for the differential diagnosis of virus groups. For this, studies on the virus particle discussed in other chapters are required. However, a definitive diagnosis usually requires some information derived from host plants.

In spite of these various difficulties, symptoms and host range on a suitable set of plant species can give at least a preliminary indication of the group to which a newly discovered virus might belong.

In making a list of suggested species for the identification of virus groups (Table 1) we followed two guiding principles (1) that every known virus in the group should have at least one susceptible species in the list; and (2) that the nonhosts listed should be nonhosts for all the known viruses in the group. Such nonhosts were sometimes difficult to find. For example, among the 50 nonhosts recorded for individual members of the *Potyvirus* group, we only found 5 species that were nonhosts for all members (Table 1).

VII. SUGGESTED SPECIES FOR IDENTIFICATION OF INDIVIDUAL VIRUSES

Table 2 lists suggested species for the identification of those individual mechanically transmissible viruses, described in the series *CMI/AAB Descriptions of Plant Viruses*[46] between 1970 and 1988, and which can be found in the paper of Hull et al.[58] and in Appendix 2 of Matthews.[10] For the 232 viruses involved we have selected 141 host species that are best known and most widely used. The selection is based on the 421 host species in the *CMI/ AAB Descriptions of Plant Viruses.*[46] A few viruses have only a single known host each (e.g., blueberry red ringspot *Caulimovirus*, ginger chlorotic fleck? *Sobemovirus*, Narcissus tip necrosis *Carmovirus*).

The following genera have proved particularly useful for virus diagnosis: *Amaranthus; Capsicum; Chenopodium,* particularly *C. amaranticolor* and *C. quinoa; Cucumis,* especially *C. sativus; Curcurbita; Datura,* especially *D. stramonium* and *D. metel; Gomphrena,* especially *G. globosa* and *G. dispersa; Lycium; Nicotiana,* especially *N. tabacum* and its cultivars, *N. clevelandii, N. glutinosa,* and *N. benthamiana; Phaseolus,* especially *P. vulgaris,* and its cultivars; *Pisum,* especially *P. sativum; Physalis,* especially *P. floridana* and *P. minima; Solanum,* especially *S. demissum* and hybrids; *Tetragonia,* especially *T. expansa; Vigna,* especially *V. sinensis* and *V. unguiculata.*

As far as nonhost species for individual viruses are concerned, very few have been recorded (see *CMI/AAB Descriptions of Plant Viruses*[46]). For this reason we have also relied on original sources, and on helpful information provided by colleagues.

We should like to emphasize here that there are differences of opinion among virologists concerning the reliability of accepting a particular species as a nonhost of a given virus. For example, Fulton[48] considers that it often depends on the adequacy of the technique whether or not a plant is considered to be nonhost for a virus. Fulton,[49] in an earlier work, studied the host range of tobacco streak *Ilarvirus,* but did not give its nonhosts at the time. Thirty years later he reexamined the same species, and found that those which earlier had been considered nonhosts were all, in fact, hosts. He considered that this was due to improved inoculating methods. Similarly, he was unable to infect two species of fern and one species of bryophyte; but the following year someone with a better method infected them. Lister[50] holds the same view, namely, that it is difficult to establish that a plant is nonhost for a virus. It must also be emphasized that the variability of strains and isolates of some viruses (e.g., cucumber mosaic *Cucumovirus*, tomato spotted wilt *Tospovirus,* or the ilarviruses) is very great, and can be reflected in the host range.[48,51-54]

The host and nonhost plants listed in Table 2 are not sufficient for the exact identification of the different viruses, but the host species are suitable for detecting the presence of a given virus. The nonhost species may allow certain viruses to be excluded when a diagnosis is being established. For example, cucumber mosaic *Cucumovirus* (CMV) and alfalfa mosaic virus (AMV) can be distinguished by the use of two test species. *Cucurbita pepo* is a host for CMV and a nonhost for AMV; while *Beta vulgaris* is a nonhost for CMV and a host for AMV. Thus, differentiation between the two viruses is possible on the basis of a limited host range study.[55-57]

Table 1. Suggested Species for Identification of Virus Groups

Group	Host plants	Nonhost plants
Alfalfa mosaic virus[a]	*Chenopodium amaranticolor*	*Cucurbita pepo*
	C. quinoa	*Paulownia fargesii*
	Nicotiana benthamiana	
	Phaseolus vulgaris	
	Vigna unguiculata	
Bromovirus	*Chenopodium amaranticolor*	*Brassica oleracea*
	C. hybridum	*Dianthus caryophyllus* cv.
	Nicotiana clevelandii	William Sim
	Pisum sativum	*Nicotiana tabacum* cv.
	Vigna unguiculata	Samsun
		Phaseolus lunatus cv.
		Henderson
Carlavirus	*Chenopodium quinoa*	*Cucumis sativus*
	Datura metel	*Datura stramonium*
	Humulus lupulus	*Nicotiana glutinosa*
	Lilium longiflorum	*N. sylvestris*
	Nicotiana clevelandii	*Phaseolus vulgaris* cv. Red
	Pisum sativum	Kidney
		Vigna unguiculata
Carmovirus	*Brassica pekinensis*	*Beta vulgaris*
	Chenopodium quinoa	*Datura metel*
	Cucumis sativus	*Lagenaria siceraria*
	Glycine max	*Nicotiana glutinosa*
	Hibiscus rosa-sinensis	*N. tabacum*
	Narcissus cv. Barnett	*Pelargonium domesticum* cv.
	Browning	Nittany Lion
	Nicotiana clevelandii	*Pisum sativum*
	Phaseolus vulgaris	*Tropaeolum majus*
	Spinacia oleracea	*Vicia faba*
	Tetragonia expansa	*Vigna sinensis*
	Vigna unguiculata	*Vigna sinensis* cv. Blackeye
Caulimovirus	*Brassica campestris*	*Chenopodium quinoa*
	Glycine max	*Cucumis sativus*
	Phaseolus vulgaris cv.	*Datura stramonium*
	Kintoki	*Gomphrena globosa*
	Saponaria vaccaria var.	*Nicotiana glutinosa*
	Pink Beauty	*Vigna sinensis*
	Vaccinium spp.	
	Zinnia elegans	
Closterovirus	*Chenopodium amaranticolor*	*Cucumis sativus*
	C. capitatum	*Datura stramonium*
	C. quinoa	*Nicotiana glutinosa*
	Dianthus caryophillus	*Vigna sinensis*
	Nicotiana clevelandii	
Comovirus	*Brassica campestris*	*Citrullus lanatus*
	Chenopodium amaranticolor	*Datura stramonium*
	C. quinoa	*Gomphrena globosa*
	Cucurbita pepo	*Nicotiana glutinosa*
	Nicotiana clevelandii	*N. tabacum* cv. Havana 425
	Phaseolus vulgaris	*N. tabacum* cv. White
	Vicia faba	Burley
	Vigna unguiculata	*N. tabacum* cv. Xanthi-nc
Cucumovirus	*Chenopodium quinoa*	*Beta vulgaris*
	Cucumis sativus	*Cucumis myriocarpus*
	Nicotiana benthamiana	*Pisum sativum*
	N. glutinosa	*Tropaeolum majus*
	Vigna sinensis	*Vicia faba*

Table 1. Suggested Species for Identification of Virus Groups (continued)

Group	Host plants	Nonhost plants
Dianthovirus	*Chenopodium quinoa*	*Datura stramonium*
	Dianthus barbatus	*Glycine max*
	Gomphrena globosa	*Petunia hybrida*
	Nicotiana clevelandii	*Pisum sativum*
	Tetragonia expansa	*Zinnia elegans*
Furovirus	*Chenopodium amaranticolor*	*Chenopodium capitatum*
	C. quinoa	*Cucumis melo*
	Nicotiana benthamiana	*Datura stramonium*
	N. clevelandii	*Gomphrena globosa*
	N. glutinosa	*Vicia faba*
	N. tabacum cv. Xanthi-nc	*Vigna sinensis*
	Tetragonia expansa	*Vigna unguiculata*
	Vicia faba	
Geminivirus	*Datura stramonium*	*Chenopodium amaranticolor*
	Nicotiana benthamiana	*C. quinoa*
	N. clevelandii	*Gomphrena globosa*
	Phaseolus vulgaris	*Tetragonia expansa*
	Vigna radiata	*Vigna unguiculata*
Idaeovirus	*Chenopodium amaranticolor*	*Cucumis sativus* cv.
	C. murale	Sperlings Mervita
	C. quinoa	l*Lycopersicon esculentum*
	Phaseolus vulvaris cv.	*Nicotiana debneyi*
	The Prince	*N. sylvestris*
		N. tabacum cv. Samsun
Ilarvirus	*Chenopodium quinoa*	*Cucumis melo*
	Cucumis sativus	*Nicotiana sylvestris*
	Nicotiana megalosiphon	*N. tabacum* cv. Samsun
	N. occidentalis	*Phaseolus mungo*
	N. tabacum	*Vicia faba*
	Phaseolus vulgaris	*Vigna sinensis* cv. Blackeye
Necrovirus	*Chenopodium amaranticolor*	*Cucumis sativus*
	Nicotiana tabacum	*Vigna radiata*
	Phaseolus vulgaris	
Nepovirus	*Chenopodium quinoa*	*Capsicum annuum*
	Cucumis sativus	*Cucurbita maxima*
	Gomphrena globosa	*Datura metel*
	Nicotiana clevelandii	*Phaseolus aureus*
	N. tabacum	*Solanum demissum*
	Phaseolus vulgaris	*S. tuberosum* spp. andigena
	Physalis floridana	CPC 1801
	Tetragonia expansa	*Vicia faba*
		Zinnia elegans
Pea enation mosaic virus[a]	*Chenopodium amaranticolor*	*Cucumis sativus*
	C. quinoa	*Datura stramonium*
	Nicotiana clevelandii	*Nicotiana glutinosa*
	Pisum sativum	
	Trifolium incarnatum	
	Vicia faba	
Potexvirus	*Chenopodium amaranticolor*	*Brassica pekinensis*
	Cucumis sativus	*Crambe abyssinica*
	Datura stramonium	*Cucurbita pepo*
	Gomphrena globosa	*Glycine max*
	Nicotiana benthamiana	*Petunia hybrida*
	N. clevelandii	*Zinnia elegans*
	N. glutinosa	
	N. tabacum	

Table 1. Suggested Species for Identification of Virus Groups (continued)

Group	Host plants	Nonhost plants
Potyvirus	*Agropyron repens*	*Cucumis sativus*
	Allium cepa	*Medicago sativa*
	Apium graveoleus var. dulce	*Phaseolus vulgaris* cv. Red
	Avena sativa	Kidney
	Belamcanda chinensis	*Pisum sativum* cv. Perfection
	Beta vulgaris	*Vigna sinensis* cv. Blackeye
	Brassica rapa	
	Capsicum frutescens cv.	
	Tabasco	
	Chenopodium amaranticolor	
	C. foetidum	
	C. quinoa	
	Cucurbita pepo	
	C. pepo cv. Small sugar	
	Dactylis glomerata	
	Datura stramonium	
	Dianthus barbatus	
	Glycine max	
	Gomphrena globosa	
	Hordeum vulgare	
	Lilium formasanum	
	Lolium multiflorum cv. S22	
	Narcissus pseudonarcissus	
	Nicotiana benthamiana	
	N. clevelandii	
	N. debneyi	
	N. glutinosa	
	N. glutinosa × *N.*	
	clevelandii	
	N. occidentalis	
	N. tabacum cvs. Havana	
	425, Kentucky 14, Samsun	
	Oryza sativa	
	Pastinaca sativa	
	Petunia hybrida	
	Phaseolus vulgaris	
	Philodendron selloum	
	Pisum sativum	
	Setaria italica	
	Solanum demissum × *S.*	
	tuberosum A6 hybrid	
	Sorghum bicolor cv. Rio	
	Tetragonia expansa	
	Triticum aestivum cv. Kent	
	Vicia faba	
	Vigna unguiculata	
Sobemovirus	*Brassica pekinensis*	*Cucumis melo*
	Chenopodium amaranticolor	*C. sativus* cv. Delicatess
	Dactylis glomerata	*Cucurbita maxima*
	Glycine max	*Datura stramonium*
	Lolium persicum	*Lolium perenne*
	Nicotiana clevelandii	*Nicotiana megalosiphon*
	Oryza sativa	*N. tabacum* cv. Xanthi-nc
	Phaseolus vulgaris	*Phaseolus vulgaris* cv. Pinto
	Trifolium subterranum	*Tetragonia expansa*
	Zea mays	*Vicia faba*
	Zingiber officinale	*Zinnia elegans*

Table 1. Suggested Species for Identification of Virus Groups (continued)

Group	Host plants	Nonhost plants
Tobamovirus	Capsicum spp.	Petunia hybrida
	Chenopodium amaranticolor	Pisum sativum
	Cucumis sativus	Nicotiana glauca
	Nicotiana clevelandii	Trifolium spp.
	N. glutinosa	Vigna sinensis
	N. tabacum cv. Samsun	V. sinensis cv. Blackeye
	Phaseolus vulgaris cv. The Prince	
Tobravirus	Chenopodium amaranticolor	Cucurbita pepo cv. capitatum
	Nicotiana clevelandii	Lycopersicon esculentum
	Phaseolus vulgaris	Medicago sativa
Tombusvirus	Chenopodium quinoa	Cucumis melo
	Cucumis sativus	Nicotiana tabacum cv. Samsun
	Gomphrena globosa	
	Nicotiana clevelandii	Pisum sativum
	Ocimum basilicum	
Tospovirus	Nicotiana clevelandii	Tetragonia expansa
	N. glutinosa	
	N. tabacum	
	Petunia hybrida	
Tymovirus	Brassica pekinensis	Amaranthus caudatus
	B. chinensis	Ammi majus
	Chenopodium amaranticolor	Capsicum annuum
	Cucumis sativus	Gomphrena globosa
	Cucurbita pepo	Nicotiana tabacum cv. Xanthi-nc
	Datura stramonium	
	Nicotiana benthamiana	Phaseolus vulgaris cv. Pinto
	N. clevelandii	P. vulgaris cv. The Prince
	N. tabacum cv. Samsun	Physalis floridana
	Phaseolus aureus	Zinnia elegans
	P. vulgaris cv. Great Northern	
	P. vulgaris cv. Long Tom	
	Vigna unguiculata subspp. unguiculata	

[a] Monotypic groups with no approved group names.

VIII. CONCLUSION

Disease symptoms and host range for natural and experimental hosts can often given an indication of what virus or viruses might be involved in a particular disease. However, for a definitive diagnosis, a selection of the other methods described in this book, which depend on the properties of the virus particle itself, are almost always required.

Back inoculation of the purified and characterized virus to the original host is essential to demonstrate that the virus studied by *in vitro* methods was, in fact, the causative agent, and the only causative agent, of the disease.

Virus diseases in the field caused by chance mixed infections with two viruses, virus strains, or a virus and a satellite RNA are not at all uncommon. In such circumstances, differential hosts that can separate the components may be essential. Host plants are even more important for diagnosis when a disease is caused by an obligatory association between two viruses (Chapter 13).

Table 2. Suggested Species for Identification of Individual Viruses

Virus name[a]	Acronym[b]	Host plants	Nonhost plants
African cassava mosaic *Geminivirus* (II)	ACMV	*Datura stramonium* *Nicotiana benthamiana* *N. clevelandii*	*Chenopodium amaranticolor* *C. quinoa* *Gomphrena globosa* *Tetragonia expansa*
Agropyron mosaic *Potyvirus* (mite)	AgMV	*Agropyron repens* *Lolium multiflorum*	*Chenopodium amaranticolor* *C. quinoa* *Datura stramonium* *Nicotiana tabacum*
Alfalfa latent *Carlavirus*	ALV	*Cassia occidentalis* *Vicia faba*	*Chenopodium amaranticolor* *C. quinoa* *Cucumis sativus* *Datura stramonium* *Gomphrena globosa*
Alfalfa mosaic virus[c]	AMV	*Chenopodium quinoa* *Nicotiana benthamiana* *Nicotiana tabacum* *Phaseolus vulgaris*	*Cucurbita pepo* *Tetragonia crystallina*
American hop latent *Carlavirus*	AHLV	*Chenopodium quinoa* *Datura stramonium*	*Nicotiana clevelandii* *N. glutinosa* *N. tabacum* cv. White Burley
American plum line pattern *Ilarvirus*	APLPV	*Cucumis sativus* *Nicotiana megalosiphon* *N. occidentalis*	*Gomphrena globosa*
Andean potato mottle *Comovirus*	APMV	*Nicandra physaloides* *Nicotiana clevelandii*	*Chenopodium amaranticolor* *C. quinoa*
Apple chlorotic leaf spot ? *Closterovirus*	ACLSLV	*Chenopodium amaranticolor* *C. quinoa* *Phaseolus vulgaris* cv. Pinto	*Cucumis sativus* *Gomphrena globosa* *Nicotiana clevelandii* *N. tabacum* *Vigna sinensis*
Apple mosaic *Ilarvirus*	ApMV	*Cucumis sativus* *Vigna sinensis*	*Capsicum annuum* *Nicotiana tabacum* *Petunia hybrida* *Vicia faba*
Arabis mosaic *Nepovirus*	ArMV	*Chenopodium quinoa* *Cucumis sativus*	*Capsicum annuum* *Vicia faba*
Arracacha A *Nepovirus*	AVA	*Chenopodium murale* *C. quinoa* *Nicotiana clevelandii* *Tetragonia expansa*	*Capsicum annuum* *Phaseolus vulgaris* *Vigna sinensis* *Zinnia elegans*
Arracacha B ? *Nepovirus*	AVB	*Chenopodium amaranticolor* *C. murale*	*Capsicum annuum* *Datura metel* *Phaseolus vulgaris*
Artichoke Italian latent *Nepovirus*	AILV	*Cucumis sativus* *Gomphrena globosa* *Nicotiana tabacum* cv. White Burley	*Capsicum annuum* *Nicotiana glutinosa* *Vicia faba* *Zinnia elegans*
Artichoke vein banding? *Nepovirus*	AVBV	*Chenopodium amaranticolor* *C. quinoa* *Phaseolus vulgaris*	*Nicotiana glutinosa* *N. tabacum* cv. Xanthinc

Table 2. Suggested Species for Identification of Individual Viruses (continued)

Virus name[a]	Acronym[b]	Host plants	Nonhost plants
Artichoke yellow ringspot *Nepovirus*	AYRSV	*Chenopodium amaranticolor* C. murale* *C. quinoa* *Cucumis sativus* *Gomphrena globosa* *Nicotiana glutinosa*	*Beta vulgaris* *Brassica rapa* *Capsicum annuum* *Cucurbita maxima* *C. pepo* *Solanum melongena*
Asparagus virus 2 *Ilarvirus*	AV2	*Chenopodium quinoa* *Gomphrena globosa* *Nicotiana tabacum* cv. Havana 425	*Ammi majus* *Cucumis melo* *C. sativus* *Datura stramonium*
Barley yellow mosaic *Potyvirus* (fungus)	BaYMV	*Hordeum vulgare*	*Chenopodium amaranticolor* *Nicotiana tabacum*
Bean common mosaic *Potyvirus* (aphid)	BCMV	*Chenopodium quinoa* *Nicotiana benthamiana* *N. clevelandii* *Phaseolus vulgaris*	*Cucumis sativus* *Datura stramonium*
Bean golden mosaic *Geminivirus* (II)	BGMV	*Phaseolus vulgaris*	*Glycine max* *Rhynchosia minima*
Bean mild mosaic ? *Carmovirus*	BMMV	*Dolichos lablab* *Phaseolus vulgaris* cv. Pinto	*Phaseolus lunatus* *Vicia* spp. *Vigna unguiculata*
Bean pod mottle *Comovirus*	BPMV	*Chenopodium quinoa* *Phaseolus vulgaris* cvs. Pinto, Bountiful	*Cucumis sativus* *Nicotiana glutinosa*
Bean rugose mosaic *Comovirus*	BRMV	*Chenopodium amaranticolor* *Phaseolus vulgaris* cv. Plentiful *Vicia faba*	*Cucumis sativus* *Nicotiana tabacum* cv. White Burley
Bean yellow mosaic *Potyvirus* (aphid)	BYMV	*Chenopodium quinoa* *Nicotiana clevelandii* *Phaseolus vulgaris* *Vicia faba*	*Cucumis sativus* *Datura stramonium* *Pisum sativum* cv. Perfection
Bearded iris mosaic *Potyvirus* (aphid)	BIMV	*Belamcanda chinensis* *Iris* spp.	*Chenopodium amaranticolor* *Cucumis sativus* *Nicotiana megalosiphon* *Vigna sinensis*
Beet mosaic *Potyvirus* (aphid)	BtMV	*Beta vulgaris* *Chenopodium quinoa* *Spinacia oleracea*	*Cucumis sativus* *Datura stramonium* *Nicotiana glutinosa*
Beet necrotic yellow vein *Furovirus*	BNYVV	*Beta macrocarpa* *Chenopodium quinoa* *Nicotiana benthamiana* *Tetragonia expansa* (syn.: *T. tetragonoides*)	*Cucumis sativus* *Datura stramonium* *Nicotiana glutinosa* *Phaseolus vulgaris* *Vicia faba*
Beet yellow *Closterovirus*	BYV	*Beta vulgaris* *Chenopodium capitatum* *C. foliosum*	*Datura stramonium* *Nicotiana glutinosa* *Phaseolus vulgaris* *Vicia faba* *Vigna sinensis*
Belladonna mottle *Tymovirus*	BeMV	*Datura stramonium* *Nicotiana tabacum* cv. Samsun	*Ammi majus* *Chenopodium amaranticolor*

Table 2. Suggested Species for Identification of Individual Viruses (continued)

Virus name[a]	Acronym[b]	Host plants	Nonhost plants
			Cucurbita pepo
			Phaseolus vulgaris cv. Pinto
Bidens mottle *Potyvirus* (aphid)	BiMoV	*Chenopodium quinoa* *Nicotiana glutinosa* × *N. clevelandii*	*Lactuca sativa* cv. Valmaine *Pisum sativum*
Black raspberry latent *Ilarvirus*	BRLV	*Chenopodium quinoa* *Gomphrena globosa* *Nicotiana tabacum* cv. White Burley	*Pisum sativum* *Vigna sesquipedalis*
Blackeye cowpea mosaic *Potyvirus* (aphid)	BlCMV	*Chenopodium quinoa* *Glycine max* *Nicotiana benthamiana* *Vigna unguiculata* subsp. unguiculata	*Cucumis sativus* *Nicotiana glutinosa* *N. tabacum*
Blackgram mottle ? *Carmovirus*	BMoV	*Phaseolus aureus* *P. vulgaris* cvs. Black Valentine, Pinto	*Chenopodium* spp. *Nicotiana clevelandii*
Blueberry leaf mottle *Nepovirus*	BLMV	*Chenopodium quinoa* *Nicotiana clevelandii*	*Cucurbita maxima* *Datura stramonium* *Petunia hybrida* *Vigna unguiculata* *Zinnia elegans*
Blueberry red ringspot *Caulimovirus*	BRRV	*Vaccinium* spp.	*Chenopodium quinoa* *Datura stramonium* *Nicotiana clevelandii* *Vigna sinensis*
Broad bean mottle *Bromovirus*	BBMV	*Chenopodium amaranticolor* *Nicotiana clevelandii* *Vicia faba*	*Nicotiana tabacum*
Broad bean necrosis? *Furovirus*	BBNV	*Chenopodium quinoa* *Nicotiana clevelandii* *Pisum sativum* *Ficia faba*	*Cucumis sativus* *Gomphrena globosa* *Nicotiana glutinosa* *Vigna radiata*
Broad bean stain *Comovirus*	BBSV	*Chenopodium quinoa* *Phaseolus vulgaris* cv. Tendergreen	*Gomphrena globosa* *Nicotiana clevelandii*
Broad bean true mosaic *Comovirus*	BBTMV	*Lathyrus odoratus* *Phaseolus vulgaris* *Vicia faba*	*Chenopodium quinoa* *Gomphrena globosa* *Nicotiana clevelandii*
Brome mosaic *Bromovirus*	BMV	*Chenopodium amaranticolor* *C. hybridum* *C. quinoa*	*Gomphrena globosa* *Nicotiana glutinosa* *Nicotiana tabacum* cv. Samsun *Petunia hybrida*
Cacao necrosis *Nepovirus*	CNV	*Chenopodium quinoa* *Phaseolus vulgaris* cv. The Prince	*Datura stramonium* *Physalis floridana* *Vicia faba*
Cacao yellow mosaic *Tymovirus*	CYMV	*Chenopodium amaranticolor* *C. quinoa* *Nicandra physaloides* *Nicotiana clevelandii*	*Datura stramonium* *Gomphrena globosa* *Lycopersicon esculentum* *Nicotiana glutinosa* *Petunia hybrida*
Cactus X *Potexvirus*	CVX	*Chenopodium quinoa* *Gomphrena globosa*	*Nicotiana* spp.

Table 2. Suggested Species for Identification of Individual Viruses (continued)

Virus name[a]	Acronym[b]	Host plants	Nonhost plants
Carnation etched ring *Caulimovirus*	CERV	*Saponaria vaccaria* var. Pink Beauty *Silene armeria*	*Nicotiana glutinosa*
Carnation latent *Carlavirus*	CLV	*Chenopodium amaranticolor* *C. quinoa*	*Nicotiana glutinosa* *Lycopersicon esculentum*
Carnation mottle *Carmovirus*	CarMV	*Chenopodium amaranticolor* *Chenopodium quinoa* *Dianthus caryophyllus* *Gomphrena globosa* *Tetragonia expansa*	*Datura stramonium* *Nicotiana glutinosa* *Pisum sativum* *Vigna sinensis* *V. sinensis* cv. Blackeye
Carnation necrotic fleck *Closterovirus*	CNFV	*Dianthus barbatus* *D. caryophyllus*	*Chenopodium amaranticolor* *Cucumis sativus* *Datura stramonium* *Nicotiana glutinosa* *Phaseolus vulgaris*
Carnation ringspot *Dianthovirus*	CRSV	*Chenopodium amaranticolor* *Chenopodium quinoa* *Dianthus barbatus* *Gomphrena globosa*	*Glycine max* *Pisum sativum*
Carnation vein mottle *Potyvirus* (aphid)	CVMV	*Chenopodium amaranticolor* *C. quinoa* *Dianthus barbatus*	*Gomphrena globosa* *Phaseolus vulgaris*
Carrot thin leaf *Potyvirus* (aphid)	CTLV	*Chenopodium amaranticolor* *C. murale* *C. quinoa* *Daucus carota* ssp. sativa *Nicotiana clevelandii*	*Cucumis sativus* *Datura stramonium* *Gomphrena globosa* *Nicotiana glutinosa* *Phaseolus vulgaris* *Tetragonia expansa* *Vigna sinensis*
Cassava common mosaic *Potexvirus*	CsCMV	*Chenopodium amaranticolor* *C. quinoa* *Manihot esculenta*	*Phaseolus vulgaris* *Vigna sinensis*
Cassia yellow blotch *Bromovirus*	CYBV	*Chenopodium amaranticolor* *Nicotiana clevelandii*	*Datura stramonium* *Pisum sativum* *Vigna unguiculata*
Cauliflower mosaic *Caulimovirus*	CaMV	*Brassica campestris* *B. rapa* *Crambe abyssinica*	*Chenopodium amaranticolor* *C. quinoa* *Cucumis sativus* *Datura stramonium* *Gomphrena globosa* *Nicotiana glutinosa* *N. tabacum* cv. Xanthi-nc *Petunia hybrida* *Tetragonia expansa*
Celery mosaic *Potyvirus* (aphid)	CeMV	*Apium graveolens* var. dulce *Pastinaca sativa*	*Datura stramonium* *Gomphrena globosa* *Nicotiana clevelandii* *Ocimum basilicum*

Table 2. Suggested Species for Identification of Individual Viruses (continued)

Virus name[a]	Acronym[b]	Host plants	Nonhost plants
Cherry leaf roll *Nepovirus*	CLRV	*Chenopodium quinoa* *Cucumis sativus* *Nicotiana tabacum* cv. White Burley	*Cucurbita pepo* *Nicotiana glutinosa* *Petunia hybrida* *Solanum rostratum*
Cherry rasp leaf? *Nepovirus*	CRLV	*Chenopodium amaranticolor* *C. quinoa* *Cucumis sativus*	*Solanum melongena* *Spinacia oleracea* *Verbascum thapsus*
Chicory yellow mottle *Nepovirus*	ChYMV	*Chenopodium quinoa* *Cucurbita pepo* *Phaseolus vulgaris*	*Pisum sativum* *Vicia faba*
Chrysanthemum B *Carlavirus*	CVB	*Nicotiana clevelandii* *Petunia hybrida* *Tetragonia expansa*	*Chenopodium amaranticolor* *C. quinoa* *Cucumis sativus* *Gomphrena globosa* *Vigna sinensis*
Citrus leaf rugose *Ilarvirus* (II)	CiLRV	*Chenopodium quinoa* *Cucumis sativus*	*Chenopodium album* *Datura stramonium* *Phaseolus lunatus*
Clitoria yellow vein *Tymovirus*	CYVV	*Nicotiana clevelandii* *Phaseolus vulgaris* cv. Long Tom	*Chenopodium amaranticolor* *C. quinoa* *Datura stramonium* *Gomphrena globosa* *Nicotiana glutinosa*
Clover yellow mosaic *Potexvirus*	ClYMV	*Chenopodium amaranticolor* *Gomphrena globosa* *Phaseolus aureus*	*Datura stramonium* *Nicotiana glutinosa* *N. tabacum*
Clover yellow vein *Potyvirus* (aphid)	ClYVV	*Chenopodium quinoa* *Nicotiana clevelandii*	*Vigna sinensis*
Cocksfoot mild mosaic ? *Sobemovirus*	CMMV	*Dactylis glomerata* *Lolium persicum*	*Festuca pratensis* *Lolium perenne* *Nicotiana glutinosa*
Cocksfoot mottle *Sobemovirus*	CoMV	*Dactylis glomerata* *Hordeum vulgare*	*Chenopodium quinoa* *Nicotiana clevelandii* *N. glutinosa*
Cocksfoot streak *Potyvirus* (aphid)	CSV	*Dactylis glomerata* *Paspalum membranaceum*	*Agropyron intermedium* *Datura stramonium* *Gomphrena globosa* *Nicotiana glutinosa*
Cowpea aphid-borne mosaic *Potyvirus* (aphid)	CABMV	*Chenopodium amaranticolor* *Glycine max* *Phaseolus vulgaris* cv. Bataaf *Petunia hybrida* *Vigna unguiculata*	*Datura stramonium* *Nicotiana glutinosa* *Vicia faba*
Cowpea chlorotic mottle *Bromovirus*	CCMV	*Chenopodium hybridum* *Vigna unguiculata*	*Lycopersicon esculentum* *Nicotiana glutinosa* *Phaseolus lunatus* cv. Henderson
Cowpea mild mottle *Carlavirus*	CPMMV	*Nicotiana clevelandii* *Phaseolus vulgaris* cv. The Prince	*Cucumis sativus* *Datura stramonium* *Nicotiana glutinosa*

Table 2. Suggested Species for Identification of Individual Viruses (continued)

Virus name[a]	Acronym[b]	Host plants	Nonhost plants
Cowpea mosaic *Comovirus*	CPMV	*Chenopodium amaranticolor* *Vigna unguiculata*	*Cucumis sativus*
Cowpea mottle ? *Carmovirus*	CPMoV	*Chenopodium quinoa* *Phaseolus vulgaris* *Vigna unguiculata*	*Cucumis sativus* *Datura stramonium* *Nicotiana clevelandii* *Vicia faba*
Cowpea severe mosaic *Comovirus*	CPSMV	*Chenopodium amaranticolor* *Vigna unguiculata*	*Medicago sativa*
Cucumber green mottle mosaic *Tobamovirus*	CGMMV	*Chenopodium amaranticolor* *Cucumis sativus* *Cucurbita pepo* convar. patissonina f. radiata	*Datura stramonium* *Nicotiana glutinosa* *Petunia hybrida* *Vigna sinensis*
Cucumber leaf spot ? *Carmovirus*	CLSV	*Chenopodium quinoa* *Cucumis sativus* *Nicotiana benthamiana*	*Datura stramonium* *Nicotiana glutinosa* *N. tabacum* cv. Samsun *Phaseolus vulgaris* *Vicia faba*
Cucumber mosaic *Cucumovirus*	CMV	*Chenopodium quinoa* *Nicotiana benthamiana* *N. clevelandii* *N. megalosiphon* *Vigna unguiculata*	*Beta vulgaris*
Cucumber necrosis *Tombusvirus*	CuNV	*Chenopodium amaranticolor* *Cucumis sativus* *Gomphrena globosa*	Differential nonhosts are not known
Cymbidium mosaic *Potexvirus*	CyMV	*Chenopodium quinoa* *Cymbidium* spp. *Datura stramonium*	*Lycopersicon esculentum* *Nicotiana glutinosa* *Phaseolus vulgaris*
Cymbidium ringspot *Tombusvirus*	CyRSV	*Chenopodium quinoa* *Nicotiana clevelandii* *N. glutinosa* *Vigna unguiculata*	*Cucumis melo* *C. sativus* *Pisum sativum* *Spinacia oleracea*
Dahlia mosaic *Caulimovirus*	DMV	*Verbesina encelioides* *Zinnia elegans*	*Chenopodium amaranticolor* *Gomphrena globosa* *Nicotiana glutinosa* *Phaseolus vulgaris* *Tetragonia expansa*
Daphne X *Potexvirus*	DVX	*Chenopodium quinoa* *Cucumis melo* *Gomphrena globosa* *Nicotiana clevelandii*	*Capsicum frutescens* *Vicia faba* *Zinnia elegans*
Dasheen mosaic *Potyvirus* (aphid)	DsMV	*Philodendron selloum*	*Capsicum annuum* *Chenopodium amaranticolor* *Cucurbita pepo* *Datura stramonium* *Gomphrena globosa* *Nicotiana tabacum* cv. Samsun *Phaseolus vulgaris* cv. Red Kidney *Vigna sinensis*

Table 2. Suggested Species for Identification of Individual Viruses (continued)

Virus name[a]	Acronym[b]	Host plants	Nonhost plants
Desmodium yellow mottle *Tymovirus*	DYMV	*Phaseolus vulgaris* cv. Great Northern *Vigna sinensis*	*Chenopodium quinoa* *Nicotiana glutinosa*
Dioscorea latent *Potexvirus*	DLV	*Dioscorea composita* *Nicotiana benthamiana* *N. megalosiphon*	*Nicotiana glutinosa* *N. tabacum*
Eggplant mosaic *Tymovirus*	EMV	*Chenopodium amaranticolor* *C. quinoa* *Nicotiana clevelandii* *Vigna sinensis*	*Phaseolus vulgaris* cv. The Prince
Elderberry carla *Carlavirus*	ECV	*Chenopodium amaranticolor* *C. quinoa* *Gomphrena globosa*	*Datura metel* *Datura stramonium* *Nicotiana benthamiana* *N. clevelandii* *N. tabacum* cv. Xanthinc
Elderberry latent ? *Carmovirus*	ELV	*Chenopodium quinoa* *Cucumis sativus* *Datura stramonium* *Gomphrena globosa* *Nicotiana benthamiana* *N. clevelandii* *N. glutinosa*	*Phaseolus vulgaris* cv. The Prince *Pisum sativum*
Elm mottle *Ilarvirus* (II)	EMoV	*Chenopodium murale* *C. quinoa* *Cucumis sativus* *Nicotiana clevelandii* *N. glutinosa* *N. megalosiphon*	*Brassica oleracea* *Citrullus lanatus* *Tropaeolum majus*
Erysimum latent *Tymovirus*	ErLV	*Brassica chinensis* *Sinapis alba*	*Chenopodium amaranticolor* *C. quinoa* *Nicotiana clevelandii* *Vigna sinensis*
Foxtail mosaic *Potexvirus*	FoMV	*Chenopodium amaranticolor* *Gomphrena globosa* *Nicotiana clevelandii* *N. megalosiphon*	*Agropyron smithii* *Commelina* spp. *Cyperus esculentus* *Sorghum halapense*
Frangipani mosaic *Tobamovirus*	FrMV	*Chenopodium quinoa* *Datura stramonium* *Nicotiana glutinosa*	*Chenopodium amaranticolor* *Cucumis sativus* *Phaseolus vulgaris*
Galinsoga mosaic *Carmovirus*	GaMV	*Chenopodium amaranticolor* *C. quinoa* *Spinacia oleracea*	*Cucumis sativus* *Datura stramonium* *Vicia faba*
Ginger chlorotic fleck? *Sobemovirus*	GCFV	*Zingiber officinale*	*Chenopodium amaranticolor* *C. quinoa* *Nicotiana glutinosa* *N. tabacum*
Grapevine Bulgarian latent *Nepovirus*	GBLV	*Chenopodium quinoa* *Gomphrena globosa* *Nicotiana clevelandii*	*Cucumis sativus* *Nicotiana tabacum* cv. Havana 423 *Phaseolus vulgaris* *Vigna unguiculata*

Table 2. Suggested Species for Identification of Individual Viruses (continued)

Virus name[a]	Acronym[b]	Host plants	Nonhost plants
Grapevine chrome mosaic *Nepovirus*	GCMV	*Chenopodium quinoa* *Gomphrena globosa* *Nicotiana clevelandii* *Phaseolus vulgaris*	*Cucurbita pepo* *Nicotiana glutinosa* *N. tabacum* cv. Samsun *Petunia hybrida*
Grapevine fanleaf *Nepovirus*	GFLV	*Chenopodium quinoa* *Cucumis sativus* *Gomphrena globosa* *Nicotiana benthamiana*	*Cucurbita pepo* *Datura stramonium* *Nicotiana glutinosa* *Petunia hybrida*
Guinea grass mosaic *Potyvirus* (aphid)	GGMV	*Setaria italica*	*Chenopodium quinoa* *Datura stramonium* *Gomphrena globosa* *Nicotiana clevelandii* *Phaseolus vulgaris* cv. Pinto *Vigna sinensis*
Helenium S *Carlavirus*	HVS	*Chenopodium amaranticolor* *C. quinoa*	*Cucumis sativus* *Nicotiana clevelandii* *Phaseolus vulgaris* cv. Red Kidney *Vigna unguiculata*
Henbane mosaic *Potyvirus* (aphid)	HMV	*Chenopodium amaranticolor* *Datura stramonium* *Nicotiana sylvestris* *N. tabacum* cv. Xanthi-nc	*Cucumis sativus* *Gomphrena globosa* *Phaseolus vulgaris* *Vigna sinensis*
Heracleum latent ? *Closterovirus*	HLV	*Chenopodium murale* *C. quinoa*	*Nicotiana clevelandii* *N. tabacum*
Hibiscus chlorotic ringspot *Carmovirus*	HCRSV	*Chenopodium quinoa* *Hibiscus cannabinus* *H. rosa-sinensis*	*Amaranthus* spp. *Capsicum* spp. *Cucumis* spp. *Datura* spp. *Nicotiana* spp. *Tetragonia expansa*
Hibiscus latent ringspot *Nepovirus*	HLRSV	*Chenopodium amaranticolor* *C. murale* *C. quinoa* *Gomphrena globosa* *Nicotiana clevelandii*	*Nicotiana sylvestris* *Phaseolus vulgaris* *Physalis floridana* *Solanum demissum* *Vigna unguiculata*
Hippeastrum mosaic *Potyvirus* (aphid)	HiMV	*Chenopodium quinoa* *Gomphrena globosa* *Hippeastrum hybridum* *Nicotiana clevelandii*	*Datura stramonium* *Nicotiana glutinosa* *Tetragonia expansa* *Vicia faba*
Honeysuckle latent *Carlavirus*	HnLV	*Chenopodium quinoa* *Nicotiana clevelandii*	*Nicotiana tabacum* cv. White Burley
Hop latent *Carlavirus*	HpLV	*Chenopodium murale* *Humulus lupulus* *Phaseolus vulgaris*	*Chenopodium amaranticolor* *C. quinoa* *Cucumis sativus* *Datura stramonium* *Gomphrena globosa* *Nicotiana clevelandii* *N. debneyi* *N. glutinosa* *N. rustica* *Pisum sativum*

Table 2. Suggested Species for Identification of Individual Viruses (continued)

Virus name[a]	Acronym[b]	Host plants	Nonhost plants
Hop mosaic *Carlavirus*	HpMV	*Chenopodium album* *C. amaranticolor* *C. murale* *C. quinoa* *Nicotiana clevelandii* *N. debneyi*	*Gomphrena globosa* *Nicotiana sylvestris* *N. tabacum* cv. White Burley *Phaseolus vulgaris* cv. Kinghorn Wax *Tetragonia expansa* *Vicia faba* *Vigna unguiculata* cv. Blackeye
Hydrangea ringspot *Potexvirus*	HRSV	*Chenopodium quinoa* *Gomphrena globosa* *Primula malacoides*	*Nicotiana glutinosa* *N. tabacum* cv. Samsun
Hypochoeris mosaic ? *Furovirus*	HyMV	*Chenopodium* *amaranticolor* *C. quinoa* *Cucumis sativus* *Gomphrena globosa* *Nicotiana clevelandii*	*Datura stramonium* *Nicotiana benthamiana* *Vigna unguiculata*
Iris fulva mosaic *Potyvirus* (aphid)	IFMV	*Belamcanda chinensis* *Chenopodium quinoa*	*Chenopodium* *amaranticolor* *Gomphrena globosa* *Nicotiana clevelandii* *Phaseolus vulgaris* cv. Bountiful *Tetragonia expansa* *Vigna unguiculata*
Iris mild mosaic *Potyvirus* (aphid)	IMMV	*Chenopodium quinoa* *Iris* × *hollandica* cv. Wedgwood *Nicotiana clevelandii*	*Datura stramonium* *Gomphrena globosa* *Petunia hybrida*
Iris severe mosaic *Potyvirus* (aphid)	ISMV	*Belamcanda chinensis* *Crocus* spp. *Iris* spp.	*Chenopodium* *amaranticolor* *C. quinoa*
Kennedya yellow mosaic *Tymovirus*	KYMV	*Datura stramonium* *Kennedya rubicunda* *Phaseolus aureus*	*Lycopersicon esculentum* *Nicotiana clevelandii*
Leck yellow stripe *Potyvirus* (aphid)	LYSV	*Allium porrum* *Chenopodium* *amaranticolor* *C. quinoa*	*Allium fistulosum* *Nicotiana clevelandii* *N. megalosiphon* *Petunia hybrida* *Phaseolus vulgaris* cv. Bataaf *Tetragonia expansa*
Lettuce mosaic *Potyvirus* (aphid)	LMV	*Carthamus tinctorius* *Chenopodium quinoa* *Gomphrena globosa*	*Datura stramonium* *Nicotiana glutinosa* *Petunia hybrida* *Phaseolus vulgaris*
Lilac chlorotic leafspot *Closterovirus*	LCLV	*Chenopodium quinoa* *Nicotiana clevelandii*	*Beta vulgaris* cv. Monobush
Lilac ring mottle *Ilarvirus*	LRMV	*Chenopodium quinoa* *Nicotiana glutinosa* *N. tabacum* cv. White Burley	*Amaranthus caudatus* *Cucumis sativus* *Lycopersicon esculentum*
Lily symptomless *Carlavirus*	LSV	*Lilium longiflorum*	*Nicotiana clevelandii* *N. glutinosa* *Phaseolus vulgaris*

Table 2. Suggested Species for Identification of Individual Viruses (continued)

Virus name[a]	Acronym[b]	Host plants	Nonhost plants
Lucerne Australian latent *Nepovirus*	LALV	*Chenopodium amaranticolor* *C. quinoa* *Gomphrena globosa*	*Cucurbita maxima* cv. Buttercrop *Vicia faba*
Lucerne transient streak *Sobemovirus*	LTSV	*Chenopodium amaranticolor* *C. quinoa* *Nicotiana clevelandii* *Pisum sativum*	*Cucumis sativus* *Gomphrena globosa* *Nicotiana glutinosa* *Phaseolus vulgaris* *Vigna unguiculata*
Maclura mosaic ? *Potyvirus* (aphid)	MacMV	*Chenopodium amaranticolor* *Nicotiana clevelandii* *Tetragonia expansa*	*Cucumis sativus* *Gomphrena globosa* *Nicotiana megalosiphon* *Petunia hybrida* *Vigna sinensis*
Maize chlorotic mottle ? *Sobemovirus*	MCMV	*Zea mays*	*Sorghum bicolor* cv. Asgrow Bugoff
Melandrium yellow fleck *Bromovirus*	MYFV	*Chenopodium quinoa* *Gomphrena globosa* *Nicotiana clevelandii* *Tetragonia expansa*	*Brassica oleracea* *Dianthus caryophyllus* cv. William Sim *Trifolium incarnatum*
Melon necrotic spot *Carmovirus*	MNSV	*Cucumis anguria* var. longipes *C. melo* *C. sativus*	*Lagenaria siceraria* *Vigna unguiculata* spp. sinensis cv. Blackeye
Mulberry ringspot *Nepovirus*	MRSV	*Chenopodium quinoa* *Nicotiana clevelandii* *Vigna sinensis* cv. Blackeye	*Chenopodium amaranticolor* *Nicotiana glutinosa*
Mung bean yellow mosaic *Geminivirus* (II)	MYMV	*Phaseolus vulgaris* cv. Top Crop *Vigna radiata*	*Lablab purpureus* *Phaseolus lunatus* *Vigna unguiculata*
Myrobalan latent ringspot *Nepovirus*	MLRSV	*Chenopodium quinoa* *Nicotiana clevelandii*	*Zinnia elegans*
Narcissus latent *Carlavirus*	NLV	*Nicotiana benthamiana* *N. clevelandii* *Tetragonia expansa*	*Nicotiana tabacum* cv. White Burley *Petunia hybrida* *Vicia faba*
Narcissus mosaic *Potexvirus*	NMV	*Chenopodium amaranticolor* *Gomphrena globosa* *Nicotiana clevelandii*	*Nicotiana glutinosa* *N. tabacum* cv. White Burley
Narcissus tip necrosis ? *Carmovirus*	NTNV	*Narcissus* cv. Barett Browning	*Chenopodium quinoa* *Nicotiana tabacum*
Narcissus yellow stripe *Potyvirus* (aphid)	NYSV	*Narcissus pseudonarcissus* *Tetragonia expansa*	Differential nonhosts are not known
Nerine X *Potexvirus*	NVX	*Chenopodium amaranticolor* *C. quinoa* *Gomphrena globosa* *Tetragonia expansa*	*Cucumis sativus* *Vicia faba* *Zinnia elegans*
Nicotiana velutina mosaic ? *Furovirus*	NVMV	*Chenopodium quinoa* *Gomphrena globosa* *Nicotiana glutinosa*	*Tetragonia expansa*
Oat mosaic *Potyvirus* (fungus)	OMV	*Avena sativa*	*Hordeum vulgare* *Triticum aestivum*

Table 2. Suggested Species for Identification of Individual Viruses (continued)

Virus name[a]	Acronym[b]	Host plants	Nonhost plants
Oat necrotic mottle *Potyvirus* (? mite)	ONMV	*Avena sativa* *Lolium multiflorum*	*Hordeum vulgare* *Secale cereale* *Triticum aestivum*
Odontoglossum ringspot *Tobamovirus*	ORSV	*Chenopodium quinoa* *Nicotiana clevelandii* *N. tabacum* cv. Xanthi- nc	*Datura stramonium*
Okra mosaic *Tymovirus*	OkMV	*Chenopodium amaranticolor* *Cucumis sativus* *Nicotiana clevelandii* *Vigna sinensis*	*Amaranthus caudatus* *Capsicum annuum* *Datura stramonium* *Nicotiana glutinosa* *N. tabacum* cv. Xanthi- nc *Phaseolus vulgaris* cv. The Prince *Zinnia elegans*
Olive latent ringspot *Nepovirus*	OLRSV	*Chenopodium amaranticolor* *C. quinoa* *Gomphrena globosa*	*Nicotiana glutinosa* *N. tabacum* cv. Samsun *N. tabacum* cv. White Burley *Phaseolus aureus*
Onion yellow dwarf *Potyvirus* (aphid)	OYDV	*Allium cepa*	*Allium porrum*
Panicum mosaic ? *Sobemovirus*	PMV	*Setaria italica* *Zea mays*	*Chenopodium amaranticolor* *C. quinoa* *Cucumis melo* *Datura stramonium* *Gomphrena globosa* *Nicotiana glutinosa* *Tetragonia expansa*
Papaya mosaic *Potexvirus*	PapMV	*Carica papaya* *Chenopodium amaranticolor* *Gomphrena globosa*	*Datura stramonium* *Nicotiana glutinosa* *Phaseolus vulgaris* *Tetragonia expansa* *Vigna sinensis*
Papaya ringspot (= watermelon mosaic virus 1) *Potyvirus* (aphid)	PRSV	*Cucurbita pepo* *Luffa acutangula*	*Nicotiana benthamiana*
Parsnip mosaic *Potyvirus* (aphid)	ParMV	*Chenopodium amaranticolor* *C. quinoa* *Pastinaca sativa*	*Cucumis melo* *C. sativus* *Datura stramonium* *Nicotiana clevelandii* *N. glutinosa* *N. tabacum* cv. Samsun *Phaseolus vulgaris* *Tetragonia expansa* *Vigna sinensis*
Passionfruit woodiness *Potyvirus* (aphid)	PWV	*Chenopodium amaranticolor* *Glycine max* *Passiflora edulis* *Phaseolus vulgaris*	*Datura stramonium* *Nicotiana glutinosa* *Vigna unguiculata*
Pea early-browning *Tobravirus*	PEBV	*Chenopodium amaranticolor*	*Lycopersicon esculentum* *Medicago sativa*

Table 2. Suggested Species for Identification of Individual Viruses (continued)

Virus name[a]	Acronym[b]	Host plants	Nonhost plants
Pea enation mosaic virus[c]	PEMV	*Nicotiana clevelandii* *Vicia faba* *Chenopodium amaranticolor* *C. quinoa* *Nicotiana clevelandii* *Pisum sativum* *Trifolium incarnatum* *Vicia faba*	*Cucumis sativus* *Datura stramonium* *Nicotiana glutinosa* *Tetragonia expansa*
Pea seed-borne mosaic *Potyvirus* (aphid)	PSbMV	*Chenopodium amaranticolor* *C. quinoa* *Pisum sativum* *Vicia faba*	*Cucumis sativus* *Datura stramonium* *Nicotiana glutinosa* *N. tabacum* cv. Xanthi-nc *Phaseolus vulgaris*
Pea streak *Carlavirus*	PeSV	*Chenopodium amaranticolor* *Gomphrena globosa* *Pisum sativum* *Vicia faba*	*Cucumis sativus* *Nicotiana glutinosa* *N. tabacum* *Petunia hybrida*
Peach rosette mosaic *Nepovirus*	PRMV	*Chenopodium amaranticolor* *C. quinoa*	*Vitis labrusca* cv. Delaware
Peanut clump *Furovirus*	PCV	*Chenopodium amaranticolor* *Nicotiana benthamiana* *N. glutinosa*	*Cucumis melo* *Gomphrena globosa* *Tetragonia expansa* *Vicia faba*
Peanut mottle *Potyvirus* (aphid)	PeMoV	*Chenopodium amaranticolor* *Nicotiana clevelandii* *Phaseolus vulgaris* cv. Prince	*Cucumis sativus* *Gomphrena globosa*
Peanut stunt *Cucumovirus*	PSV	*Chenopodium quinoa* *Gomphrena globosa* *Nicotiana megalosiphon* *Phaseolus vulgaris* *Vigna sinensis*	*Brassica napus* *Glycine max* *Medicago sativa*
Pelargonium flower-break *Carmovirus*	PFBV	*Chenopodium quinoa* *Gomphrena globosa* *Nicotiana clevelandii*	*Pelargonium domesticum* cv. Nittany Lion
Pepper mild mottle *Tobamovirus*	PMMV	*Capsicum* spp. *Chenopodium amaranticolor* *C. quinoa* *Datura stramonium* *Nicotiana sylvestris*	*Lycopersicon esculentum* *Nicotiana glauca*
Pepper mottle *Potyvirus* (aphid)	PepMoV	*Capsicum frutescens* cv. Tabasco *Chenopodium amaranticolor* *Nicotiana tabacum* cv. Xanthi-nc	*Nicotina tabacum* cv. V-20
Pepper veinal mottle *Potyvirus* (aphid)	PVMV	*Chenopodium amaranticolor* *C. quinoa* *Nicotiana clevelandii* *N. megalosiphon*	*Cucumis sativus* *Gomphrena globosa* *Phaseolus vulgaris* *Vicia faba*

Table 2. Suggested Species for Identification of Individual Viruses (continued)

Virus name[a]	Acronym[b]	Host plants	Nonhost plants
Peru tomato *Potyvirus* (aphid)	PTV	*Chenopodium quinoa* *N. occidentalis* *N. tabacum* cv. Burley 21	*Solanum demissum* × *S. tuberosum* A6 hybrid
Plantain X *Potexvirus*	P1VX	*Nicotiana benthamiana* *N. clevelandii* *Nicotiana* × *edwarsonii*	*Chenopodium quinoa* *Gomphrena globosa* *Nicotiana glutinosa*
Plum pox *Potyvirus* (aphid)	PPV	*Chenopodium foetidum* *Nicotiana clevelandii*	*Cucumis sativus* cv. Lange Gele Tros
Poinsettia mosaic ? *Tymovirus*	PnMV	*Nicotiana benthamiana* *N. megalosiphon*	*Chenopodium amaranticolor* *C. quinoa* *Datura stramonium* *Gomphrena globosa* *Nicotiana clevelandii* *N. glutinosa*
Pokeweed mosaic *Potyvirus* (aphid)	PkMV	*Chenopodium quinoa* *Gomphrena globosa* *Phytolacca americana*	*Cucumis sativus* *Datura stramonium* *Nicotiana clevelandii* *N. glutinosa* *Phaseolus vulgaris* *Vigna sinensis*
Poplar mosaic *Carlavirus*	PopMV	*Nicotiana glutinosa* *N. megalosiphon* *Vigna sinensis*	*Chenopodium amaranticolor* *C. quinoa* *Phaseolus vulgaris*
Potato A *Potyvirus* (aphid)	PVA	*Lycopersicon pimpinellifolium* *Nicandra physaloides* *Nicotiana tabacum* cv. Samsun *Solanum demissum* × *S. tuberosum* A6 hybrid	*Datura metel* *Vigna sinensis*
Potato aucuba mosaic ? *Potexvirus*	PAMV	*Capsicum annuum* *Nicotiana glutinosa*	*Cucumis sativus* *Vigna sinensis* *Zinnia elegans*
Potato black ringspot *Nepovirus*	PBRSV	*Chenopodium amaranticolor* *C. quinoa* *Cucumis sativus* *Datura stramonium* *Gomphrena globosa* *Nicotiana benthamiana* *N. clevelandii* *N. rustica* *N. tabacum* cv. Xanthi-nc *Phaseolus vulgaris* cv. The Prince	*Solanum phureja* CPC 4110 *S. tuberosum* spp. andigena CPC 1801
Potato M *Carlavirus*	PVM	*Chenopodium quinoa* *Datura metel* *Nicotiana debneyi* *Solanum rostratum*	*Cucurbita pepo* *Nicotiana tabacum* cv. Xanthi-nc *Petunia hybrida* *Tinantia erecta*
Potato mop-top *Furovirus*	PMTV	*Chenopodium amaranticolor* *C. quinoa*	*Chenopodium capitatum* *Cucumis sativus* *Gomphrena globosa*

Table 2. Suggested Species for Identification of Individual Viruses (continued)

Virus name[a]	Acronym[b]	Host plants	Nonhost plants
		Datura stramonium	*Phaseolus vulgaris*
		Nicotiana benthamiana	*Vicia faba*
		N. clevelandii	*Vigna sinensis*
		N. glutinosa	
		N. megalosiphon	
		N. sylvestris	
		N. tabacum cv. Xanthi-nc	
		Tetragonia expansa	
Potato S *Carlavirus*	PVS	*Chenopodium amaranticolor*	*Cucurbita pepo*
			Nicotiana tabacum cv. Xanthi-nc
		C. quinoa	
		Nicotiana debneyi	*Petunia hybrida*
Potato V *Potyvirus* (aphid)	PVV	*Nicotiana clevelandii*	*Chenopodium amaranticolor*
		N. glutinosa	
		N. occidentalis	*C. quinoa*
		N. tabacum cv. White Burley	*Cucumis sativus*
			Datura stramonium
		Solanum tuberosum cv. Maris Piper	*Gomphrena globosa*
			Phaseolus vulgaris cv. Pinto
			Vigna unguiculata
Potato X *Potexvirus*	PVX	*Datura stramonium*	*Crambe abyssinica*
		Gomphrena globosa	*Cucurbita pepo*
			Cucumis sativus
			Glycine max
Potato Y *Potyvirus* (aphid)	PVY	*Chenopodium amaranticolor*	*Datura stramonium*
		C. quinoa	
		Lycium spp.	
		Nicotiana tabacum cv. Xanthi-nc	
		Physalis floridana	
		Solanum demissum × *S. tuberosum* A6 hybrid	
Prune dwarf *Ilarvirus* (III)	PDV	*Cucumis sativus*	*Chenopodium quinoa*
		Cucurbita maxima	*N. tabacum*
Prunus necrotic ringspot *Ilarvirus* (III)	PNRSV	*Chenopodium quinoa*	*Nicotiana langsdorfii*
		Cucumis sativus	*N. tabacum*
		Cucurbita maxima	*Solanum sysimbrifolium*
Quail pea mosaic *Comovirus*	QPMV	*Phaseolus vulgaris*	*Gomphrena globosa*
Radish mosaic *Comovirus*	RaMV	*Brassica campestris*	*Nicotiana tabacum* cv. Havana 425
		Chenopodium amaranticolor	
Raspberry bushy dwarf *Idaeovirus*	RBDV	*Chenopodium amaranticolor*	*Cucumis sativus* cv. Sperlings Mervita
		C.murale	*Lycopersicon esculentum*
		C. quinoa	*Nicotiana debneyi*
		Phaseolus vulgaris cv. The Prince	*N. sylvestris*
			N. tabacum cv. Samsun
			Vigna sinensis cv. Blackeye
Raspberry ringspot *Nepovirus*	RRSV	*Chenopodium amaranticolor*	*Vicia faba*
		C. quinoa	
		Nicotiana clevelandii	
		Petunia hybrida	

Table 2. Suggested Species for Identification of Individual Viruses (continued)

Virus name[a]	Acronym[b]	Host plants	Nonhost plants
Red clover mottle *Comovirus*	RCMV	*Chenopodium amaranticolor* *C. quinoa* *Gomphrena globosa* *Phaseolus vulgaris* *Vicia faba*	*Cucumis sativus* *Nicotiana glutinosa*
Red clover necrotic mosaic *Dianthovirus*	RCNMV	*Chenopodium quinoa* *Gomphrena globosa* *Nicotiana clevelandii* *Phaseolus vulgaris*	*Petunia hybrida*
Red clover vein mosaic *Carlavirus*	RCVMV	*Chenopodium amaranticolor* *C. quinoa* *Gomphrena globosa*	*Cucumis sativus* *Datura stramonium* *Nicotiana glutinosa* *Vigna radiata*
Ribgrass mosaic *Tobamovirus*	RMV	*Chenopodium amaranticolor* *Nicotiana glutinosa* *N. tabacum* cv. Turkish	*Phaseolus vulgaris*
Rice necrosis mosaic *Potyvirus* (fungus)	RNMV	*Oryza sativa*	*Chenopodium amaranticolor* *C. quinoa* *Nicotiana tabacum*
Rice yellow mottle *Sobemovirus*	RYMV	*Oryza sativa*	*Avena sativa* *Nicotiana glutinosa* *Secale cereale* *Zea mays*
Robinia mosaic *Cucumovirus*	RbMV	*Chenopodium murale* *C. quinoa* *Nicotiana glutinosa* *Vigna sinensis*	*Amaranthus retroflexus* *Ocimum basilicum*
Ryegrass mosaic *Potyvirus* (mite)	RGMV	*Lolium multiflorum* cv. S22	*Triticum aestivum*
Saguaro cactus *Carmovirus*	SCV	*Chenopodium amaranticolor* *C. capitatum* *C. quinoa* *Gomphrena globosa*	*Datura metel* *Nicotiana glutinosa* *Vigna sinensis* cv. Blackeye
Satsuma dwarf ? *Nepovirus*	SDV	*Chenopodium quinoa* *Physalis floridana*	*Citrus unshiu* cv. Mexican lime *C. unshiu* cv. Rough lemon
Scrophularia mottle *Tymovirus*	ScrMV	*Chenopodium quinoa* *Datura stramonium* *Vicia faba*	*Zea mays*
Shallot latent *Carlavirus*	SLV	*Allium porrum* *Chenopodium amaranticolor* *C. quinoa*	*Allium neapolitanum*
Soilborne wheat mosaic *Furovirus*	SBWMV	*Chenopodium amaranticolor* *C. quinoa* *Triticum aestivum* cv. Michigan Amber	*Nicotiana tabacum* *Phaseolus vulgaris*
Solanum nodiflorum mottle *Sobemovirus*	SNMV	*Nicotiana clevelandii* *N. debneyi* *N. velutina*	*Chenopodium amaranticolor* *C. quinoa*

Table 2. Suggested Species for Identification of Individual Viruses (continued)

Virus name[a]	Acronym[b]	Host plants	Nonhost plants
		Solanum nodiflorum	*Datura stramonium*
			Nicotiana glutinosa
			Phaseolus vulgaris
			Vicia faba
			Vigna unguiculata
Southern bean mosaic	SBMV	*Glycine max*	*Cassia tora*
Sobemovirus		*Phaseolus vulgaris*	*Cicer arietinum*
		Vigna sinensis	*Lupinus albus*
			Melilotus alba
Sowbane mosaic	SoMV	*Chenopodium*	*Cucumis sativus* cv.
Sobemovirus		*amaranticolor*	Delicatess
		C. quinoa	*Cucurbita maxima*
			Datura stramonium
			Nicotiana glutinosa
			N. megalosiphon
			N. tabacum cv. Xanthi-nc
			Phaseolus vulgaris cv. Pinto
			Vigna sinensis
Soybean chlorotic mottle	SbCMV	*Dolichos lablab*	*Chenopodium*
Caulimovirus		*Glycine max*	*amaranticolor*
		Phaseolus vulgaris	*C. quinoa*
		P. vulgaris cv. Kintoki	*Cucumis sativus*
			Datura stramonium
			Gomphrena globosa
			Nicotiana clevelandii
			N. tabacum
			Tetragonia expansa
Soybean mosaic	SbMV	*Chenopodium quinoa*	*Cucumis sativus*
Potyvirus (aphid)		*Glycine max*	*Datura stramonium*
		Phaseolus vulgaris	*Gomphrena globosa*
			Nicotiana tabacum
			Vicia faba
Spinach latent *Ilarvirus*	SpLV	*Chenopodium*	*Cucumis sativus*
		amaranticolor	*Datura stramonium*
		C. quinoa	*Lycopersicon esculentum*
			Petunia hybrida
			Phaseolus vulgaris cv. The Prince
Squash mosaic	SMV	*Cucumis sativus*	*Citrullus lanatus*
Comovirus		*Cucurbita pepo*	
Strawberry latent	SLRSV	*Chenopodium*	*Amaranthus caudatus*
ringspot ? *Nepovirus*		*amaranticolor*	*Cucurbita ficifolia*
		C. murale	
		C. quinoa	
		Cucumis melo	
		C. sativus	
		Datura stramonium	
		Tetragonia tetragonoides	
Subterranean clover	SCMoV	*Pisum sativum*	*Chenopodium*
mottle *Sobemovirus*		*Trifolium subterraneum*	*amaranticolor*
			C. quinoa
			Nicotiana clevelandii
Sugarcane mosaic	SCMV	*Sorghum bicolor* cv. Rio	*Chenopodium*
Potyvirus (aphid)			*amaranticolor*
			C. quinoa
			Datura stramonium

Table 2. Suggested Species for Identification of Individual Viruses (continued)

Virus name[a]	Acronym[b]	Host plants	Nonhost plants
			Nicotiana glutinosa
			Vigna sinensis
Sunnhemp mosaic *Tobamovirus*	SHMV	*Nicotiana glutinosa* *Phaseolus vulgaris* cv. The Prince	*Cucumis sativus*
Sweet clover necrotic mosaic *Dianthovirus*	SCNMV	*Chenopodium amaranticolor* *C. quinoa* *Nicotiana clevelandii* *N. glutinosa*	*Datura stramonium* *Petunia hybrida* *Zinnia elegans*
Sweet potato mild mottle *Potyvirus* (whitefly)	SPMMV	*Chenopodium murale* *C. quinoa* *Datura stramonium* *Gomphrena globosa* *Nicotiana benthamiana* *N. glutinosa* *N. rustica* *N. tabacum* *N. tabacum* cv. White Burley *Tetragonia expansa*	*Chenopodium captitatum* *Cucumis sativus* *Phaseolus vulgaris* cv. Pinto *Physalis floridana* *Vicia faba* *Vigna sinensis* cv. Blackeye
Tephrosia symptomless ? *Carmovirus*	TeSV	*Glycine max*	*Chenopodium quinoa* *Gomphrena globosa*
Tobacco etch *Potyvirus* (aphid)	TEV	*Chenopodium quinoa* *Datura stramonium* *Nicotiana tabacum* cv. Havana 425	*Pisum sativum* *Spinacia oleracea*
Tobacco mosaic *Tobamovirus*	TMV	*Chenopodium amaranticolor* *Nicotiana glutinosa* *N. sylvestris* *N. tabacum* cv. Samsun	*Pisum sativum* *Trifolium* spp.
Tobacco necrosis *Necrovirus*	TNV	*Chenopodium amaranticolor* *Phaseolus aureus* *Phaseolus vulgaris* cv. The Prince	*Cucumis sativus* *Pisum sativum* *Vigna radiata*
Tobacco rattle *Tobravirus*	TRV	*Chenopodium amaranticolor* *Nicotiana clevelandii* *Phaseolus vulgaris*	*Chenopodium capitatum* *Cucurbita pepo* *Datura arborea*
Tobacco ringspot *Nepovirus*	TRSV	*Chenopodium quinoa* *Cucumis sativus* *Nicotiana clevelandii*	*Lycopersicon esculentum*
Tobacco streak *Ilarvirus* (I)	TSV	*Chenopodium quinoa* *Gomphrena globosa* *Nicotiana tabacum* *Phaseolus vulgaris*	*Nicotiana glutinosa* *Petunia hybrida* *Phaseolus mungo* *Spinacia oleracea* *Zinnia elegans*
Tobacco vein mottling *Potyvirus* (aphid)	TVMW	*Nicotiana tabacum* cv. Burley 21 *N. tabacum* cv. Kentucky 14	*Capsicum annuum* *C. frutescens*
Tomato aspermy *Cucumovirus*	TAV	*Chenopodium amaranticolor* *Nicotiana clevelandii* *N. glutinosa*	*Brassica oleracea* var. botrytis *Cucurbita pepo* *Phaseolus vulgaris* cv.

Table 2. Suggested Species for Identification of Individual Viruses (continued)

Virus name[a]	Acronym[b]	Host plants	Nonhost plants
		Vigna sinensis	Pinto
			Pisum sativum
			Tropaeolum majus
			Vicia faba
Tomato black ring	TBRV	*Chenopodium quinoa*	*Medicago sativa*
Nepovirus		*Nicotiana clevelandii*	
		Petunia hybrida	
Tomato bushy stunt	TBSV	*Chenopodium*	*Brassica pekinensis*
Tombusvirus		*amaranticolor*	*Glycine max*
		C. quinoa	*Pisum sativum*
		Gomphrena globosa	*Tropaeolum majus*
		Nicotiana clevelandii	*Solanum nigrum*
		N. glutinosa	
Tomato golden mosaic	TGMV	*Datura stramonium*	*Solanum nigrum*
Geminivirus (II)		*Nicotiana benthamiana*	*S. tuberosum* spp.
		N. glutinosa	tuberosum cv. Maris
			Bard
Tomato mosaic	ToMV	*Chenopodium*	*Phaseolus vulgaris* cv.
Tobamovirus		*amaranticolor*	The Prince
		C. quinoa	*Vigna sinensis* cv.
		Datura stramonium	Blackeye
		Nicotiana clevelandii	
		N. glutinosa	
Tomato ringspot	ToRSV	*Chenopodium quinoa*	*Tetragonia expansa*
Nepovirus		*Cucumis sativus*	
		Nicotiana clevelandii	
		Vigna sinensis	
Tomato spotted wilt	TSWV	*Nicotiana clevelandii*	*Tetragonia expansa*
Tospovirus		*Petunia hybrida*	
		Tropaeolum majus	
Tulare apple mosaic	TAMV	*Nicotiana tabacum*	*Chenopodium quinoa*
Ilarvirus (II)		*Phaseolus vulgaris* cv.	
		Bountiful	
Tulip breaking *Potyvirus*	TBV	*Chenopodium*	*Nicotiana tabacum* cv.
(aphid)		*amaranticolor*	White Burley
		Lilium formosanum	*Phaseolus vulgaris*
		Nicotiana benthamiana	*Vicia faba*
		Nicotiana clevelandii	
Tulip X *Potexvirus*	TVX	*Chenopodium*	*Nicotiana clevelandii*
		amaranticolor	
		C. quinoa	
Turnip crinkle	TCV	*Brassica pekinensis*	*Beta vulgaris*
Carmovirus		*Chenopodium album*	*Nicandra physaloides*
		C. amaranticolor	*Nicotiana glutinosa*
		C. quinoa	*N. tabacum*
		Datura stramonium	*Phaseolus vulgaris*
		Gomphrena globosa	*Pisum sativum*
		Sinapis alba	*Tropaeolum majus*
		Tetragonia expansa	*Vicia faba*
Turnip mosaic *Potyvirus*	TuMV	*Brassica rapa*	*Phaseolus vulgaris*
(aphid)		*Chenopodium*	*Vicia faba*
		amaranticolor	
		C. quinoa	
		Nicotiana tabacum	
Turnip rosette	TRoV	*Brassica pekinensis*	*Nicotiana tabacum*
Sobemovirus		*Nicotiana clevelandii*	

Table 2. Suggested Species for Identification of Individual Viruses (continued)

Virus name[a]	Acronym[b]	Host plants	Nonhost plants
Turnip yellow mosaic *Tymovirus*	TYMV	*Brassica pekinensis* *B. rapa* *Crambe abyssinica*	*Chenopodium amaranticolor* *Cucumis sativus* *Nicotiana chinensis* *N. glutinosa* *N. tabacum* *Phaseolus vulgaris* *Physalis floridana* *Tinantia erecta*
Ullucus C *Comovirus*	UVC	*Chenopodium amaranticolor* *C. murale* *C. quinoa* *Tetragonia expansa*	*Cucumis sativus* *Datura stramonium* *Nicotiana glutinosa* *N. tabacum* cv. Xanthi-nc
Velvet tobacco mottle *Sobemovirus*	VTMoV	*Nicotiana clevelandii* *N. glutinosa* *N. velutina*	*Chenopodium amaranticolor* *C. quinoa* *Cucumis sativus* *Phaseolus vulgaris* *Vigna sinensis* *Zinnia elegans*
Viola mottle *Potexvirus*	VMV	*Chenopodium amaranticolor* *C. quinoa* *Gomphrena globosa* *Nicotiana clevelandii*	*Cucumis sativus* *Nicotiana glutinosa* *Petunia hybrida* *Phaseolus vulgaris*
Voandzeia necrotic mosaic *Tymovirus*	VNMV	*Chenopodium amaranticolor* *Vigna unguiculata* subspp. *unguiculata* *Voandzeia subterranea*	*Chenopodium quinoa* *Cucumis sativus* *Gomphrena globosa* *Nicotiana glutinosa*
Watermelon mosaic virus 2 *Potyvirus* (aphid)	WMV2	*Chenopodium amaranticolor* *C. quinoa* *Cucurbita pepo* cv. Small Sugar	*Datura stramonium* *Nicotiana glutinosa* *Petunia hybrida* *Ranunculus sardous*
Wheat spindle streak mosaic *Potyvirus* (fungus)	WSSMV	*Triticum aestivum* cv. Kent	*Amaranthus retroflexus* *Chenopodium album*
Wheat streak mosaic *Potyvirus* (mite)	WSMV	*Avena sativa* *Hordeum vulgare* *Triticum aestivum*	*Agropyron repens* *Bromus inermis* *Sorghum bicolor*
White clover mosaic *Potexvirus*	WClMV	*Cucumis sativus* *Phaseolus vulgaris* *Vigna sinensis*	*Antirrhinum majus* *Chenopodium amaranticolor* *Gomphrena globosa*
Wild cucumber mosaic *Tymovirus*	WCMV	*Cucumis melo* *C. sativus* *Cucurbita pepo*	*Chenopodium amaranticolor* *N. tabacum* cv. Samsun *Vigna sinensis*
Wineberry latent *Potexvirus*	WLV	*Chenopodium amaranticolor* *C. quinoa* *Gomphrena globosa*	*Nicotiana* spp.
Yam mosaic *Potyvirus* (aphid)	YMV	*Dioscorea cayenensis* *Nicotiana benthamiana* *N. megalosiphon*	*Dioscorea composita* *D. floribunda*

Table 2. Suggested Species for Identification of Individual Viruses (continued)

Virus name[a]	Acronym[b]	Host plants	Nonhost plants
Zucchini yellow mosaic *Potyvirus* (aphid)	ZYMV	*Chenopodium amaranticolor* *C. quinoa* *Cucurbita pepo* *Gomphrena globosa*	*Lavatera trimestris*

[a] Question marks indicate uncertainty of the taxonomic status of the viruses.[58]
[b] Proposed standard acronyms.[58]
[c] Monotypic groups with no approved group names.

ACKNOWLEDGMENTS

I should like to express my sincere appreciation to Drs. T. C. Allen, Corvallis (U.S.), L. Bos, Wageningen (The Netherlands), K. F. Boswell, Canberra (Australia), A. A. Brunt, Littlehampton (U.K.), R. G. Christie, Gainesville (U.S.), A. F. L. M. Derks, Lisse (The Netherlands), J. Dunez, Villenave (France), J. R. Edwardson, Gainesville (U.S.), R. W. Fulton, Middleton (U.S.), A. J. Gibbs, Canberra (Australia), M. Kameya-Iwaki, Yoshida (Japan), R. H. Lawson, Beltsville (U.S.), D.-H. Lesemann, Braunschweig (Germany), R. M. Lister, West Lafayette (U.S.), A. F. Murant, Invergowrie, Dundee (Scotland), M. Németh, Budapest (Hungary), F. Nienhaus, Bonn (Germany), H. E. Schmidt, Aschersleben (Germany), R. J. Shepherd, Lexington (U.S.), M. Verhoyen, Louvain (Belgium), H.-L. Weidemann, Braunschweig (Germany) for providing pertinent reprints, series, periodicals, books, unpublished results, and valuable information on host range and nonhost range of some viruses.

I am indebted to Professor R. E. F. Matthews, Auckland, New Zealand, for his helpful critical reading of the manuscript.

Thanks are due to Miss K. Molnár and Mrs. M. Vágusz for their technical assistance, and to Mrs. G. Vizmathy for the English translation.

REFERENCES

1. **Kassanis, B. and Selman, I. W.,** Variations in the reaction of White Burley tobacco to the tomato aucuba mosaic virus and some other strains of tobacco mosaic virus, *J. Pomol. Hortic. Sci.,* 23, 167, 1947.
2. **De Bokx, J. A.,** Reactions of various plant species to inoculation with potato virus S, *Neth. J. Plant Pathol.,* 76, 70, 1970.
3. **Van der Want, J. P. H., Boerjan, M. L., and Peters D.,** Variability of some plant species from different origins and their suitability for virus work, *Neth. J. Plant Pathol.,* 81, 205, 1975.
4. **Van Dijk, P., Van der Meer, F. A., and Piron, P. G. M.,** Accessions of Australian *Nicotiana* species suitable as indicator hosts in the diagnosis of plant virus diseases, *Neth. J. Plant Pathol.,* 93, 73, 1987.

5. **Hoekstra, R. and Seidewitz, L., Eds.,** *Evaluation data on tuber-bearing Solanum species,* German-Dutch Curatorium Plant Genetic Resources, Braunschweig and Wageningen, 1987.
6. **Horváth, J.,** Susceptibility and hypersensitivity to tobacco mosaic virus in wild species of potatoes, *Acta Phytophathol. Hung.,* 3, 35, 1968.
7. **Horváth, J.,** Susceptibility, hypersensitivity and immunity to potato virus Y in wild species of potatoes, *Acta Phytopathol. Hung.,* 3, 199, 1968.
8. **Gibbs, A. and Harrison, B.,** *Plant Virology, the Principles,* Edward Arnold, London, 1976.
9. **Matthews, R. E. F.,** Host plant responses to virus infection, in *Comprehensive Virology,* Vol. 16, *Virus-Host Interaction, Viral Invasion, Persistence, and Diagnosis,* Fraenkel-Conrat, H. and Wagner, R. R., Eds., Plenum Press, New York, 1980, 297.
10. **Matthews, R. E. F.,** *Plant Virology,* 3rd ed., Academic Press, New York, 1991.
11. **Ross, A. F.,** Identification of plant viruses, in *Plant Virology,* Corbett, M. K. and Sisler, H. D., Eds., University of Florida Press, Gainesville, 1964, chap. 68.
12. **Fulton, R. W.,** Transmission of plant viruses by grafting, dodder, seed, and mechanical inoculation, in *Plant Virology,* Corbett, M. K. and Sisler, H. D., Eds., University of Florida Press, Gainesville, 1964, chap. 39.
13. **Holmes, F. O.,** Symptomatology of viral diseases in plants, in *Plant Virology,* Corbett, M. K. and Sisler, H. D., Eds., University of Florida Press, Gainesville, 1964, chap. 17.
14. **Roberts, D. A.,** Local-lesion assay of plant viruses, in *Plant Virology,* Corbett, M. K. and Sisler, H. D., Eds., University of Florida Press, Gainesville, 1964, chap. 194.
15. **Bos, L.,** Methods of studying plants as virus hosts, in *Methods in Virology,* Maramorosch, K. and Koprowski, H., Eds., Academic Press, New York, 1967, 129.
16. **Schmelzer, K.,** Symptomatologie, a/ Äussere Symptome, in *Pflanzliche Virologie,* Band 1, Klinkowski, M., Ed., Akademie Verlag, Berlin, 1980, 21.
17. **Noordam, D.,** *Identification of Plant Viruses, Methods and Experiments,* Centre for Agr. Publ. Doc., Wageningen, 1973.
18. **Horváth, J.,** *Plant Viruses, Vectors and Virus Transmission,* Akadémiai Kiadó, Budapest, 1972.
19. **Costa, A. S. and Kitajima, E. W.,** Cassava common mosaic virus, *CMI/AAB Descriptions of Plant Viruses,* No. 90, Commonwealth Mycological Institute, Association of Applied Biologists, 1972.
20. **Schmelzer, K., Schmidt, H. E., and Beczner, L.,** Spontane Wirtspflanzen des Luzernemosaik-Virus, *Biol. Zentralbl.,* 92, 211, 1973.
21. **Lisa, V., Boccardo, G., and Milne, R. G.,** Viola mottle virus, *CMI/AAB Descriptions of Plant Viruses,* No. 247, Commonwealth Mycological Institute, Association of Applied Biologists, 1982.
22. **Horváth, J.,** New artificial hosts and non-hosts of plant viruses and their role in the identification and separation of viruses. XVIII. Concluding remarks, *Acta Phytopathol. Hung.,* 18, 121, 1983.
23. **Kahn, R. P., Monroe, R. L., Hewitt, W. B., Goheen, A. C., Wallace, J. M., Roistacher, C. N., Neuer, E. M., Ackerman, W. L., Winters, H. F., Seaton, G. A., and Pifer, W. A.,** Incidence of virus detection in vegetatively propagated plant introductions under quarantine in the United States, 1957–1976, *Plant Dis. Rep.,* 51, 715, 1967.
24. **Matthews, R. E. F.,** *Plant Virology,* 2nd ed., Academic Press, New York, 1981.
25. **Bos, L.,** *Introduction to Plant Virology,* Centre for Agr. Publ. Doc., Wageningen, 1983.
26. **Jaspars, E. M. J. and Bos, L.,** Alfalfa mosaic virus, *CMI/AAB Descriptions of Plant Viruses,* No. 229, Commonwealth Mycological Institute, Association of Applied Biologists, 1980.
27. **Bos, L.,** *Symptoms of Virus Diseases in Plants,* Centre for Agr. Publ. Doc., Wageningen, 1970.

28. **Francki, R. I. B., Mossop, D. W., and Hatta, T.,** Cucumber mosaic virus, *CMI/AAB Descriptions of Plant Viruses,* No. 213, Commonwealth Mycological Institute, Association of Applied Biologists, 1979.

29. **Horváth, J., Juretić, N., Besada, W. H., and Mamula, D.,** Natural occurrence of turnip mosaic virus in Hungary, *Acta Phytopathol. Hung.,* 10, 77, 1975.

30. **Pawley, R. R.,** Year-round production of test plants for virology, *Rep. Glasshouse Crops Res. Inst.,* Littlehampton, 1973, 149.

31. **Yarwood, C. E.,** Procedures to increase virus yield from infected plants, in *Methods in Virology,* Vol. 5, Maramorosch, K. and Koprowski, H., Eds., Academic Press, New York, 1971, chap. 14.

32. **Hill, S. A.,** *Methods in Plant Virology,* Blackwell Scientific, Oxford, 1984.

33. **Horváth, J.,** Die Anfälligkeit von *Chenopodium amaranticolor* Coste et Reyn. gegenüber dem Kartoffel-Y-Virus im Hinblick auf die Blattsequenz, *Acta Bot. Hung.,* 15, 71, 1969.

34. **Horváth, J.,** Die Anfälligkeit und Symptomausprägung der Blätter von *Nicotiana tabacum* L. cv. Hicks Fixed A2-426 nach Infektion mit dem tobacco mosaic virus in Abhängigkeit von ihrer Sequenz am Spross, *Acta Phytopathol. Hung.,* 4, 131, 1969.

35. **Hansen, A. J.,** Antiviral chemicals for plant disease control, *Crit. Rev. Plant Sci.,* 8, 45, 1989.

36. **Van der Meer, F. A., Maat, D. Z., and Vink, J.,** *Lonicera* latent virus, a new carlavirus serologically related to poplar mosaic virus: some properties and inactivation in vivo by heat treatment, *Neth. J. Plant Pathol.,* 86, 69, 1980.

37. **Hamilton, R. L., Edwardson, J. R., Francki, R. I. B., Hsu, H. T., Hull, R., Koenig, R., and Milne, R. H.,** Guidelines for the identification and characterization of plant viruses, *J. Gen. Virol.,* 54, 223, 1981.

38. **Matthews, R. E. F.,** Studies on potato virus X. II. Criteria of relationships between strains, *Ann. Appl. Biol.,* 36, 460, 1949.

39. **Schmelzer, K., Bartels, R., and Klinkowski, M.,** Interferenzen zwischen den Viren der Tabakätzmosaik-Gruppe, *Phytopathol. Z.,* 40, 52, 1960.

40. **Köhler, E.,** Über einige Probleme der Virusinterferenz bei Pflanzen, *Zentralb. Bakteriol. Abt. 2,* 616, 1962.

41. **Kassanis, B.,** Interactions of viruses in plants, *Adv. Virus Res.,* 10, 219, 1963.

42. **Horváth, J.,** Cross protection test with four strains of potato virus Y in *Nicotiana tabacum* L. cv. Samsun, *Zentralbl. Bakteriol. Abt. 2,* 123, 249, 1969.

43. **Hamilton, R. I.,** Defenses triggered by previous invaders: viruses, in *Plant Disease,* Vol. 5, *How Plants Defend Themselves,* Horsfall, J. G. and Cowling, E. B., Eds., Academic Press, New York, 1980, 279.

44. **Dodds, J. A.,** Cross-protection and interference between electrophoretically distinct strains of cucumber mosaic virus in tomato, *Virology,* 118, 235, 1982.

45. **Bos, L.,** A hundred years of Koch's postulates and the history of etiology in plant virus research, *Neth. J. Plant Pathol.,* 87, 91, 1981.

46. *CMI/AAB Descriptions of Plant Viruses,* No. 1–339, Commonwealth Mycological Institute, Association of Applied Biologists, 1970–1988.

47. **Kurstak, E., Ed.,** *Handbook of Plant Virus Infection, Comparative Diagnosis,* Elsevier/North-Holland, Amsterdam, 1981.

48. **Fulton, R. W.,** personal communication, 1991.

49. **Fulton, R. W.,** Hosts of the tobacco streak virus, *Phytopathology,* 38, 429, 1948.

50. **Lister, R. M.,** personal communication, 1991.

51. **Horváth, J.,** Host Ranges of Viruses and Virus Differentiation, D.Sc. Thesis, Budapest and Rostock, 1976.

52. **Fulton, R. W.,** Ilarviruses, in *Plant Virus Infections, Comparative Diagnosis,* Kurstak, E., Ed., Elsevier/North-Holland, Amsterdam, 1981, chap 13.

53. **Kaper, J. M. and Waterworth, H. E.,** Cucumoviruses, in *Plant Virus Infections, Comparative Diagnosis,* Kurstak, E., Ed., Elsevier/North-Holland, Amsterdam, 1981, chap. 11.

54. **Francki, R. I. B. and Hatta, T.,** Tomato spotted wilt virus, in *Plant Virus Infections, Comparative Diagnosis,* Kurstak, E., Ed., Elsevier/North-Holland, Amsterdam, 1981, chap. 17.

55. **Horváth, J.,** Viruses of lettuce. II. Host ranges of lettuce mosaic virus and cucumber mosaic virus, *Acta Agric. Hung.,* 29, 333, 1980.

56. **Quacquarelli, A. and Avgelis, A.,** *Nicotiana benthamiana* Domin, as host for plant virus, *Phytopathol. Medit.,* 14, 36, 1975.

57. **Crill, P., Hagedorn, D. J., and Hanson, E. W.,** Alfalfa mosaic, the disease and its virus incitant, *Univ. Wisconsin Res. Bull.,* 280, 1970.

58. **Hull, R., Milne, R. G., and van Regenmortel, M. H. W.,** A list of proposed standard acronyms for plant viruses and viroids, *Arch. Virol.,* 120, 151, 1991.

Chapter 3

EXPERIMENTAL TRANSMISSION OF VIRUSES IN DIAGNOSIS

A. T. Jones

TABLE OF CONTENTS

0-8493-4284-8/93/$0.00 + $.50

I. INTRODUCTION

The ability to be transmitted between plants is a fundamental property of plant viruses that is necessary for their survival and spread. Additionally, the experimental transmission of viruses to host plants is a basic practice for the study of such viruses and is one of the main ways of distinguishing symptoms in plants induced by viruses and other transmissible diseae agents from those induced by nontransmissible causes, such as nutrient deficiency, and physiological and genetic disorders. Unaided, plant viruses are unable to pass through the surface cuticle of host plants to infect cells. Thus, for transmission and infection to occur, viruses must enter cells through sublethal wounds in the plant surface.

In early experimental work, transmission was accomplished by graft-inoculation, whereby the freshly cut surfaces of infected and healthy plant tissue are maintained in contact, usually until union between the tissues occurs; most viruses are able to move from infected to healthy tissues across the graft union. In some later virus work, the parasitic plant dodder (*Cuscuta* spp.) was used to establish vascular bridges between plants of different species and genera, thus extending the possibilities of transmitting virus across tissue unions. With few exceptions, e.g., cryptoviruses, almost all plant viruses are transmissible by these means, although it may prove to be technically difficult with some species, particularly with monocotyledonous plants. However, it should be remembered that other disease-inducing agents, e.g., mycoplasma-like organisms and bacteria, can also be transmitted in these ways. Many viruses can be transmitted experimentally by gently abrading the leaf surface of plants with sap from infected plants, a process often referred to as "sap inoculation" or "mechanical inoculation".

Despite the ease with which many viruses can be transmitted experimentally by graft-inoculation and/or mechanical inoculation, transmission in nature by grafting (usually of roots) or by the rubbing together of leaves or stems is comparatively rare. Almost all plant viruses make use of an invertebrate or fungal vector to insert them through the outer plant surface barrier and deposit them within a cell in which they can replicate, and from which they can spread to other cells within the plant. Transmission between plants by these various means is sometimes termed "horizontal transmission" as distinct from "vertical transmission", which is used to refer to passage of virus from a plant to its progeny, either vegetatively or through seed or pollen, without the aid of a vector.

For diagnostic purposes, the advent of serological and molecular techniques for rapidly identifying infection with specific viruses has tended towards a decrease in the use of transmission studies. Nevertheless, the inoculation of plant viruses to different experimental plants, determining whether infection occurs and whether it becomes systemic or not, and careful observation of symptom development, if any, remains a simple and most useful tool in plant virology and is usually essential for studying new viruses or virus strais. Experimental transmission therefore continues to be used for (1) detecting a virus and showing that it causes a particular disease (i.e., fulfilling Koch's postulates as they can be applied to viruses), (2) differentiating viruses in mixed infections, (3) distinguishing and characterizing new viruses and virus strains and experimentally induced mutants, and (4) determining the occurrence of satellite RNA, or defective interfering particles. In the characterization of a previously unknown or poorly described virus, the transmission or lack of transmission by specific means and the induction of particular symptoms in plants can often provide useful clues to the identify of the virus, its affinities with other, better-characterized viruses, and its probable mode of spread in nature.

This chapter considers the experimental transmission of virus without the aid of natural vectors. The use of vector transmission for diagnosis is considered in Chapters 4 and 13.

II. TRANSMISSION BY GRAFT INOCULATION

A. GENERAL FEATURES OF GRAFT TRANSMISSION

Grafting is a standard horticultural practice. It involves the bringing together of the cut surfaces of tissues of two plants so that they can establish a union. For centuries, horticulturists were aware that by grafting, certain types of leaf variegation could be transferred to nonvariegated clones or species. Many kinds of leaf variegation in ornamental plants are now known to be caused by infection with viruses or virus-like agents. Provided that the union between healthy and infected tissue is successful, most kinds of viruses are graft transmissible. The cryptoviruses are an exception, in that they are not transmitted, or at least not efficiently transmitted, by either grafting or dodder.[1]

Successful graft union is dependent on the presence of meristematic cells, usually cambial tissue, at the contact surfaces between the plant tissues being grafted, and is usually most efficient when the two plants involved are from the same or related species. However, successful union does not guarantee transmission and there are several instances in which the efficiency of graft transmission is relatively low despite successful unions.[2,3] (A. T. Jones, unpublished data) Some of these failures may be due to the erratic distribution of the virus within the plant (resulting in the absence of virus from the source tissue used for graft inoculation) and some to inherent resistance to infection. However, other factors, as yet unknown, are probably also involved. Instances are also known of transmission occurring despite the failure of the tissues to unite. Thus, Cadman and co-workers[4] successfully graft-transmitted grapevine fanleaf *Nepovirus* from *Chenopodium amaranticolor* to grapevine even though callus tissue and not true tissue union developed at the grafted surfaces. There may be several reasons why graft union does not occur between plant tissues. Incompatibility of tissues is common where unrelated plant species are used, but can also occur between some closely related species.[5] The presence of virus may itself make graft union more difficult. Graft inoculation of monocotyledonous plants is commonly regarded as being impossible, except with bulb and corm tissue. This is due to the absence of the ring of cambial tissue that is present in dicotyledonous plants. While this might preclude successful graft union, it may not necessarily prevent virus transmission. However, Muzik and LaRue[6] have reported grafts with grasses and have indicated that it is the presence of meristematic cells at the graft union, rather than cambial tissue itself, that is important to successful grafts.

B. METHODS OF GRAFTING

Graft-inoculation was the first widely used technique for the transmission and detection of viruses, especially of woody plants. It became, and remains, the basis of a reliable diagnostic method for many viruses of crop plants in which tissue from the plant to be tested is grafted to a plant species or clone sensitive to infection with the virus in question. Many of the viruses detected in this traditional way can now be detected by more rapid tests, such as enzyme-linked immunosorbent assay (ELISA) or nucleic acid hybridization. However, for some viruses and virus strains, especially in woody plants, graft-inoculation to indicator plants is still the most sensitive detection method, and for several others it remains the only method available. Thus, graft-inoculation continues to be a major component of national and international schemes for virus-indexing of nursery stock of tree, vine, and bush fruit crops.

Many different forms of graft-inoculation have been devised, but they can be broadly considered to be of two main types:

1.   Grafting tissues (usually stems or stolons) of two intact rooted plants; this is often referred to as *approach grafting* (Figure 1)
2.   Grafting detached tissue (called the *scion*) from one plant onto an intact, rooted second plant, called the *stock* plant (Figures 2 to 7)

Approach graft

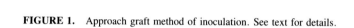

Stock plant

FIGURE 1. Approach graft method of inoculation. See text for details.

Tongue graft   Bottle graft

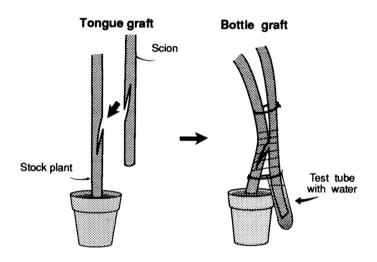

Scion

Stock plant

Test tube
with water

FIGURE 2. (a) Tongue graft; (b) bottle graft.

Approach grafting has the advantage that the plant partners are grown on their own roots, thereby minimizing graft failure due to dehydration of tissues. For methods involving detached tissues (scions) it is essential to minimize dehydration of tissues while the graft union develops. Several simple procedures or combinations of procedures can help achieve this: removal of all excess leaf material from scions to decrease the transpiration area; bottle grafting (Figure 2b) to provide a source of water for the scion until graft union is complete; humidity covers, e.g., plastic bags or tubes, to decrease loss by

Bud grafting

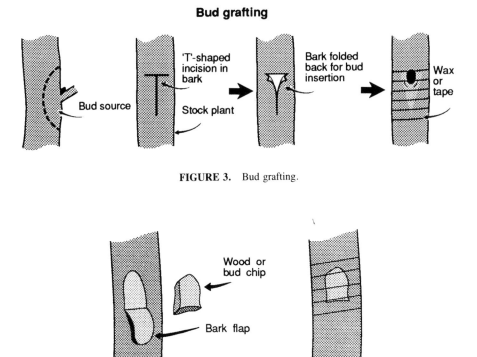

FIGURE 3.   Bud grafting.

FIGURE 4.   Chip graft.

transpiration; and, where dormant tissue (e.g., buds, bark) is the scion source, covering with wax or tape (Figures 3 and 4).

When grafts involving a stock and scion are used, it is preferable to graft an infected scion to a healthy stock, rather than the other way around, because any symptoms in the stock plant due to nonviral causes, e.g., nutritional deficiency, may become evident in the scion and thus confuse the interpretation of the results.

In all forms of grafting it is essential that contact surfaces between plant partners fit neatly and closely and are bound firmly together. Binding material should be flexible and/or stretchable, stable to ultraviolet light if used outdoors, and nontoxic to plants. Materials that have been used include raffia tape, rubber tubing, self-adhesive medical bandage, and Parafilm membrane. Tissues used for grafting can be dormant, e.g., buds, bulbs, corms, tubers, or actively growing, e.g., leaves, petioles, stems, stolons.

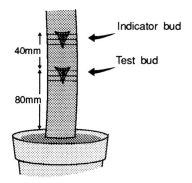

FIGURE 5. Double budding.

The technique of grafting will vary with the tissues used. Some techniques used for grafting scions are

1. Tongue graft — If soft stem or stolon tissue is used, the *tongue* graft is usually the preferred technique. A downward slit of about 15 to 20 mm extending only part way into the center of the stem of the stock plant is made with a razor or sharp scalpel blade and a corresponding, similar-sized upward slit into the stem of the scion (Figure 2a and b). The epidermis of each "tongue" produced by the incisions in the plant partners is sometimes removed before the tongue of the scion is fitted into the incision of the stock plant. The grafted tissues are bound tightly together to cover tissues to at least 10 mm above and below the graft incision (Figures 1 and 2).

2. Bud and shield graft — Grafting of dormant buds is commonly used for graft-inoculation of woody plants. In this technique, sometimes referred to as *shield grafting,* a T-shaped cut is made in the bark of the stock plant and the two bark flaps are lifted up to receive the bud (Figure 3). The bud is sliced from a twig so that the cut surface beneath the bud is wider than the bud but smaller than the T-shaped opening in the bark of the stock plant. The bud is inserted in the opening, the bark flaps folded over to hold it in place, and the grafted area either sealed with wax or bound tightly to hold the tissues together, but leaving the bud exposed (Figures 3 and 5). In addition to buds, wood chips, leaves, and even pollen have been used successfully as inoculation sources in shield grafts (Figure 4). An extension of this technique, often used for virus-indexing of fruit trees, is *double budding*. In this instance, two buds are grafted on a rootstock plant, one directly above the other, about 40 to 60 mm apart (Figure 5). The upper bud is from a healthy virus indicator plant and the lower one from a virus source plant or from a plant under test. Virus (if present) moves across the grafted

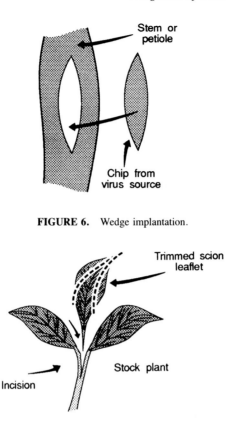

FIGURE 6.   Wedge implantation.

FIGURE 7.   Leaflet graft.

surfaces from the lower bud to the upper virus-indicator bud during the
dormant season. When the upper bud grows out in the spring, symptoms
appear on the young growth.

3. Wedge graft — In this instance, a slit is made completely through the
    stem of the stock plant and into this is placed a piece of similar-sized
    stem or petiole tissue, tapered at both ends, from the virus donor plant
    (Figure 6). Grafting tape is then used to cover the grafted area com-
    pletely, so preventing dehydration of tissues.
4. Leaflet graft — This technique, initially reported by Bringhurst and
    Voth,[7] has become widely used for graft-transmission and virus-index-
    ing in strawberry.[8] The central (terminal) leaflet of healthy plant is
    removed and an incision made down the center of the petiole between
    the two remaining leaflets. A leaflet from the donor plant with its petiole
    trimmed into a wedge shape is inserted into the petiole slit of the stock
    plant and bound with grafting tape (Figure 7). The scion leaflet is
    trimmed of excess leaf and the grafted plant maintained under a humidity

cover until graft union occurs. Usually two or three such grafts are made to each stock plant and, in strawberry, over 90% of grafted plants usually have at least one successful graft.

5. Tuber graft — In early work on viruses of potato, graft-inoculation of tubers was done by binding the cut surfaces of tuber partners together. In 1926 a simpler method was introduced by Murphy and McKay[9] and Goss.[10] A cork borer was used to extract a tissue core from the tuber used to produce the stock plant. A similar-sized cork borer was used to extract a core from the donor tuber and this was inserted into the hole in the "stock tuber". Similar procedures can be used for bulbs and corms.

The speed of movement of virus across the graft junction can vary enormously depending on the virus, the plant partners involved, the grafting technique used, and the time of year when grafting takes place. Generally, however, viruses that reach a high concentration and are widely distributed in plant tissues, tend to move very quickly, in actively growing tissue perhaps within 2 to 4 days. Viruses that reach only low concentrations and/or have a restricted distribution in plants take much longer, perhaps 7 to 30 days. Bennett[11] found that following graft inoculation, tobacco ringspot *Nepovirus* and cucumber mosaic *Cucumovirus,* which are both mechanically transmissible, passed the graft union within 2 or 3 days, whereas beet curly top *Geminivirus,* which is not mechanically transmissible, required more than 5 to 6 days.

C. CONCLUSIONS

In relation to diagnosis, the speed of transmission across a grafting interface may give some indication of the type of virus involved[12] and, in multiple infections, this may provide a means of separating viruses. For woody plants in particular, graft inoculation to sensitivie indicators is the only way of identifying some viruses and some virus strains,[13,14] and remains a widely used technique for transmission and diagnosis. Furthermore, graft inoculation to new species and indicator cultivars continues to detect apparently new viruses and virus-like agents in plants (A. T. Jones, unpublished data). Failure to be graft-transmitted seems to be a unique feature of cryptoviruses.

Despite its usefulness in these specific examples, graft-inoculation has several disadvantages. First, it is laborious and, for some techniques and crop species, requires considerable skill and dexterity. Second, in some crop species, especially of woody plants, symptom development can take many months and in a few instances 2 to 3 years. Third, it is not generally applicable to monocotyledonous plants. While this may not be true for all species, technically it is much more difficult to achieve satisfactory results with monocotyledonous plants. Finally, viruses are not the only agents to be transmitted by grafting. Indeed, many of the diseases once attributed to viruses on the basis of, among other criteria, graft-inoculation, are now known to be caused by mycoplasma-like organisms. Grafting therefore remains a somewhat crude detection system but nevertheless a vital one for some virus/host combinations.

III. TRANSMISSION BY DODDER

A. GENERAL FEATURES OF DODDER TRANSMISSION

Dodder (*Cuscuta* spp.) is a parasitic vine, lacking leaves and chlorophyll, belonging to the family Convolvulaceae. It parasitizes higher plants by twining its thin slender stems around them and, where it makes contact with the host plant, forming special organs called haustoria that penetrate the host stem and connect with its vascular tissue. A single dodder plant can embrace two or more plants and establish vascular bridges between them. Different *Cuscuta* species have different host ranges but some, such as *C. campestris* and *C. subinclusa,* have wide host ranges, so enabling vascular bridges to be made between two unrelated host plant species. These species therefore offer an advantage over graft-inoculation in permitting transmission of virus from one plant species to another.

Bennett[15] was the first to report that some viruses move through a dodder bridge from one plant to another. He also showed that for some viruses, dodder acts only as a "pipeline" through which they travel passively; for others it also acts as a host and they multiply within its tissues. Indeed, some viruses multiply and induce disease symptoms in dodder.[16] Not all viruses are transmitted by dodder and not all dodder species are able to transmit the same viruses. In general, viruses that multiply in the dodder species are more efficiently transmitted than those which merely pass passively through the dodder. While it is known that the sap of some dodder species contains substances that interfere with the transmission of some viruses,[17,18] there is no evidence that such inhibitors are involved in the specificity of virus transmission by dodder.

B. METHODS USED FOR DODDER TRANSMISSION

The dodder species used most frequently for virus transmission have been *C. campestris* and *C. subinclusa,* mainly because they have wide host ranges. However, other species such as, *C. americana, C. californica, C. epilinium, C. epithymum, C. europaea, C. gronovii, C. japonica, C. lupuliformis,* and *C. repens* have also been used with success for some viruses.[19] For transmission work, dodder cultures can be produced either by growing plants directly from seed or by using pieces of dodder stem from a culture maintained on plants.

Dodder seed is viable for many years and can be sown on the soil surface of pots containing a vigorously growing host plant. Alternatively, dodder seedlings can be germinated on moist filter paper until they are about 20 to 30 mm long and then transferred to the soil around the host plant or attached to host plant stems and sustained with water in a small container until the host plant is parasitized. To establish a new culture from one already existing on plants, dodder stems are placed in the leaf axils of the new plant host. For virus transmission work, it is usual to establish the dodder culture on an infected plant and, when sufficient dodder growth has occurred, to transfer dodder stems to the healthy plant and allow it to colonize this second plant while still attached to its initial host. Although viruses have been transmitted by transferring stem pieces from dodder established on a virus-infected plant to healthy plants, this is a less reliable method, especially for viruses that do not multiply in the dodder. Some workers have obtained increased frequencies of dodder transmission by adjusting various conditions. For example, Cochran[20] achieved 75% passive transmission of tobacco mosaic *Tobamovirus* when the dodder was first pruned back and the recipient healthy plant was shaded; without these conditions he obtained no virus transmission.

C. CONCLUSIONS

Although dodder transmission was much favored by early workers on plant viruses, it is now rarely used. The only advantage it offers is the ability to transmit some viruses that for technical reasons cannot be transmitted by grafting. However, cryptoviruses seem not to be transmitted by either method. While it is possible to separate some viruses from mixtures in plants using dodder, it has rarely been used for this purpose. However, it has been used to transfer virus-like agents from woody plants, which are difficult hosts in which to study the agents, into herbaceous hosts, which are more amenable to study. In this regard it has been especially useful for mycoplasma-like organsims and it has been possible to compare such organisms by introducing them into a common host, e.g., *Vinca* spp. Some of the disadvantages of graft-inoculation for the transmission of plant viruses, indicated earlier, are shared by dodder inoculation.

IV. TRANSMISSION BY MECHANICAL INOCULATION

A. INTRODUCTION

The classical work of Holmes[21,22] on the use of local lesions as a quantitative assay for plant viruses made it possible for the first time to compare the relative effectiveness of different methods of mechanically inoculating plants with virus. From such assays, it was clear that rubbing leaves of plants with infective sap using the finger, brush, or glass rod, was a much more effective way of inoculating plants than injecting or pricking them with needles, as first described by Mayer.[23] This finding by Holmes laid the foundation for the many refinements made over the years to maximize the efficiency of mechanical transmission of different viruses to plants, refinements which have enabled the mechanical transmission of viruses thought previously not to be transmitted in this way. Nevertheless, it remains true that about 20% of the viruses assigned to recognized plant virus groups are not mechanically transmissible to plants. The ability, or lack of ability, to be transmitted mechanically to plants may therefore be a useful pointer in identifying a virus. However, for many viruses, mechanical transmissibility may depend critically on optimizing several different factors, and it is important to consider these factors before concluding that a virus is not mechanically transmissible. Some of the more important of these factors are highlighted below and more detailed information on these aspects can be found elsewhere.[24-26]

B. IMPORTANT FACTORS INFLUENCING TRANSMISSION BY MECHANICAL INOCULATION

1. Virus Host Range

Some viruses have restricted host ranges, necessitating the use of representatives of as wide a range of plant families as possible in transmission studies. The growing conditions of such plants, before and after inoculation, may greatly affect their infectibility with virus. The host range of viruses is discussed in Chapter 2.

2. Inoculum Source

For some plant virus/host combinations, the choice of the plant parts used as a source of inoculum may determine whether transmission is successful or not. There are at least two possible reasons for this. First, virus distribution and concentration may vary greatly within plants, especially woody plants, and it is not uncommon for some buds, branches, or whole limbs to be free from infection. In other instances, e.g., with tobacco necrosis *Necrovirus,* the virus is confined to the roots of most naturally infected plants, though it may be capable of causing local lesions when inoculated mechanically to the leaves of the same plant species. In *Brassica* spp. infected with turnip yellow mosaic *Tymovirus,* the concentration of virus can be much less in normal green areas of infected leaves than in chlorotic areas[27] and this may be a common feature of mosaic diseases.

Second, the saps of some plant species contain substances that interfere with virus transmission, and the concentration and distribution of these substances in plant parts can vary. For example, polyphenols (tannins) that can bind to virus particles and precipitate them from solution are often present in high concentrations in woody plant species. However, as a general rule in actively growing plants of this kind, such materials are scarce in the tip leaves and flower parts, and sometimes also in the root tips.

3. Effect of Additives in the Inoculum

As indicated at the beginning of this chapter, viruses must penetrate the nonliving surface of plants to infect cells. Transmission by mechanical inoculation is therefore greatly aided by the presence of an abrasive in the inoculum to induce sublethal wounds for entry of virus into cells. Aluminum oxide, carborundum, corundum, or celite are commonly used for this purpose.

As mentioned earlier, the saps of many plant species contain materials that can interfere with virus transmission. It is therefore necessary to minimize the effect of such substances. Dilution of the virus-containing sap is often very effective but, with those viruses that occur in low concentration in plants, there is a delicate balance between diluting out the inhibitory effects of interfering substances and maintaining sufficient virus concentration to be infective. Most virus inhibitors in plants are enzymes, other proteins, or polyphenols; various chemicals added to the inoculum can diminish the interference of these substances with virus transmission. Polyphenols exert their inhibitory effect by binding to the virus particle protein (tanning) and this can be minimized by adding other proteins to act as "decoy" agents. Materials such as hide powder, caffeine, or insoluble polyvinyl pyrrolidone (PVP) have also been used for this purpose.[28] As tanning of proteins is greatly decreased under alkaline conditions, another approach to the problem is to use high pH buffer or 2% nicotine solution (pH 9) for the extraction medium.[4] Oxidation of polyphenols can also interfere with virus transmission but the effects can be minimized by the presence of the copper chelating agent, sodium diethyl-dithiocarbamate (DIECA) in the extraction medium.[29] More general oxidation in plant sap can have a dramatic influence on virus infectivity. This is especially noticeable with most ilarviruses, which can lose much infectivity within a few minutes of extraction from infected plants unless a strong reducing agent such as dithiothreitol or 2-mercaptoethanol is present.[30] Ribonucleases pose a problem for the infectivity of some viruses and especially for those that lack a protein coat to encapsulate their nucleic acid. Alkaline buffer (pH 8) decreases the activity of nucleases and, when combined with the presence of bentonite, which absorbs nucleases, can be especially effective (see below). Such procedures have also been effective in the mechanical transmission of "umbraviruses" (see Chapter 13).

4. Inoculation Method

Different workers tend to have their own method for mechanical inoculation of plants. However, this is usually for historical or personal reasons rather than scientific ones. Nevertheless, rubbing sap from infected plants on leaves with the finger or muslin pad is an effective way of transmitting most mechanically transmissible viruses to herbaceous plants. However, for some viruses and/or some hosts, e.g., monocotyledonous plants and woody plants, this method is ineffective and other methods have been devised. For example, nepoviruses are transmitted in this way readily to herbaceous plants but very poorly or not at all to woody plants, despite the fact that they commonly infect such plants in nature. However, Bitterlin et al.[31] succeeded in transmitting tomato ringspot *Nepovirus* to 50 to 100% of *Prunus* seedlings by repeated (>100 times) slashing of stems with a razor contaminated with the virus. A similar method has been used for transmitting citrus tristeza *Closterovirus* and viroids to citrus seedlings.[32] Severe wounding by heavy rubbing with infective sap seems necessary to transmit onion yellow dwarf *Potyvirus* to onion.[33]

The use of equipment that sprays inoculum under pressure onto leaves has given considerable improvements in the transmission of some viruses. Toler and Hebert[34] found that an artist's air brush, used for the mass inoculation of a large number of plants, produced a 20-fold increase in the transmission of oat mosaic virus to oats when compared to inoculation with the finger. This approach has been refined even further by Laidlaw,[35] through the use of two spray guns — one to spray carborundum to wound the plants, the other to spray the inoculum. Using this system, with careful attention to, among other factors, carborundum particle size, gun distance from leaves, and air pressure, he obtained an estimated 64-fold improvement in transmission efficiency compared with traditional inoculation with the finger. With further refinements, the claimed potential improvement was up to 500-fold.

Finally, using electroendosmosis to introduce virus into the cut surfaces of leaves, Polson and von Wechmar[36] reported the transmission to maize plants of maize streak *Geminivirus,* a virus shown previously to be transmitted only by leafhoppers.

C. SPECIFIC INFECTIVITY ASSAYS FOR DETERMINING VIRUS PROPERTIES AND AFFINITIES

1. Dilution Curves

Where assay is possible with a local lesion host of the virus, plotting the dilution curve of infectivity may provide useful information on the virus involved. Fulton[37] found that for two ilarviruses, infectivity decreased severalfold more rapidly than the dilution factor. He concluded that this was because these viruses had their genome parts divided among different particle types and that for infection to occur at an inoculation site, the presence of more than one particle type was necessary. Viruses in several other virus groups that are known to have divided genomes, with the genome parts packaged in different virus particles, have been shown to produce steep infectivity curves.[38,39] In plotting dilution curves it should be noted that other factors can influence the curve. For example, if assays are made with plant sap, the effects of any inhibitors will be greatly decreased by dilution, causing an increase in apparent infectivity with dilution over part of the dilution range. Additionally, the particles of some viruses tend to aggregate, especially when purified, but disaggregate on dilution, causing an apparent increase in infectivity.

2. Cross-Protection Tests

Cross-protection is the name given to a phenomenon in which systemic infection of a plant with one strain of a virus may protect it from developing additional symptoms when inoculated with a closely related strain of the same virus. This type of test, which involves mechanical inoculation, is discussed briefly in Chapter 2, Section IV, and in Matthews.[26]

3. Infectivity of Nucleic Acid Extracts

For some mechanically transmissible viruses, variants are known in which the gene(s) for the virus coat protein is either absent or defective. Mechanical transmission of such variants in plant sap is difficult or impossible because the uncoated virus nucleic acid is degraded by nucleases released from plant cells by extraction of sap. However, transmission can be effected in most instances by the use of alkaline extraction buffer containing bentonite, to adsorb nucleases. Alternatively, measures can be taken to extract the nucleic acid while protecting it from degradation. This is usually done by grinding leaves in a mixture of alkaline buffer and water-saturated phenol containing sodium dodecyl sulfate (SDS), at 4°C, and centrifuging the slurry to separate the nucleic acid-containing aqueous phase from the denser phenol phase containing protein and plant debris. The nucleic acid can then be precipitated from the aqueous phase with ethanol overnight at $-15°C$ before resuspending in buffer containing bentonite to inoculate to plants. In this way, coat protein-deficient variants of, among others, tobacco mosaic *Tobamovirus,* tobacco necrosis *Necrovirus,* and tobacco rattle *Tobravirus,* have been transmitted to plants. Such an approach also led to the finding that the agent of potato spindle tuber disease is a naked nucleic acid without a coat protein; this was the first to be described of a number of similar pathogens now known as viroids[40] (Chapter 12). However, it should be noted that increased infectivity of phenol extracts compared to sap extracts may have causes other than the protection of naked virus nucleic acid. For example, it could be brought about by the removal of proteinaceous inhibitors, or by the release of virus particles attached to cell organelles and other structures.

D. CONCLUSIONS

The ability of a virus to be mechanically transmitted allows the study of its host range, symptomatology, and many other properties with relative ease when compared with transmission by grafting or dodder (Chapter 2). Mechanical transmission tests may give valuable clues for diagnosis and enable the separation of virus mixtures. Bearing in mind the factors necessary for the mechanical transmission of some viruses (mentioned above), the ability, lack of ability, and even the relative ability to transmit a virus by this means, may in itself suggest affinity or lack of affinity with viruses of recognized plant virus groups (Table 1). In general, viruses transmitted in a persistent manner by leafhoppers, plant hoppers, or aphids are transmitted poorly or not at all by mechanical inoculation to plants. Some notable exceptions to this general rule are pea enation mosaic virus and lettuce necrotic yellows *Rhabdovirus,* each of which is transmitted in a persistent manner by aphids but is also mechanically transmissible.

Table 1. Virus Groups, Most of Whose Members are Poorly Transmitted, or Not Transmitted, to Plants by Mechanical Inoculation

Virus group	Main vector	Mode of transmission
Poorly Transmitted		
"Barley yellow mosaic virus"	Fungus	Persistent
"Commelina yellow mosaic virus"	Aphids, leafhoppers	Semipersistent
Furovirus	Fungus	Persistent
Geminivirus — subgroup B	Whitefly	Semipersistent
Rhabdoviruses	Leafhoppers and aphids	Propagative
Not Transmitted		
Cryptovirus	None	
Fijivirus	Planthoppers	Propagative
Geminivirus — subgroup A	Leafhoppers	Persistent
Luteovirus	Aphids	Persistent
"Maize chlorotic dwarf virus"	Leafhoppers	Semipersistent
Marafivirus	Leafhoppers	Persistent
Phytoreovirus	Leafhoppers	Propagative
"Rice ragged stunt virus" (Reoviridae)	Planthoppers	Propagative
Tenuivirus	Planthoppers	Propagative

Some of the factors necessary to achieve transmission of a virus may also be a useful pointer to identification. For example, the instability of virus in sap without the presence of a strong reducing agent may suggest a possible affinity with ilarviruses. However, it should be noted that some ilarviruses, like elm mottle,[41] are very stable in plant sap. If the infectivity of nucleic acid extracts from leaves is greater than that of plain sap extracts, a defective or deletion virus variant, a viroid or a possible "umbravirus" (see Chapters 12 and 13) may be involved, although other unrelated factors may be the cause. Assays on plants to plot a dilution curve of infectivity may indicate the presence of a multicomponent virus.

In addition to the above kinds of experiment, mechanical inoculation to plants, especially if a good local lesion host of the virus is available, can form the basis of assays to assess the effects of various treatments on the virus in plant sap. Such experiments are usually necessary to determine the most effective ways of purifying the virus and can, for example, determine the isoelectric point of the virus, its response to organic solvents, chelating agents, salts, etc. Taken together, the results of such experiments can help to build up a useful picture of the kind of virus being studied and lead to accurate diagnosis; they may also provide a suitable purification schedule for the virus. Infectivity assays can also be used to determine the location of virus following centrifugation in sucrose density gradients and/or isopycnic gradients in cesium salts. Thus, the ability to transmit a virus mechanically to plants, although usually not diagnostic in itself, is a major step forward in that it can provide a sensitive assay tool for other diagnostic tests.

V. TRANSMISSION THROUGH SEED

A. GENERAL FEATURES OF SEED TRANSMISSION

Transmission of viruses from host plant to progeny through seed plays a major role in the ecology and epidemiology of many plant viruses. It occurs, at least in experimental work, more frequently and in a wider range of viruses than was once thought, being recorded for at least one member in more than half of the established groups of plant viruses. Nevertheless, in nature, viruses transmitted most frequently in the seed of host plants tend to be confined to a few plant virus groups, most notably *Cryptovirus, Hordeivirus, Ilarvirus, Nepovirus, Potyvirus, Tobamovirus, Tobravirus,* and viroids. All cryptoviruses are seed-transmitted with high frequency and, apart from vegetative propagation of the host plant, this seems to be their only mode of transmission from one generation to another. Conversely, most members of some other groups and genera, such as the "barley yellow mosaic virus" *Carmovirus, Caulimovirus,* "commelina yellow mosaic virus", *Dianthovirus, Fijivirus, Luteovirus, Marafivirus, Phytoreovirus,* and *Tenuivirus* groups, seem not to be seed-transmitted, while for most members of other groups, seed transmission is rare, e.g., *Rhabdovirus, Closterovirus, Carlavirus,* and *Geminivirus* groups. The reasons why some viruses are seed-transmitted while others are not, is not understood, although the restricted distribution within tissues of the host plant may provide an obvious reason why some viruses, such as luteoviruses, are not seed-transmitted. Luteoviruses are confined to the phloem tissue of their plant hosts and no direct vascular connections exist with the embryo. Nevertheless, seed-transmissiblity of any individual virus is often host specific, sometimes specific to certain strains of the virus, and can be temperature dependent. Furthermore, for those viruses that are seed-borne, the time of infection is critical in determining whether transmission occurs or not, and the extent of transmission to progeny seedlings.

B. FACTORS INFLUENCING THE FREQUENCY OF TRANSMISSION

Many factors may influence the frequency of seed transmission, and indeed, whether such transmission occurs at all. This variability may limit the value of the presence or absence of seed transmission as a diagnostic criterion.

For example, the temperature at which plants are grown can have a significant effect on the incidence of seed transmission in some virus/host combinations; extremes of temperature (high or low) sometimes preventing seed transmission altogether. Thus, cherry leaf roll *Nepovirus* was transmitted to 100 and 0% of seed from *Nicotiana rustica* plants grown at 20 and 30°C, respectively.[42] Seed transmission of southern bean mosaic *Sobemovirus* in bean was 95 and 55% in plants grown at 16 to 20°C and 28 to 30°C, respectively.[43]

Seed transmission may be restricted to certain plant species or even cultivars of a species.[19,26] Peanut stripe *Potyvirus* is seed-borne in peanut but not soybean.[44] Some strains of a virus may be more readily seed-transmitted than others.[45] For example, strains of barley stripe mosaic *Hordeivirus* varied in their transmission through seed of wheat and barley from 0 to 53%.[46]

Finally, location of seeds on the seed-bearing plant may affect seed transmission. Jones[47] found that the frequency of seed transmission of cucumber mosaic *Cucumovirus* in *Lupinus angustifolius* decreased in those pods that matured earliest, indicating that the time of infection prior to seed set is an important factor for seed transmission to occur.

C. DETERMINING SEED TRANSMISSION

It is impossible to prove that seed transmission never occurs in any particular virus/host combination, because a negative result cannot preclude the possibility of a rare instance. Nevertheless, if a large enough sample is assayed, the probability of seed transmission can be assessed. Testing progeny seedlings is the only certain way of determining if seed transmission occurs. However, assays (infectivity or serological) based either on individual seeds or on homogenates of bulked seed samples are often able to predict the likelihood of transmission. Slack and Shepherd[48] were able to detect barley stripe mosaic *Hordeivirus* (BSMV) directly in individual barley seeds by placing them in agar containing BSMV antibodies (radial immunodiffusion) and observing for virus-specific precipitation lines. This method was also used for detecting brome mosaic *Bromovirus* in small grain cereals.[49] Other serological tests such as ELISA and immunoelectroblotting have extended the sensitivity of detection, allowing detection of one infected seed in homogenates of 30 to more than 400 healthy seeds, depending on the virus and on the quality of virus antiserum used.[49-51] Using such techniques, mathematical models with confidence limits to predict seed-borne transmission have been devised.[52] However, in developing these models or assessing results of seed assays, it is important to recognize that many viruses carried externally on the seed coat, or internally in endosperm or embryo, may not ultimately infect the developing seedling. Using ELISA on only small portions of seed parts, Culver and Sherwood,[51] studying peanut stripe *Potyvirus* in peanut seed, and Nolan and Campbell[50] studying squash mosaic *Potyvirus* in cucurbit seed, found that no cotyledon or embryo, respectively, free from virus ever produced virus-infected seedlings. However, some seedlings grown from such sources infected with virus also remained uninfected.

D. CONCLUSIONS

Bennett,[53] in a perceptive review of seed transmission of viruses, noted some features common to viruses that are seed-borne. In particular, he observed that viruses that are not readily transmitted by inoculation of plant sap are usually not seed-borne. This general rule is still true, with one major exception: cryptoviruses are not transmitted mechanically or by grafting, yet are seed-transmitted with high frequency, often >90%. At the time of Bennett's writing, most plant viruses had not been classified into groups. With the benefit of such grouping it is now possible to make some general comments on features about virus groups and seed transmission.[54]

1.   There is no apparent relation between seed transmission and particle morphology or nucleic acid type.
2.   Viruses transmitted by leafhoppers or planthoppers (fijiviruses, geminiviruses — subgroup A, "maize chlorotic dwarf virus", marafiviruses, *Phytoreovirus,* tenuiviruses, and some rhabdoviruses) or, by aphids in a persistent manner (luteoviruses and some rhabdoviruses), are not seed-transmitted. A major exception to this is pea enation mosaic virus which is transmitted in a persistent manner by aphids but can also be transmitted mechanically. Another, more complex example is the "umbraviruses" which are mechanically transmitted and also aphid-transmitted in a persistent manner, although only when present in plants with a luteovirus (see Chapter 13).
3.   Viruses that belong to virus groups in which some members are transmitted by nematodes (*Nepovirus, Tobravirus*), beetles (*Bromovirus, Comovirus, Sobemovirus, Tymovirus*), thrips (*Tospovirus*), or by aphids in a nonpersistent manner (alfalfa mosaic virus, *Carlavirus, Cucumovirus, Potyvirus*) are often seed-borne.
4.   Viruses in groups that tend to reach high concentrations in host plants (*Capillovirus, Ilarvirus, Hordeivirus, Necrovirus, Potexvirus, Tobamovirus, Tombusvirus, Tymovirus*) are often seed-borne in some hosts.

Detection of seed transmission in particular hosts of the virus under study may therefore be helpful in eliminating the likelihood of the virus belonging to certain groups. However failure to detect seed transmission is much less helpful because of the many possibilities that restrict such transmission even in viruses known to be seed-borne.

VI. TRANSMISSION BY POLLINATION

A. INTRODUCTION

Several viruses are transmitted vertically through infected pollen to seedling progeny. For a few of these, evidence exists that the virus in infected pollen is also able to infect the plant pollinated, i.e., it is transmitted horizontally. Although pollination was postulated in 1918[55] as a mechanism of horizontal transmission, Das and Milbrath[56] were the first to demonstrate its occurrence. When pollen from *Cucurbita maxima* plants infected with Prunus necrotic ringspot *Ilarvirus* was used to pollinate healthy plants, about 10% of the pollinated plants became infected with the virus. Subsequently, work has indicated that this mechanism of transmission occurs for several other ilarviruses,[57-59] raspberry bushy dwarf virus,[60] and some nepoviruses.[61-64]

B. DETECTION OF HORIZONTAL TRANSMISSION BY POLLEN AND POSSIBLE MECHANISMS OF TRANSMISSION

Alvarez and Campbell[65] showed that infectivity of squash mosaic *Potyvirus* was removed from pollen grains of infected melon by thorough washing, indicating that virus was located on the outside and not the inside of the pollen grains. Subsequently, however, virus particles and virus antigen have been identified on the surface and/or within pollen grains of several virus/host combinations.[64-67] Conceivably, therefore, virus could be transmitted horizontally to plants from infected pollen either through (1) the escape of virus into the pollinated plant following fertilization of the ovule, or (2) mechanical inoculation of virus attached to the external surface of pollen grains. It is difficult to prove the former possibility except by circumstantial evidence, e.g., absence of transmission in the absence of flowers. By contrast, convincing evidence is now being accumulated with some viruses for the latter possibility. The infectivity of virus-carrying pollen is well documented[64] and, in association with virus-infected pollen, thrips have been shown to transmit tobacco streak and Prunus necrotic ringspot ilarviruses.[68,69] In the light of this evidence, it may be necessary to re-examine earlier evidence of pollen transmission, as it is possible that brushes used to pollinate stigmas with virus-infected pollen may have caused sufficient abrasion for mechanical inoculation to occur. Certainly, trees and shrubs, in which horizontal transmission of some viruses by pollen is believed to occur naturally, support a wide variety of insects as well as vast volumes of pollen during the flowering season. It is very plausible, therefore, that insects, through their feeding and foraging activities on foliage and flowers covered in virus-carrying pollen, could mechanically inoculate the plants.

To determine whether virus is carried externally and/or internally, pollen should be washed several times to remove any virus particles attached to the exterior and the washings then assayed separately from homogenates of the washed pollen grains. Some viruses can be held very strongly to surfaces. For example, cherry leaf roll *Nepovirus* on birch pollen required five or more washes with phosphate-buffered saline containing the detergent Tween 20 before most of the detectable virus antigen was removed from the pollen surface.[70] Provided that the sample size is large enough, failure to detect virus in, or on, pollen, by the methods described above, almost certainly indicates that no pollen transmission will occur. Furthermore, as all viruses showing horizontal transmission by pollen in a given host are also known to be transmitted vertically through its seed, failure to detect seed transmission also indicates that horizontal transmission by pollen is most unlikely to occur. However detection of virus in such tests indicates only an association of virus or, for serological tests, only an association of virus antigen with these plant parts; it does not show that transmission occurs through pollen to the plant pollinated, or to seed and to progeny seedlings.

C. CONCLUSIONS

Only a few plant viruses are known to be transmitted horizontally through pollen. Potentially therefore, the finding that a virus is transmitted in this way is a major step not only in virus diagnosis, but also virus identification. However, in practice, the frequency of horizontal transmission is often low, even using hand pollination with virus-infected pollen. Unequivocal evidence of horizontal transmission by pollen is therefore not easy to obtain.

All the viruses reported to be transmitted horizontally to plants through pollen are also seed-borne and transmissible by mechanical inoculation of sap. These additional properties therefore offer easier approaches for virus diagnosis.

REFERENCES

1. **Boccardo, G., Lisa, V., and Milne, R. G.,** Cryptic viruses in plants, in *Double-Stranded RNA Viruses,* Compans, R. W. and Bishop, D. H. L., Eds., Elsevier, New York, 1983, 425.
2. **Davidson, T. R.,** Tuber graft transmission of potato leaf roll, *Can. J. Agric. Sci.,* 35, 238, 1955.
3. **Cropley, R.,** Erratic transmission of strawberry viruses by grafting excised leaves, *Annu. Rep. East Malling Res. Stat. (Kent) 1957,* 1958, p. 124.
4. **Cadman, C. H., Dias, H. F., and Harrison, B. D.,** Sap-transmissible viruses associated with diseases of grape vines in Europe and North America, *Nature (London),* 187, 577, 1960.
5. **Santamour, F. S.,** Cambial peroxidase enzymes related to graft incompatibility in red oak, *J. Environ. Hortic.,* 6, 87, 1988.
6. **Muzik, T. J. and LaRue, C. D.,** Further studies on the grafting of monocotyledonous plants, *Am. J. Bot.,* 41, 448, 1954.
7. **Bringhurst, R. S. and Voth, V.,** Strawberry virus transmission by grafting excised leaves, *Plant Dis. Rep.,* 40, 596, 1956.
8. **Converse, R. H.,** Detection and elimination of virus and virus-like diseases in strawberry, in *Virus Diseases of Small Fruits,* Converse, R. H., Ed., USDA Agric. Handbook No. 631, U.S. Department of Agriculture, Washington, D.C., 1987, 4.
9. **Murphy, P. A. and McKay, R.,** Methods for investigating the virus diseases of the potato, and some results obtained by their use, *Sci. Proc. R. Dublin Soc.,* 18, 169, 1926.
10. **Goss, R. W.,** Transmission of potato spindle-tuber by cutting knives and seed piece contact, *Phytopathology,* 16, 299, 1926.
11. **Bennett, C. W.,** Influence of contact period on the passage of viruses from scion to stock in Turkish tobacco, *Phytopathology,* 33, 818, 1943.
12. **Kunkel, L. O.,** Contact periods in graft transmission of peach viruses, *Phytopathology,* 28, 491, 1938.
13. **Barbara, D. J., Jones, A. T., Henderson, S. J., Wilson, S. C., and Knight, V. H.,** Isolates of raspberry bushy dwarf virus differing in *Rubus* host range, *Ann. Appl. Biol.,* 105, 49, 1984.

14. **Jones, A. T., Mitchell, M. J., and Brown, D. J. F.,** Infectibility of some new raspberry cultivars with arabis mosaic and raspberry ringspot viruses and further evidence for variation in British isolates of these two nepoviruses, *Ann. Appl. Biol.,* 115, 57, 1989.

15. **Bennett, C. W.,** Acquisition and transmission of plant viruses by dodder *(Cuscuta subinclusa), Phytopathology,* 30, 2, 1940.

16. **Costa, A. S.,** Multiplication of viruses in dodder, *Cuscuta campestris, Phytopathology,* 34, 151, 1944.

17. **Miyakawa, T. and Yoshii, H.,** Transmission of tobacco mosaic virus by the dodder *(Cuscuta japonica)* and its inhibitory effect on the virus activity, *Sci. Bull. Fac. Agric. Kyushu Univ.,* 12, 143, 1951.

18. **Schmelzer, K.,** Beitrage zur Kenntris der Ubertragbarkeit von Vinen durch Cuscuta-Arten, *Phytopathol. Z.,* 28, 1, 1956.

19. **Fulton, R. W.,** Transmission of plant viruses by grafting, dodder, seed, and mechanical inoculation, in *Plant Virology,* Corbett, M. K. and Sisler, H. D., Eds., University of Florida Press, Gainsville, 1964, 39.

20. **Cochran, G. W.,** Effect of shading techniques on transmission of tobacco mosaic virus through dodder, *Phytopathology,* 36, 396, 1946.

21. **Holmes, F. O.,** Local lesions in tobacco mosaic, *Bot. Gaz.,* 87, 39, 1929.

22. **Holmes, F. O.,** Local lesions of mosaic in *Nicotiana tabacum* L., *Contrib. Boyce Thomp. Inst.,* 3, 163, 1931.

23. **Mayer, A. E.,** Uber die Masaikkrankheit des Tabaks, *Die Landwirtsch. Versuchst.,* 32, 451, 1886.

24. **Kado, C. I.,** Mechanical and biological inoculation principles, in *Principles and Techniques in Plant Virology,* Kado, C. I. and Agrawal, H. O., Eds., van Nostrand Reinhold, New York, 1972, 3.

25. **Yarwood, C. E. and Fulton, R. W.,** Mechanical transmission of plant viruses, in *Methods in Virology,* Vol. 1, Maramorosch, K. and Koprowski, H., Eds., Academic Press, New York, 1967, 237.

26. **Matthews, R. E. F.,** *Plant Virology,* 3rd ed., Academic Press, New York, 1991.

27. **Reid, M. S. and Matthews, R. E. F.,** On the origin of the mosaic induced by turnip yellow mosaic virus, *Virology,* 28, 563, 1966.

28. **Brunt, A. A. and Kenten, R. H.,** Mechanical transmission of cocoa swollen-shoot virus, *Virology,* 12, 328, 1960.

29. **Harrison, B. D. and Pierpoint, W. S.,** The relation of polyphenoloxidase in leaf extracts to the instability of cucumber mosaic and other viruses, *J. Gen. Microbiol.,* 32, 417, 1963.

30. **Fulton, R. W.,** Purification and some properties of tobacco streak and Tulare apple mosaic viruses, *Virology,* 32, 153, 1967.

31. **Bitterlin, M. W., Gonsalves, D., and Scorza, R.,** Improved mechanical transmission of tomato ringspot virus to *Prunus* seedlings, *Phytopathology,* 77, 560, 1987.

32. **Garnsey, S. M., Gonsalves, D., and Purcifull, D. E.,** Mechanical transmissionn of citrus tristeza virus, *Phytopathology,* 67, 965, 1977.

33. **Louie, R. and Lorbeer, J. W.,** Mechanical transmission of onion yellow dwarf virus, *Phytopathology,* 56, 1020, 1966.

34. **Toler, R. W. and Hebert, T. T.,** Properties and transmission of soil-borne oat mosaic virus, *Phytopathology,* 54, 428, 1964.

35. **Laidlaw, W. M. R.,** Mechanical aids to improve the speed and sensitivity of plant virus diagnosis by the biological test method, *Ann. Appl. Biol.,* 108, 309, 1986.

36. **Polson, A. and von Wechmar, M. B.,** A novel way to transmit plant viruses, *J. Gen. Virol.,* 51, 179, 1980.

37. **Fulton, R. W.,** The effect of dilution on necrotic ringspot virus infectivity and the enhancement of infectivity by noninfective virus, *Virology,* 18, 477, 1962.

38. **Kammen, van A.,** The relationship between the components of cowpea mosaic virus. I. Two ribonucleoprotein particles necessary for the infectivity of CPMV, *Virology,* 34, 312, 1968.

39. **van Vloten-Doting, L. and Jaspars, E. M. J.,** Enhancement of infectivity by combination of two ribonucleic acid components from alfalfa mosaic virus, *Virology,* 33, 684, 1967.

40. **Diener, T. O.,** Viroids, *Adv. Virus Res.,* 17, 295, 1972.

41. **Jones, A. T. and Mayo, M. A.,** Purification and properties of elm mottle virus, *Ann. Appl. Biol.,* 75, 347, 1973.

42. **Copper, V. C.,** The Seed Transmission of Cherry Leaf Roll Virus, Ph.D. Thesis, University of Birmingham, 1976, p. 184.

43. **Crowley, N. C.,** Studies on the time of embryo infection by seed transmitted viruses, *Virology,* 8, 116, 1959.

44. **Warwick, D. and Demski, J. W.,** Susceptibility and resistance of soybeans to peanut stripe virus, *Plant Dis.,* 72, 19, 1988.

45. **Tu, J. C.,** Effect of different strains of soybean mosaic virus on growth, maturity, yield, seed mottling and seed transmission in several soybean cultivars, *J. Phytopathol.,* 126, 231, 1989.

46. **McKinney, H. H. and Greeley, L. W.,** Biological characteristics of barley stripe mosaic virus strains and their evolution, *USDA Tech. Bull.,* 1324, 1, 1965.

47. **Jones, R. A. C.,** Seed-borne cucumber mosaic virus infection of narrow-leafed lupin (*Lupinus angustifolius*) in Western Australia, *Ann. Appl. Biol.,* 113, 507, 1988.

48. **Slack, S. A. and Shepherd, R. J.,** Serological detection of seed-borne barley stripe mosaic virus by simplified radial diffusion technique, *Phytopathology,* 65, 948, 1975.

49. **von Wechmar, M. B., Kaufmann, A., Desmarais, F., and Rybicki, E. P.,** Detection of seed-transmitted brome mosaic virus by ELISA, radial immunodiffusion and immunoelectroblotting tests, *Phytopathol. Z.,* 109, 341, 1984.

50. **Nolan, P. A. and Campbell, R. N.,** Squash mosaic virus detection in individual seeds and seed lots of cucurbits by enzyme-linked immunosorbent assay, *Plant Dis.,* 68, 971, 1984.

51. **Culver, J. N. and Sherwood, J. L.,** Detection of peanut stripe virus in peanut seed by an indirect enzyme-linked immunosorbent assay using a monoclonal antibody, *Plant Dis.,* 72, 676, 1988.

52. **Maury, Y., Duby, C., Bossennec, J.-M., and Bouduzin, G.,** Group analysis using ELISA: determination of the level of transmission of soybean mosaic virus in soybean seed, *Agronomie,* 5, 405, 1985.

53. **Bennett, C. W.,** Seed transmission of plant viruses, *Adv. Virus Res.,* 14, 221, 1969.

54. **Stace-Smith, R. and Hamilton, R. I.,** Inoculum thresholds of seedborne pathogens. Viruses, *Phytopathology,* 78, 875, 1988.

55. **Reddick, D. and Stewart, V. B.,** Varieties of beans susceptible to mosaic, *Phytopathology,* 8, 530, 1918.

56. **Das, C. R. and Milbrath, J. A.,** Plant-to-plant transfer of stone fruit ringspot virus in squash by pollination, *Phytopathology,* 51, 489, 1961.

57. **Converse, R. H. and Lister, R. M.,** The occurrence and some properties of black raspberry latent virus, *Phytopathology,* 59, 325, 1969.

58. **George, J. A. and Davidson, T. R.,** Pollen transmission of necrotic ring spot and sour cherry yellows viruses from tree to tree, *Can. J. Plant Sci.,* 43, 276, 1963.

59. **Cameron, H. R., Milbrath, J. A., and Tate, L. A.,** Pollen transmission of prunus necrotic ringspot virus in prune and sour cherry orchards, *Plant Dis. Rep.,* 57, 241, 1973.

60. **Murant, A. F., Chambers, J., and Jones, A. T.,** Spread of raspberry bushy dwarf virus by pollination, its association with crumbly fruit, and problems of control, *Ann. Appl. Biol.,* 77, 271, 1974.

61. **Mircetich, S. M., Sanborn, R. R., and Ramos, D. E.**, Natural spread, graft-transmission, and possible etiology of walnut blackline disease, *Phytopathology*, 70, 962, 1980.
62. **Kyriakopoulou, P. E., Rana, G. L., and Roca, F.**, Geographic distribution, natural host range, pollen and seed transmissibility of artichoke yellow ringspot virus, *Ann. Inst. Phytopathol. Benaki (N.S.)*, 14, 137, 1985.
63. **Childress, A. M. and Ramsdell, D. C.**, Detection of blueberry leaf mottle in highbush blueberry pollen and seed, *Phytopathology*, 76, 1333, 1986.
64. **Cooper, J. I., Kelley, S. E., and Massalski, P. R.**, Virus-pollen interactions, *Adv. Dis. Vector Res.*, 5, 221, 1988.
65. **Alvarez, M. and Campbell, R. N.**, Transmission and distribution of squash mosaic virus in seeds of cantaloupe, *Phytopathology*, 68, 257, 1978.
66. **Hamilton, R. I., Leung, E., and Nichols, C.**, Surface contamination of pollen by plant viruses, *Phytopathology*, 67, 395, 1977.
67. **Cole, A., Mink, G. I., and Regev, S.**, Prunus necrotic ringspot virus located on the surface of pollen from infected almond and cherry trees, *Phytopathology*, 72, 1542, 1982.
68. **Sdoodee, R. and Teakle, D. S.**, Transmission of tobacco streak virus by *Thrips tabaci*: a new method of plant virus transmission, *Plant Pathol.*, 36, 377, 1987.
69. **Greber, R. S., Klose, M. J., Milne, J. R., and Teakle, D. S.**, Transmission of prunus necrotic ringspot virus using plum pollen and thrips, *Ann. Appl. Biol.*, 118, 589, 1991.
70. **Massalski, P. R. and Cooper, J. I.**, The location of virus-like particles in the male gametophyte of birch, walnut and cherry naturally infected with cherry leaf roll virus and its relevance to vertical transmission of the virus, *Plant Pathol.*, 33, 255, 1984.

Chapter 4

VIRUS TRANSMISSION THROUGH SOIL AND BY SOIL-INHABITING ORGANISMS IN DIAGNOSIS

A. T. Jones

TABLE OF CONTENTS

0-8493-4284-8/93/$0.00 + $.50

I. GENERAL INTRODUCTION

Early indications that a virus causing disease in crops is transmitted through the soil are often given by the pattern of distribution of the disease. Typically, soil-borne infections tend to occur in patches that may vary in size from only a few to many meters in diameter; in exceptional instances whole fields may be affected. The infection of healthy plants grown in soil taken from such diseased areas provides evidence that the agent is probably soil-borne. This evidence is strongly supported if similar plants remain healthy when grown in soil from unaffected areas of the same field. Beijerinck[1] was probably the first to provide such evidence when he showed that healthy tobacco plants became infected with tobacco mosaic *Tobamovirus* (TMV) when they were grown in soil taken from around the roots of TMV-infected plants. Several viruses described later were also shown to be soil-borne but, although it was considered by a few workers that invertebrates in the soil might act as vectors for some of these viruses, it was not until 1958[2] that the role of nematodes as virus vectors was first demonstrated. About the same time, evidence was accumulating to indicate that the fungus, *Olpidium brassicae,* was a vector of the lettuce big vein agent[3,4] and of tobacco necrosis *Necrovirus.*[5] Subsequently, many soil-borne viruses have been found to have either nematode (Table 2) or fungus (Table 3) vectors. Nevertheless, for some soil-borne viruses (Table 1), extensive studies over more than 30 years have failed to implicate any vector in their spread, despite some earlier claims. In diagnosis, therefore, it is important to distinguish between these three different modes of transmission through soil.

Initial tests to determine transmission of virus through soil usually involve planting healthy virus-infectible "bait plant" seedlings in soil from affected areas of the crop. After several weeks the bait plants are either assessed for virus symptoms or carefully removed from the soil, their roots washed free of adhering soil particles and debris, and the roots and tops assessed for virus infection separately, usually by infectivity or serological tests. If infection occurs it suggests that the disease agent is soil-borne. It is then necessary to determine the mode of soil transmission. As nematodes are killed by air drying soil and/or by rubbing soil particles together to form fine particles, loss of soil infectivity following these treatments suggests a nematode vector may be involved. While the infectivity of some soil-water-borne viruses can also be greatly decreased by such treatments, it is not usually eliminated. However, viruliferous fungal resting spores are not greatly inactivated by these treatments and can remain viable under dry conditions for many years.

Table 1. Viruses Reported to be Transmitted to Plants Through Soil in the Absence of Vectors

Virus	Particle shape	Virus group	Ref.
Carnation mottle	Isometric	*Carmovirus*	100
Galinsoga mosaic	Isometric	*Carmovirus*	101
Carnation ringspot	Isometric	*Dianthovirus*	45
Red clover necrotic mosaic	Isometric	*Dianthovirus*	11
Potato X	Flexuous rod	*Potexvirus*	102
Southern bean mosaic	Isometric	*Sobemovirus*	16
Sowbane mosaic	Isometric	*Sobemovirus*	100
Cucumber green mottle mosaic	Rod	*Tobamovirus*	103
Tobacco mosaic	Rod	*Tobamovirus*	102
Tomato mosaic	Rod	*Tobamovirus*	104
Cucumber leaf spot	Isometric	*Tombusvirus*	15
Cymbidium ringspot	Isometric	*Tombusvirus*	10
Tomato bushy stunt	Isometric	*Tombusvirus*	14

II. SOIL TRANSMISSION WITHOUT THE APPARENT NEED FOR VECTORS

A. THE VIRUSES AND POSSIBLE MECHANISMS OF TRANSMISSION

At one time, only TMV was regarded as being spread through soil without the need for a vector, but currently at least 13 viruses appear to share this property (Table 1). Recent studies of river water in many parts of the world have shown they contain relatively high concentrations of several of these viruses, together with apparently previously undescribed potexviruses, tobamoviruses, and tombusviruses.[6] For many of these viruses, the significance of their abiotic transmission to plants in soil or soil water for their ecology and epidemiology, is not well studied. However, it is clear that abiotic transmission to plants occurs more frequently and for a wider range of viruses than was once thought.

Almost all the viruses reported to be transmitted in this way reach high concentrations in plants, are released from plant roots, are very stable, infect a wide range of plant species and genera, are readily transmitted to plants by mechanical inoculation, and, with only a few exceptions, appear not to be transmitted by air-borne vectors.

The precise mechanism(s) whereby plants become infected with virus abiotically in soil is not clear. However, it is known that most of the viruses listed in Table 1 are released in significant amounts from roots of infected plants, whereas other plant viruses that have been tested are not.[7-11] Hence these soil-borne viruses are readily detected in drainage water from soils containing infected plants and, as noted above, many have been found commonly, in relatively large concentrations, in river water.[6] The inoculum level of these viruses in soils can therefore be very high. Furthermore, their infectivity in soils can be maintained for many months after the virus-infected plants have died.[8,9,12] Virus particles are known to be adsorbed to clay particles and organic plant debris[13,14] and this considerably increases their longevity in soils and their stability to extremes of environmental conditions.[8,9,12,13]

Kegler and Kegler[15] found that tomato bushy stunt *Tombusvirus* (TBSV) and carnation ringspot *Dianthovirus* (CaRSV) were transmitted more efficiently to plants grown in sterile sand than those in sterile nutrient solution. This suggested that wounds to roots, caused by rubbing against sand particles during growth, may provide viruses with a means of entry into plants. However, studies by Teakle[16] indicated that transmission of southern bean mosaic *Sobemovirus* (SBMV) in soil was not significantly affected by the presence of the root damaging nematode, *Meloidogyne* spp. This result, and the fact that workers have reported virus infections in plants in nutrient solutions,[15,17,18] may indicate that mechanisms of virus entry into plant roots other than through wounds may occur.

B. TESTS FOR ABIOTIC TRANSMISSION

To determine if abiotic transmission in soil occurs, it is necessary to eliminate, as far as is possible, all biotic factors (apart from the test plants) that may be present in the soil/compost used for culturing plants and in the close vicinity of these plants, e.g., pots, benches, and glasshouse/laboratory structures. As viruses most likely to be transmitted abiotically occur in very high concentrations in plants and are very stable, strict hygiene and isolation from likely sources of virus contamination are also essential. For most experiments this has usually been achieved by using sterile (autoclaved) pots and growing medium (soil, compost, or sand), isolating plant roots from benches (for example by placing experimental pots inside sterile boxes or pots to prevent plant roots from contacting benches), watering plants carefully from the top and avoiding splashing, and providing adequate separation between pots.

Inoculum sources for such tests have included sterile compost mixed with leaves from virus-infected plants, growing plants in sterile compost and mechanically inoculating their leaves with virus, and watering sterile soil with suspensions of virus from infected plants.

Test systems used have included (1) growing bait plants from seed in sterile compost and watering the soil with virus suspensions, being careful not to contaminate the aerial parts of the plants;[16] (2) mechanically inoculating virus to leaves of plants growing in sterile compost and then sowing seed of bait plants in the same pot. As the germinating bait plants emerge above the soil, their aerial parts are prevented from touching the virus source plants by using screens of glass or plastic. (3) To minimize the risk of bait plant infection through wounds or damage in seed coats, Teakle[16] planted seed in a sterile layer of compost that covered virus-contaminated compost which was the inoculum source. In this way only actively growing roots extended into the inoculum zone. Obviously, as with any experiment, adequate numbers of directly comparable control plants, free from virus inoculum, are essential, especially as the numbers of virus transmissions that occur may be small.

Where virus transmitted through soil to bait plant roots infects the tops of these plants, these parts (usually leaves) can be assayed with a minimum risk of contamination to determine if virus transmission has occurred. However, for many soil-borne virus/host combinations, including viruses transmitted by nematodes and fungi, virus that enters plants through natural infection of the roots frequently fails to enter the tops of plants even though it may readily infect all plant parts when mechanically inoculated to leaves. In these situations, and in tests where virus is added to the soil as plant debris or in suspensions, and where viruses are exuded from roots, it is essential to eliminate contaminating virus present in or on soil particles and debris attached to bait plant roots before testing roots for the presence of virus within them. Teakle,[16] using SBMV, overcame this problem by immersing washed bait plant roots in 10% (w/v) trisodium phosphate solution for 1 min and then thoroughly rinsing them under running tap water before extracting sap for infectivity assays. He claimed that such treatment removed 99 to 100% of the surface contamination of roots.

III. TRANSMISSION BY SOIL-INHABITING NEMATODES

A. INTRODUCTION — THE VIRUSES AND THE VECTORS

To date, all known plant virus vector nematodes are ectoparasites restricted to species from four genera in the order Dorylaimida. Strangely, no virus vectors are reported in the order Tylenchida which contains several plant-parasitic species. The viruses transmitted by nematodes belong to only two plant virus groups, *Tobravirus* and *Nepovirus*, although some debate exists about the possibility of subdividing the *Nepovirus* group. Tobraviruses have rigid rod-shaped particles and are transmitted by species of *Trichodorus* and *Paratrichodorus* nematodes in the order Trichodoridae, whereas nepoviruses have isometric particles and are transmitted by species of *Longidorous* and *Xiphinema* nematodes in the order Dorylaimidae (Table 2). All recognized tobraviruses are known to be nematode-transmitted but, of nearly 40 viruses recognized as members or tentative members of the *Nepovirus* group, only 12 have been clearly demonstrated to be transmitted by nematodes (Table 2). Extensive tests indicate that at least two nepoviruses, cherry leaf roll and blueberry leaf mottle, are probably not nematode-borne.[19,20]

Virus vector nematodes usually feed on young tissue near root tips, using their stylets to penetrate cells to feed on their contents. Trichodorid vectors tend to feed on epidermal cells, using their comparatively short, curved, toothlike stylet. By contrast, Dorylaimid vectors use their longer stylets to penetrate and feed in vascular tissue. Virus-vector relationships are similar in principle for each of these vector types: virus is acquired from plants within 1 h, is

Table 2. Viruses with Known Nematode Vectors

Virus	Vector species	Ref.[a]
Trichodorid vectors (tobraviruses)		
Pea early browning ⎤	*Paratrichodorus* and *Trichodorus*	120
Pepper ringspot ⎬	spp.	347
Tobacco rattle ⎦		346
Longidorid vectors (nepoviruses[b])		
Artichoke Italian latent	*Longidorus apulus[c], L. fasciatus*	176
Mulberry ringspot	*L. martini*	142
Raspberry ringspot	*L. elongatus, L. macrosoma*	198
Tomato black ring (TBRV)	*L. elongatus, L. attenuatus*	38
Arabis mosaic	*Xiphinema diversicaudatum*	16
Cherry rasp leaf	*X. americanum sensu lato*	159
Grapevine chrome mosaic	*X. index*?	103
Grapevine fanleaf	*X. index, X. italiae*	28
Peach rosette mosaic	*X. americanum sensu lato, L. diadecturus*	150
Strawberry latent ringspot (SLRV)	*X. diversicaudatum*	126
Tobacco ringspot (TobRSV)	*X. americanum sensu lato*	309
Tomato ringspot	*X. americanum sensu lato X. californicum, X. rivesi*	290

[a] Numbers given are those in *AAB Descriptions of Plant Viruses,* Association of Applied Biologists, Wellesbourne.

[b] Other viruses recognized as nepoviruses or as tentative nepoviruses, but for which little or no evidence exists for transmission by nematodes, are arracacha A, arracacha B, artichoke veinbanding, artichoke yellow ringspot, blueberry leaf mottle, cassava green mottle, cassava American latent, cherry leaf roll, chicory yellow mottle, cocoa necrosis,[a] crimson clover latent, cycas necrotic stunt, grapevine Bulgarian latent, hibiscus latent ringspot, lucerne Australian latent, lucerne Australian symptomless, myrobalan latent ringspot,[b] olive latent rinspot, potato black ringspot,[c] potato U, rubus Chinese seed-borne,[d] satsuma dwarf, and tomato top necrosis. Superscripts indicate viruses serologically related to viruses known to be transmitted by nematodes: [a,b] = TBRV, [c] = TobRSV, [d] = SLRV.

[c] Species originally identified as *L. attenuatus*[105] but redescribed as *L. apulus.*[106]

retained for many weeks (Trichodorids and *Longidorus* spp.) to many months (*Xiphinema* spp.) but is lost after moulting, and is transmitted to plants in access feeds of 1 h. Both adults and larvae can transmit, but virus is not transmitted transovarially, nor does it multiply in the nematode. Despite these similarities, there is strong vector specificity not only between different viruses (Table 2) but also between virus strains[21,22] and variation exists between vector populations in their abilities to transmit virus.[23-25]

B. SAMPLING, ISOLATION, HANDLING, AND CULTURE OF VECTORS

1. Sampling

Natural outbreaks of nematode-borne viruses are usually characterized by the occurrence in the crop of patches of infection that increase in size only slowly and which persist in these discrete areas for long periods, even after fallowing. Having implicated a virus with such disease outbreaks and shown that it is soil-borne, initial studies may begin with sampling soil around the roots of plants in affected and unaffected areas to determine the kinds of nematodes present and their relative abundance. Later, more detailed studies may involve mapping nematode distribution and relative abundance within the crop and this can be determined both laterally and vertically in soil. Various soil sampling methods for detecting and mapping vector nematodes in field soil have been described.[26-28] More recently, the application to sampling of geostatistical methods, that take account of the patchy distribution of viruliferous nematodes, has shown improvements over earlier approaches.[29]

For all sampling procedures it should be noted that the largest numbers of nematodes are usually found around young roots of host plants, but nematode species differ in their vertical distribution and abundance in different crops, soil types, and under different moisture conditions.[30,31] Thus, most of the preferred hosts of virus vector *Longidorus* spp. are surface-rooting and, as might be expected, these nematodes are found most frequently in the top 20 cm of soil.[32] By contrast, *Xiphinema diversicaudatum* feeds on a wide variety of plants, including deep-rooted perennials and, depending on the plant species, the nematode is found at 20 to 60 cm depths. *X. index* on grapevines occurs at 30 to 60 cm in stony, shallow soils but at up to 3.6 m in deep fertile soils.[33] The distribution of *Trichodorus* and *Paratrichodorus* spp. is very variable in soils but they tend to occur in greater abundance in sandy, light soils. Following prolonged dry conditions, all vector nematodes tend to be found at greater soil depths.

Nematodes are prone to injury and can be killed by drying, overheating, or suffocating. Samples should therefore be collected and transported with the minimum disturbance of the soil. The top few centimeters of field soil are usually scraped away and the soil beneath removed with a trowel or spade; at greater depths an auger is usually necessary. For holding soil samples, plastic bags or containers are preferred to moisture-absorbing paper ones. Samples should be kept cool until use.

2. Isolation

Of many methods described for isolating nematodes from soil, those based on the wet-sieving method of Cobb[34] are the simplest and most commonly used. The technique is essentially one of elutriation and flotation. Sample size and nematode numbers can affect the efficiency of recovery;[35] for most purposes 200 to 300 g of soil/sample is suitable. A modified form of the technique routinely used at the Scottish Crop Research Institute (SCRI)[118] is illustrated diagrammatically in Figure 1. The soil sample is soaked in 2 to 3 mm of water for 1 to 2 h (Figure 1a) and then 400 to 800 ml of water is added and the suspension agitated vigorously to separate nematodes from soil particles and debris (Figure 1b). The mixture is allowed to settle for 15 to 20 s before decanting the supernatant into a second container through a 2-

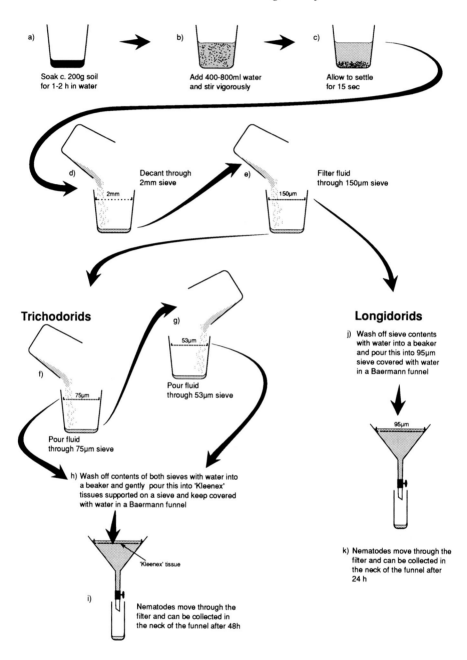

FIGURE 1. Outline of a method for isolating nematodes from soil.

mm sieve to remove the larger soil particles, debris, and small stones (Figure 1c, d). The fluid is then sieved progressively through sieves of decreasing mesh size, (mesh aperture diameter 150, 75, and 53 μm; Figure 1e, f, g). The 150-μm mesh sieve traps mainly the larger nematodes such as *Longidorus* and *Xiphinema* spp., including the juveniles, while those retained by the other two sieves are the smaller nematodes. *Trichodorus* and *Paratrichodorus*. Material on the sieves must not be allowed to dry and should be washed immediately into a beaker with water.

Further clarification is usually necessary to remove particulate matter and dead and fragmented nematodes. This is usually done on a Baermann funnel. The suspended contents from the sieves is stirred, allowed to settle for 15 s and the supernatants poured onto a 95-μm mesh sieve (large [Longidorid] nematodes) (Figure 1j, k) or a layer of two-ply Kleenex™ tissue (small [Trichodorid] nematodes) (Figure 1h, i), ensuring that the surface of these filters is submerged in water at all times. The sieves are placed on the top of a glass funnel filled with aerated water at about 20°C. Living nematodes actively move through the filters into the water beneath and settle by gravity in the narrow neck of the filter funnel that is sealed with a clip (Figure 1k, i). Because of the confined space in this part of the apparatus, it is essential to use aerated water to preserve nematode activity. Most Longidorid and Trichodorid nematodes collect in this area within 24 and 48 h, respectively, but they can be collected for use at 2 to 3 h intervals.

3. Handling and Identification

Nematodes should be concentrated into a minimum volume of water, free from other contaminating material, and examined in a glass observation dish. A binocular microscope with a magnification of $\times 20$ to $\times 40$ is required for nematode identification and handling. Nematodes can be manipulated with a fine metal tool or eyelash. For hand picking species-specific nematodes, the animals are gently raised up with a fine tool to lie just beneath the surface of the water and then in a smooth but rapid upward motion removed from the fluid and deposited in a separate dish. Various keys exist for identification of *Xiphinema* spp.,[36,37] *Longidorus* spp.,[38] and *Trichodorus* spp.[39]

Specimens can be preserved for permanent records or for further identification by gently heating in water to 65°C for 2 min to kill them and then fixing them. Various fixatives and procedures are described by Hooper.[40]

4. Culture

Ideally, a virus-free population of a specific nematode species is desirable, but in practice this may be difficult to achieve. Frequently, field populations of nematodes are the only sources available, or those available with sufficient numbers of nematodes. Such field sources may contain more than one nematode species and some of these may be viruliferous. Furthermore, most nematode-borne viruses are also seed transmitted, often to a high frequency in common weed species. The germination of such infected seed following disturbance of the soil may unwittingly provide an inoculum source for initially virus-free nematodes. When field soil is used as a source for culturing nematodes it is therefore important to determine the nematode species present, whether they are viruliferous, and also whether other soil-borne viruses, e.g., tobacco necrosis *Necrovirus,* which is a common soil contaminant, are present. The use of bait plants can determine the presence of infective virus and representative samples of soil can be used for nematode extraction.

Some nematode species are relatively easy to culture in the glasshouse and reach large numbers very quickly, e.g., *X. index* can increase a hundredfold or more within 4 to 6 months when cultured on vine.[41] However, the multiplication rate of most other virus vector species is very slow, often requiring several years to double in size. Temperature, soil moisture conditions, and host plant species greatly influence nematode survival and reproduction in soil and the optimum conditions for maximum reproduction vary with the nematode species. Fig and grapevine are commonly used to culture *X. index,* perennial ryegrass for several *Longidorus* species and Trichodorids, and raspberry, rose, and strawberry for *X. diversicaudatum* and *L. macrosoma.*

Temperatures of about 20°C have usually been used for culturing nematodes in the glasshouse; higher temperatures should be avoided. Most nematodes die following extended periods of very wet conditions and all prefer relatively low moisture content.[42] Thus, field soil or sterile soil used for nematode culture should have a high proportion of sand added, especially for Trichodorids.

C. DETECTION OF TRANSMISSION

As indicated in the Introduction, loss of infectivity from soils following air drying and/or rubbing soil particles together is usually an indication that nematodes are involved in transmission. However, any loss of infectivity of soils achieved following treatment with chemicals must be interpreted with care, as some chemical have a broad spectrum of activity. For example, some fungicides are capable of killing or inactivating nematodes as well as fungi.[43]

Once evidence from such tests suggest nematode transmission is involved, much more detailed experiments are necessary to prove that this is so and to identify the vector nematode involved. In the past, failure to design suitable experiments for this purpose has led to several erroneous claims of nematode transmission; subsequent work has shown that contamination and not transmission was involved. The following procedure, briefly illustrated diagrammatically in Figure 2, has been used successfully for many years at SCRI. It is designed to eliminate as far as is practicable, criticisms of earlier experimental systems.[22,44] It seeks to determine (1) by using hand-picked nematodes, the correct identification of the only vector used, (2) by counting the galls on roots induced by nematode feeding, that nematodes have fed both on the

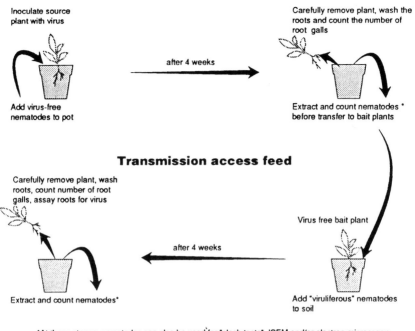

Acquisition access feed

Inoculate source
plant with virus

Carefully remove plant, wash the
roots and count the number of
root galls

after 4 weeks

Add virus-free
nematodes to pot

Extract and count nematodes *
before transfer to bait plants

Transmission access feed

Carefully remove plant, wash
roots, count number of root
galls, assay roots for virus

Virus free bait plant

after 4 weeks

Extract and count nematodes*

Add "viruliferous" nematodes
to soil

*At these stages, nematodes can also be used for "slash tests", ISEM and/or electron microscopy
of sections of the head to detect virus particles

FIGURE 2. Outline of a procedure for detecting virus transmission by nematodes.

virus source and bait plants, (3) by using slash tests, immunosorbent electron microscopy (ISEM) and/or electron microscopy, whether nematodes have acquired virus, and (4) by using adequate numbers of nematode-free replicates, whether nematodes specifically transmitted the virus. It is also necessary to include adequate numbers of virus-free replicates to ensure that the nematodes used were not carrying virus prior to the test, and to show that any virus recovered from test plants is the same as that used in the virus source plants and not a contaminant from another source. Using this and other less critical test systems, and 10 nematode species, Jones et al.[19] found that transmission of cherry leaf roll *Nepovirus* (CLRV) appeared to occur only in tests where virus source and virus-free bait plants were grown simultaneously in the same pot. No transmission was detected when the stringent system described above was used. This, together with other data, has provided strong evidence that earlier reports of transmission of CLRV by three different nematode species were due to contamination. Similar studies have indicated that reports of nematode transmission of CaRSV are also due to contamination.[45]

The system outlined above works best with micropots (25 ml for Longidorids, 1 to 2 ml for Trichodorids) without drain holes, containing a sterile medium (sand for Trichodorids; 3:1 mix sand and loam for Longidorids), sieved to provide soil particles <1500 μm to > 150 μm. Pots are partly filled with the growing medium and single small seedling plants added. Plants are allowed to settle for 2 to 3 days before virus inoculation or addition of nematodes. Once the nematodes are added, a further amount of the growing medium is used to nearly fill the pots. The micropots are then sunk in moist sand and careful attention given to maintaining the optimum conditions for good plant growth and nematode feeding, usually 20°C, a day length of about 16 h, and high humidity. Humid conditions minimize transpiration, causing a lower requirement for frequent watering and a decreased risk of widely fluctuating soil water content. Any water used should be fresh and aerated and added carefully to minimize disturbance and flooding. The virus source and bait plants must be suitable for both virus multiplication and nematode feeding; *Chenopodium quinoa, Nicotiana tabacum,* and *Petunia hybrida* have been used successfully for these purposes with a wide range of viruses and nematode species. However, specific requirements are sometimes necessary, for example, while *X. index* acquires grapevine fanleaf *Nepovirus* (GFLV) from *C. quinoa,* it will transmit the virus only to grapevine.[46] It should also be noted that different populations of the same nematode species vary in their efficiency to transmit virus isolates.[25] Ideally, therefore, the nematode population and the virus isolate used should be those obtained from the same field location. Adult nematodes and not juveniles should be used where possible, as virus specifically attached to the feeding apparatus is lost on moulting.

Ideally, nematodes are used singly, especially for Trichodorids whose identification to species level is often difficult without high magnification. Where the incidence of transmission is low, as often occurs with viruses transmitted by *Longidorus* species, groups of ten or more nematodes have been used. It is essential to have as many replicates as possible, but at least ten, and to include sufficient replicates of the necessary control treatments noted above.

The lengths of acquisition and transmission access feeds vary. For Longidorids, 3 to 4 weeks is usual but for Trichodorids, because of the smaller pot size, 7 to 14 days is recommended. In this latter instance, after nematodes have been extracted, bait plants need to be repotted into larger-sized micropots containing sterile soil and allowed to grow on for a further 2 to 3 weeks before testing in order to allow any virus present to replicate to detectable levels.

Extraction of nematodes and removal of test plants (Figure 2) are best achieved by totally immersing the micropots in water, gently dislodging the plant roots, and shaking them free of soil particles. Plant roots are washed further in another container of water and the water from this second wash pooled with that from the first and used for extracting the nematodes as described earlier. It is especially important to wash the roots of bait plants thoroughly, as nematodes and nematode feces, that contain virus, may be attached to roots or debris. The erroneous report of *L. leptocephalus* and *L. caespiticola* as vectors of arabis mosaic *Nepovirus* (AMV)[47] is now believed to be due to contamination with nematode feces.[48] After washing, plant roots can be examined with a hand lens or binocular microscope for root galls, caused by nematode feeding.

In tests for virus transmission, it is customary to test the roots and tops of bait plants separately. Virus detected in tops is the best evidence of virus transmission, but this occurs only infrequently in plants inoculated by nematodes through roots. Detection of virus infection in the different plant parts can be done by infectivity tests of sap to herbaceous test plants, or by enzyme-linked immunosorbent assay (ELISA). All positive results from infectivity tests must be tested serologically to confirm the identity of the virus present. When using ELISA on roots, the use of horseradish peroxidase and the substrate 3,3',5,5'-tetramethylbenzidine should be avoided, as nonspecific positives are commonly obtained.[49]

To determine the presence of virus in nematodes by the "slash" test, nematodes are ground up in a small volume of water or buffer between two glass slides. The extract can be either manually inoculated to indicator test plants or used in ISEM.[50] As most viruses transmitted by *Xiphinema* are inactivated in the gut of this nematode, detection by ISEM is more sensitive than infectivity tests for these viruses. ELISA is not a recommended procedure for testing nematodes for virus as most studies indicate that 20 to 50 nematodes are required to achieve sufficient sensitivity. Detection of virus in nematode extracts indicates only that the nematode ingested virus and it cannot be concluded that it transmitted the virus. Furthermore, although electron microscopy of thin sections of the feeding apparatus of nematodes has located specific sites where virus is retained,[51,52] this in itself is not evidence that virus is also released and transmitted to plants during feeding; only infection of bait plants demonstrates transmission.

IV. TRANSMISSION BY SOIL-INHABITING FUNGI

A. INTRODUCTION — THE VECTORS AND THE VIRUSES

Viruses and virus-like agents transmitted by soil-inhabiting fungi cause serious and economically important diseases of cereals, high-value horticultural crops, and root and tuber crops worldwide. Several such diseases have been known for many years. The discovery of *Olpidium brassicae* as the first recorded virus vector fungus followed observations by different workers of an association between this fungus and lettuce big vein disease (LBV).[3,4] The graft transmission of the agent of LBV to *Olpidium*-free plants provided critical evidence that the causal agent could exist independently of the fungus but that it failed to spread without it.[53,54] By contrast, Teakle[55] found that tobacco necrosis *Necrovirus* (TNV), which is very infective in plant sap, spread to plants in the absence of *Olpidium* but that spread was much more effective when the fungus was present. Although these findings on fungi and those identifying nematodes as virus vectors occurred about the same time, until very recently, our knowledge of fungal vectors, the disease agents they transmit, and the mechanisms of virus transmission, was much less than for nematode-borne viruses. Two of the main reasons for this are that the fungal vectors are obligate parasites and are not easy to culture and study. In addition, with few exceptions, the viruses involved are not easy to transmit mechanically to plants. Although much still remains to be discovered about the interactions and processes involved in the transmission of viruses by fungi, our present knowledge allows some general observations on the vectors, the viruses, and their associations.

Despite some earlier claims for *Pythium ultimum* and *Synchytrium endobioticum* as virus vectors,[56,57] only five species of soil-inhabiting fungi have been demonstrated to transmit viruses and these belong to two fungal classes. They are *Olpidium brassicae* and *O. radicale* (formerly known as *O. cucurbitacearum* Barr[58]) belonging to the Chytridiales, and *Polymyxa betae, P. graminis,* and *Spongospora subterranea,* belonging to the Plasmodiophorales (Table 3). Of these, only *S. subterranea* is a serious pathogen in its own right, inducing potato powdery scab in potato and crook root disease in watercress; the others assume significance only because of their role as virus vectors. The vectors occur worldwide and several have very wide host ranges but others are more restricted in the species they colonize. For example, *O. brassicae* infects species in more than 50 genera while *S. subterranea* is mostly restricted to Solanaceae. Most vectors are active at relatively low temperatures. None of the vectors have a true mycelium and each has a relatively naked developmental stage in their host plant. The two fungal classes are readily distinguished by the morphology of their zoospores; chytrid zoospores have a single long flagellum, while those of plasmodiophorids contain two flagella of unequal lengths. Fungal zoospores are involved in the virus transmission process and it is necessary therefore to understand the life cycle of these fungi, although for several vectors, some stages of their life history are still a matter of debate. There are differences in detail between the vector species, but the principal features of reproduction are the same.

The fungi tend to infect epidermal cells or root hairs near the root tips and in them produce zoosporangia containing many mobile zoospores. Under suitable conditions the zoosporangia produce exit tubes through which the flagellate zoospores are released into the soil water around the infected root. These zoospores drift or swim to the surface of another root, withdraw their cilia and may then do one of two things. First, they may produce thin-walled zoospore cysts and, after a few hours, these cysts produce an infection tube through the root cell wall by which the zoospore protoplasm enters the cell. After a few days, the zoospore contents develop into a thin-walled zoosporangium producing many more zoospores which can be released into the root/

Table 3. Viruses Known, or Believed, to Be Transmitted by Fungi

Virus	Particle shape	Vector species	Ref.[a]
Chytrid vectors — *Olpidium* spp.			
Cucumber leaf spot virus	Isometric ⎫	*O. radicale*[b]	#319
Cucumber necrosis *Tombusvirus*	Isometric ⎬		#82
Melon necrotic spot *Carmovirus*	Isometric		64
Freesia leaf necrosis virus	Unknown ⎫		107
Lettuce big vein virus	Rod	*O. brassicae*	81
Pepper yellow vein virus	Unknown ⎬		118
Tobacco necrosis *Necrovirus* and its			#14
satellites	Isometric		#15
Tobacco stunt virus	Rod ⎭		#313
Plasmodiophorid vectors — *Polymyxa* spp.			
Beet necrotic yellow vein *Furovirus*	Rod ⎫	*P. betae*	#144
Beet soilborne *Furovirus*	Rod ⎬		109
Barley mild mosaic ''Baymovirus''[c]	Filament ⎫		110
Barley yellow mosaic ''Baymovirus''	Filament		#143
Broad bean necrosis *Furovirus*	Rod		#223
Indian peanut clump *Furovirus*	Rod		111
Oat golden stripe *Furovirus*	Rod	*P. graminis*	112
Oat mosaic ''Baymovirus''	Filament		#145
Peanut clump *Furovirus*	Rod		#235
Rice necrosis mosaic ''Baymovirus''	Filament		#172
Rice stripe necrosis *Furovirus*	Rod		113
Wheat soil-borne mosaic *Furovirus*	Rod ⎭		#77
Wheat spindle streak ''Baymovirus''	Filament		#167
Wheat yellow mosaic ''Baymovirus''	Filament		114
Plasmodiophorid vectors — *Spongospora* spp.			
Potato mop top *Furovirus*	Rod ⎫	*S. subterranea*	#138
Watercress chlorotic leaf spot virus	Unknown ⎬		115
No fungus species identified			
Cucumber soil-borne virus	Sphere		116
Hypocheris mosaic *Furovirus*	Rod		#273
Nicotiana velutina mosaic *Furovirus*	Rod		189
Petunia asteroid mosaic *Tombusvirus*	Spheree		83
Squash necrosis *Necrovirus*	Sphere		

[a] # Indicates numbers of references in *AAB Descriptions of Plant Viruses,* Association of Applied Biologists, Wellesbourne.

[b] *O. radicale* Schwartz & Cook, formally *O. cucurbitacearum* Barr.[58]

[c] ''Baymovirus'' is a virus group not yet approved by the International Committee for the Taxonomy of Viruses.

soil environment through exit tubes. Second, instead of a thin-walled zoo-sporangium, zoospores may produce a thick-walled resting zoosporangium. These resting spores are very resistant to drying and may remain viable in decaying root material for long periods, even many years. Under suitable conditions, these spores germinate to release new zoospores. It was once assumed that all resting spores were produced following the fusion of two zoospores to form a zygote prior to infecting cells, but this is now open to question.[59]

Virus transmission occurs by zoospores, but in one of two different ways. Those viruses transmitted by *O. radicale* and TNV and its satellites (Table

3) are acquired from soil water on the outer surface membrane of zoospores. The virus particles appear to enter the zoospore through the infection canal but they do not enter the resting spore during its formation (nonpersistent transmission). By contrast, all of the remaining viruses in Table 3 known to be transmitted by fungi are acquired from plant cells and enter the resting spores of vectors and are released with zoospores on germination (persistent transmission). These different associations of the virus with the vector have implications for the epidemiology of these viruses and for their transmission experimentally.

More than 20 viruses are now known to be transmitted by fungi and others have suspected but unproven soil-borne fungal vectors (Table 3). The viruses show very much more diversity than those transmitted by nematodes; they include members of four recognized plant virus groups (*Carmovirus, Tombusvirus, Necrovirus, Furovirus*), one proposed group ("Baymovirus") and several others are ungrouped. Most have either flexuous filamentous particles or stiff rod-shaped particles. The few viruses with isometric particles are all transmitted by chytrid vectors (Table 3) and all are nonpersistently transmitted. For a few agents (presumably viruses) no particle morphology has been determined (Table 3). Some viruses, such as potato mop top, have very restricted host ranges, while others, such as TNV, infect a wide range of species in several monocotyledonous and dicotyledonous genera. Apart from TNV and the viruses transmitted by *O. radicale,* few are transmitted readily between plants by mechanical inoculation of plant sap; however, most can be transmitted in this way, but only with varying levels of difficulty. Unlike most nematode-borne viruses, soil-borne viruses transmitted by fungi are not readily transmitted through seed. Some exceptions to this are Indian peanut clump *Furovirus* (PCV)[60] and cucumber leaf spot virus[61] but this latter virus is reported also to spread abiotically in soil.[15] The general lack of seed transmissibility means that virus spread is dependent on the movement of the vector. Thus, the occurrence of these viruses in crops is also indicative of the presence of the vector. Virus movement can occur over short distances by the mobility of viruliferous zoospores in soil water[62] and over much greater distances by movement of contaminated soil and/or plant debris in drainage and irrigation channels and river water,[6] and on farm machinery or even by wind errosion; contaminated planting stock is another possible means of spread over long distances.

As noted earlier, an important feature affecting the epidemiology of the viruses is their relationship with their fungal vector. With the exception of TNV, its satellites, and viruses transmitted by *O. radicale,* all the viruses appear to have a persistent relationship with their vectors and are retained in its resting spores. These resting spores and the virus they contain can remain viable for many years, even in air-dried soils. By contrast, the remaining viruses have a nonpersistent relationship with the vector, and virus is acquired from soil water and possibly plant roots on the exterior plasma membrane of zoospores; it is not retained within the resting zoospore. While this transient (nonpersistent) relationship with the vector would tend to limit virus survival in nature, this is compensated for by the stability of the particles of such viruses, their relatively high concentration in plant roots, and, especially with TNV, the wide host range of the virus and of its vector, *O. brassicae.*

There is marked vector specificity between fungal species. For example, *O. radicale* will not transmit TNV and *O. brassicae* will not transmit any of the viruses transmitted by *O. radicale* (Table 3).[63,64] Furthermore, vector specificity exists among strains of these fungi. Thus, some strains of *O. radicale* that transmit cucumber necrosis *Tombusvirus* (CNV) and melon necrotic spot virus (MNSV), will not transmit cucumber leaf spot virus (CLSV)[64] and for TNV, vector and nonvector strains of *O. brassicae* are well documented.[65-67] In addition, host biotypes of fungal vectors are known. For example, *S. subterranea* that colonizes potato and watercress, are distinguished as *S. subterranea* f. sp. *subterranea* on potato and *S. subterranea* f.sp. *nasturtii* on watercress because they will not infect the heterologous plant species, yet they are morphologically indistinguishable.[68] More remarkable still is the report by van Dorst and Peters[69] that *O. brassicae* from freesia did not infect freesia after passage on lettuce. Despite these restrictions, the ability of a fungus to transmit virus to plants is not always dependent on its ability to multiply in that plant, as noted for the transmission of TNV to tulip[70,71] and IPCV to groundnut.[59]

B. ISOLATION, CULTURE, MAINTENANCE, AND USE OF FUNGAL VECTORS

While the general principles of reproduction for the five known vector species are similar, differences in their relationship with their respective viruses (persistent or nonpersistent), their production and release of zoospores under different environmental conditions, and their host range, need to be considered in relation to their culture, maintenance, and use in experiments.

1. Isolation of Fungal Vectors

Once a possible source of infection is identified in the field and evidence indicates a soil-borne virus may be involved, soil samples can be collected and bait plants used to recover both the virus and the possible fungal vector. Soil samples can be collected as for nematode-borne viruses but, if a fungal vector is suspected, it is normally collected from around the roots of infected plants. Soil samples are usually mixed with sterile sand and placed in sterile pots using the precautions mentioned earlier to minimize abiotic transmission. The infectivity of soils can vary greatly. For example, wheat yellow mosaic virus was recovered from infected soil diluted 1:1000,[72] while only a small proportion of *P. graminis* spores were infective with barley yellow mosaic virus.[73] In some instances, notably with *P. graminis* and *S. subterranea*, infectivity of soil is usually greater if soil is sampled after overwintering and/

or air drying before testing, due to the increased release of zoospores from sporangia following these treatments.[74-76] The plant species used for baiting soils is important and will obviously be determined by the host range of both the fungal vector and the virus. Usually, the host plant found infected in the field would be an obvious choice but in some instances this may not be suitable. For example, Mowat[70] found that *O. brassicae* transmitted TNV to tulip but that the fungus died soon after penetration of the plant; a similar situation occurs when *P. graminis* transmits IPCV to groundnut.[60] In these instances, although transmission occurs and the virus can therefore be cultured, the fungus vector cannot. If possible, bait plants are best planted as seed, so minimizing the risk of abiotic infection when seedling roots are damaged during transplanting in infested soil.

2. Establishment and Maintenance of Vector Cultures

After 3 to 4 weeks, the roots of bait plants are washed free from soil and they can then be used for establishing further cultures and for identification of the fungi and other microorganisms present. The identification of fungal species present in soil cultures can be difficult and the classification of some fungal vectors is still the subject of controversy. A $\times 100$ magnification microscope is necessary for studying zoospores and a phase-contrast microscope is recommended.[77] Details on the taxonomy of fungal vectors are given by Barr[59] and Braselton.[78] Subcultures may be derived from small root portions or from zoospore suspensions obtained from root washings. However, it should be noted that at this stage the inoculum contains a mixture of organisms, fungal species, and strains, and it is not unusual for soil samples to contain several different fungal vector species.[79] Isolation of specific fungal vector species and vector (or nonvector) strains is more difficult and may require several different approaches. One approach is to maintain cultures at different temperatures to try to eliminate or decrease unwanted organisms at nonpermissive temperatures. This has been used successfully to establish unifungal cultures of *O. brassicae*[80,81] and *P. betae*.[82] Other approaches are to establish cultures on selective hosts and to dissect out root pieces containing the desired fungal species. Ultimately, the establishment from a single sporangium is desirable, especially as vector and nonvector strains of some fungal species are known.[65,83] Again, one or more possible approaches can be used, such as inoculating to plants a dilution series of zoospore or sporangia suspensions, or dissecting out individual zoosporangia for inoculation.[63,77,84] However, it is important to remember that even after using these more refined procedures, the resulting unifungal cultures will still contain other microorganisms, e.g., bacteria and algae. This situation will remain until ways can be found for vector fungi to be grown *in vitro*.

Once unifungal cultures have been obtained, they can be maintained by transferring zoospore suspensions to fresh bait seedlings. The preferred growing media are sterile sand, Molochite, or small glass beads. Such media allow active root growth (necessary for good fungal growth), provide ample movement for zoospores in the large pore spaces and hence fungal spred, and eliminate clay and other soil particles that can adsorb virus. Active plant growth under these conditions is maintained by watering with nutrient solution. This procedure provides a convenient and effective way of obtaining zoospores for experimental work without disturbing the plant roots; plants are watered and the leachate from the pots, containing zoospores, collected. Such a collection process minimizes the risk of release of the virus that could occur when roots are damaged during extraction from pots and washing.

If plants are removed from pots to harvest large numbers of zoospores, a soilless medium is also convenient because the sand or other inert material is easily removed from roots by rinsing in water. Care should be taken to do this rinsing quickly as zoozpores can be released after 3 to 4 min when roots are in water. Once clean of contaminating materials, roots are immersed in water at 10 to 16°C for 10 to 20 min, when zoospores are released from zoosporangia.[85] Using this procedure $>10^6$ zoospores/ml have been obtained from *O. brassicae*[77] and *P. graminis*.[86] Water purity is important and it should be free from certain metal ions and from chlorine. For these reasons a weak nutrient solution or neutral buffer is often used instead of tap water. If stock plants can be kept turgid, e.g., by keeping them in a plastic bag, zoospores can be successfully re-extracted from roots after a few days, or even 1 to 2 weeks if plants are kept in a refrigerator. However, this is much less successful if plants are allowed to wilt.

In many instances, perhaps most, where fungal vectors are cultured from a natural outbreak of virus disease, the cultures will also contain the causal virus or agent. A virus-free culture of the fungus is necessary to demonstrate virus transmission. Such a culture may be obtained from soil samples taken from outside the disease area. Alternatively, virus-infected cultures may be freed from infection by attempting one or more of the following procedures. Where the association of virus with the vector is nonpersistent, i.e., virus is carried only on the outside of zoospores, virus can usually be eradicated by chemical treatments such as hydrochloric acid or detergents,[87] ribavirin,[88] or incubation with virus antiserum.[89] Alternatively, serially culturing the fungus on a nonhost of the virus for some weeks has been effective.[90,91] For example, *O. brassicae* has been freed from the persistently transmitted LBVV by serial transfer on either sugar beet, *Plantago major* or *Veronica persica*[81,90] and from the nonpersistently transmitted TNV by culture on turnip.[89] *P. graminis* was freed from soil-borne wheat mosaic virus (SBWMV) by serial culture on clover.[92] Oddly, cultures of *P. betae* were reported to be freed from beet necrotic yellow vein *Furovirus* (BNYVV) by culturing on *Chenopodium ficifolium,* a host of BNYVV.[93]

Once fungal cultures are obtained it is wise to preserve them for reference, and against the event of their loss or contamination. All the known fungal vectors can be readily preserved because they each produce stable resting spores. Root pieces infected with the fungus, when air-dired and stored in sealed containers, remain viable for many years. Indeed, after storing in this way, for those viruses that are able to infect resting spores (persistently transmitted), both virus and vector have remained viable for more than 20 years.[93-95]

C. VECTOR TRANSMISSION STUDIES

Zoospores are the natural means of virus transmission in the field and are a convenient and effective method of transmission experimentally. Zoospore suspensions can usually be obtained readily from cultures of fungal vectors as described earlier. For *O. brassicae,* a flush of zoospores is released 3 to 10 min after the roots are immersed in water. For some other vectors, e.g., *S. subterranea* and *P. graminis,* more efficient release of zoospores occurs when roots are immersed in water following air-drying.[75] Zoospores require several hours to establish infection in plants[96] and zoospore activity (mobility) is usually lost after 10 to 60 min in water. However, activity can be prolonged for many hours and up to 2 days in low-molarity buffers, or dilute nutrient solutions of neutral pH. For this purpose, 20% Knop's solution,[84] 5% Hoagland's solution,[66] 0.01 M phosphate or 0.05 M glycine buffer,[83] and bovine serum albumin[96] have been used with success.

Once a suspension of virus-free zoospores is obtained, it is poured around the roots of virus source plants, infected with virus by either mechanical inoculation of virus-containing sap or by graft-inoculation, e.g., for LBVV, and growing in sterile medium. The isolate of the virus used should be carefully considered as cultures of some viruses maintained by repeated mechanical inoculation to plants have been shown to lose their ability to be transmitted by fungal vectors.[73] The plants are incubated for 3 to 4 weeks to allow time for fungal colonization of the roots and acquisition of virus. A zoospore suspension is then collected from the roots of these virus-source plants and used to inoculate healthy virus-free seedling bait plants growing in sterile medium in pots. Where a virus is suspected of a nonpersistent association with a fungal vector and is therefore likely to be adsorbed to the exterior of zoospores, a more rapid method of virus acquisition is possible. Zoospore suspensions can be mixed *in vitro* with preparations of partially purified virus particles or virus in sap extracts from infected plants. After a few hours, this mixture can then be added directly to healthy bait plant roots.

The bait plants are kept for 3 to 8 weeks at about 20°C, depending on the virus and bait plant species involved, to await the development of virus symptoms in leaves and/or virus multiplication in roots. As with other soilborne virus studies, detection of virus in the tops of such bait plants provides more convincing evidence of transmission than detection in the roots. However, roots must be tested because most fungal-transmitted viruses attain higher concentrations in the roots than the tops of plants, many viruses become systemic in only a proportion of plants, and others, like TNV, rarely infect the tops of plants when infected via roots. Tests on the roots of bait plants suffer from the same hazards of contamination as discussed earlier in this chapter.

D. DETECTION OF TRANSMISSION

The method(s) of detecting virus in bait plants is determined by the nature of the virus involved. For those viruses listed in Table 3 with unknown particle morphology and for a few others that are difficult to transmit mechanically, the expression of disease symptoms in suitable indicator bait plants is the only reliable detection system available. For some of these viruses, the expression of symptoms is greatly affected by environmental conditions, especially temperature. With few exceptions, symptom expression is inhibited by high temperatures ($>20°C$) and temperatures of 14 to 18°C are considered optimum for most viruses. However, temperatures of 10°C or less are best for symptom expression of wheat spindle streak mosaic virus and above 15°C, symptoms are not expressed.[97]

For most other viruses listed in Table 3 detection can be by infectivity tests using sap from bait plant parts to mechanically inoculate indicator plants and/or, by serology. It should be noted, however, that some viruses are transmitted inefficiently by mechanical inoculation and others require additives to the inoculum to preserve infectivity, e.g., tobacco stunt virus and LBVV.[98,99]

Detection of virus in bait plants by these different means is not in itself proof of fungal transmission. As pointed out earlier, the risks of contamination and abiotic transmission are real and experimenters must seek to eliminate the possiblity of these occuring. Transmission tests must therefore always include adequate numbers of control treatments and, as a routine, include the following: plants inoculated with (1) zoospore suspensions with no access to virus to confirm that they were initially virus-free and that fungal infection alone is not the cause of disease symptoms in plants; (2) virus suspensions free from zoospores to test that no abiotic transmission occurs; and (3) untreated bait plants to show that they were themselves initially virus-free.

V. CONCLUSIONS

The upsurge in the last two decades of the number of reports of, and studies on, the transmission through soil of plant viruses, indicates the importance now attached to viruses transmitted in this way. Despite an extension in our knowledge of the properties of many of the viruses involved, proof of the transmission of viruses through soil can still be technically difficult to achieve, and identifying the precise mode of transmission can be even more difficult. Much of this difficulty is due to the technical problems of studying pathogens and virus vectors in the soil environment, the presence in soils of more than one soil-borne virus (e.g., TNV is present in many field soils as well as in glasshouse soils and debris), and the problem of contamination, especially with viruses transmitted abiotically in soil. This latter problem is undoubtedly the reason why some viruses, now known to be transmitted abiotically, were erroneously reported to be transmitted by nematodes.[45,100] This serves to further emphasize the need, already stressed in this chapter, for experimenters to show great care and precision in experiments to determine the mode of transmission of plant viruses through soil, and to use adequate numbers of all the necessary control plants.

REFERENCES

1. **Beijerinck, M. W.**, Uber ein contagium vivum fluidum als Ursache der Fleckenkrankheit der Tabaksbeatter Verhandee, *Koninke. Acad. Witenschap. Amsterdam*, 6, 1, 1898.
2. **Hewitt, W. B., Raski, D. J., and Goheen, A. C.**, Nematode vector of soil-borne fan leaf virus of grapevines, *Phytopathology*, 48, 586, 1958.
3. **Fry, P. R.**, The relationship of *Olpidium brassicae* (Wor.) Dang. to the big-vein disease of lettuce, *N.Z. J. Agric. Res.*, 1, 301, 1958.
4. **Grogan, R. G., Zink, F. W., Hewitt, W. B., and Kimble, K. A.**, The association of *Olpidium* with the big-vein disease of lettuce, *Phytopathology*, 48, 292, 1958.
5. **Teakle, D. S.**, Transmission of tobacco necrosis virus by a fungus, *Olpidium brassicae*, *Virology*, 18, 224, 1962.
6. **Koenig, R.**, Detection in surface waters of plant viruses with known and unknown natural hosts, in *Developments in Applied Biology 2: Viruses with Fungal Vectors*, Cooper, J. I. and Asher, M.J. C., Eds., Association of Applied Biologists, Wellesbourne, 1988, 305.
7. **Yarwood, C. E.**, Release and preservation of virus by roots, *Phytopathology*, 50, 111, 1960.
8. **Lovisolo, O., Bode, O., and Volk, J.**, Preliminary studies on the soil transmission of petunia asteroid mosaic virus (= 'Petunia' strain of tomato bushy stunt virus), *Phytopathol. Z.*, 53, 323, 1965.
9. **Smith, P. R., Campbell, R. N., and Fry, P. R.**, Root discharge and soil survival of viruses, *Phytopathology*, 59, 1678, 1969.
10. **Hollings, M., Stone, O. M., and Barton, R. J.**, Pathology, soil transmission and characterization of cymbidium ringspot, a virus from cymbidium orchids and white clover (*Trifolium repens*), *Ann. Appl. Biol.*, 85, 233, 1977.
11. **Gerhardson, B. and Insunza, V.**, Soil transmission of red clover necrotic mosaic virus. *Phytopathol. Z.*, 94, 67, 1979.
12. **Kegler, G., Kleinhempel, H., and Kegler, H.**, Untersuchung zur Bodenburtigkeit des tomato bushy stunt virus, *Arch. Phytopathol. Pflanzenschutz*, 16, 73, 1980.
13. **Miyamoto, Y.**, The nature of soil transmission in soil-borne plant viruses, *Virology*, 7, 250, 1959.
14. **Kleinhempel, H. and Kegler, G.**, Transmission of tomato bushy stunt virus without vectors, *Acta Phytopathol. Acad. Sci. Hung.*, 17, 17, 1982.
15. **Kegler, G. and Kegler, H.**, Beitrage zur Kenntnis der vektorlosen Ubertragung pflanzenpathogener Viren, *Arch. Phytopathol. Pflanzenschutz*, 17, 307, 1981.
16. **Teakle, D. S.**, Abiotic transmission of southern bean mosaic virus in soil, *Aust. J. Biol. Sci.*, 39, 353, 1986.

17. **Kegler, H., Griesbach, E., Skadow, K., Fritzsche, R., and Weber, I.,** Ausbreitung von Krankheitserregern und Schadlingen der Tomate in NFT-Kultur und ihre Vorbeugung, *Nachrichtenbl. Pfanzenschutz DDR,* 37, 28, 1983.

18. **Paludan, N.,** Spread of viruses by recirculated nutrient solutions in soilless cultures, *Tidsskr. Planteavl,* 89, 467, 1985.

19. **Jones, A. T., McElroy, F. D., and Brown, D. J. F.,** Tests for transmission of cherry leaf roll virus using *Longidorus, Paralongidorus* and *Xiphinema* nematodes, *Ann. Appl. Biol.,* 99, 143, 1981.

20. **Childress, A. M. and Ramsdell, D. C.,** Detection of blueberry leaf mottle virus in highbush blueberry pollen and seed, *Phytopathology,* 76, 1333, 1986.

21. **Harrison, B. D.,** Specific nematode vectors for serologically distinctive forms of raspberry ringspot and tomato black ring viruses, *Virology,* 22, 544, 1964.

22. **Brown, D. J. F., Ploeg, A. T., and Robinson, D. J.,** A review of reported associations between *Trichodorus* and *Paratrichodorus* species (Nematoda: Trichodoridae) and tobraviruses with a description of laboratory methods for examining virus transmission by trichodorids, *Rev. Nematol.,* 12, 235, 1989.

23. **Brown, D. J. F.,** The transmission of two strains of strawberry latent ringspot virus by populations of *Xiphinema diversicaudatum,* (Nematoda: Dorylaimoidea), *Nematol. Mediterr.,* 13, 217, 1985.

24. **Brown, D. J. F.,** The transmission of two strains of arabis mosaic virus from England by populations of *Xiphinema diversicaudatum* (Nematoda: Dorylaimoidea) from ten countries, *Rev. Nematol.,* 9, 83, 1986.

25. **Brown, D. J. F. and Trudgill, D. L.,** Differential transmissibility of arabis mosaic and strains of strawberry latent ringspot viruses by three populations of *Xiphinema diversicaudatum* (Nematoda: Dorylaimoidea) from Scotland, Italy and France, *Rev. Nematol.,* 6, 229, 1983.

26. **Cotton, J.,** The effectiveness of soil sampling for virus-vector nematodes in MAFF certification schemes for fruit and hops, *Plant Pathol.,* 28, 40, 1987.

27. **Mass, P. W. T. and Brinkman, I.,** Sampling of soil for nematode vectors of plant viruses in the Netherlands, *Med. Fac. Landbouw. Rijksuniv. Gent,* 45, 769, 1980.

28. **Boag, B., Brown, D. J. F., and Banck, A.,** Optimising sampling strategies for nematode-transmitted viruses and their vectors, *EPPO Bull.,* 19, 491, 1989.

29. **Brown, D. J. F., Boag, B., Jones, A. T., and Topham, P. B.,** An assessment of the soil-sampling density and spatial distribution required to detect viruliferous nematodes (Nematoda: Longidoridae and Trichodoridae) in fields, *Nematol. Mediterr.,* 18, 153, 1990.

30. **Taylor, C. E. and Brown, D. J. F.,** The geographical distribution of *Xiphinema* and *Longidorus* nematodes in the British Isles and Ireland, *Ann. Appl. Biol.,* 84, 383, 1976.

31. **Alphey, T. J. W. and Boag, B.,** Distribution of trichodorid nematodes in Great Britain, *Ann. Appl. Biol.,* 84, 371, 1976.

32. **Taylor, C. E.,** The multiplication of *Longidorus elongatus* (de Man) on different host plants with reference to virus transmission, *Ann. Appl. Biol.,* 59, 275, 1967.

33. **Taylor, C. E.,** Transmission of viruses by nematodes, in *Principles and Techniques in Plant Virology,* Kado, E. I. and Agrawal, H. O., Eds., Van Nostrand Reinhold, New York, 1972, 226.

34. **Cobb, N. A.,** Estimating the Nema Population of Soil, U.S.D.A. Technical Circ. No. 1, U.S. Department of Agriculture, Washington, D.C., 1918.

35. **Towshend, J.L.,** Plant parasitic nematodes in grape and raspberry soils of Ontario and a comparison of extraction techniques, *Can. Plant Dis. Surv.,* 47, 83, 1967.

36. **Loof, P. A. A. and Luc, M.,** A revised polytomous key for the identification of species of the genus *Xiphinema* Cobb, 1913 (Nematoda: Longidoridae) with exclusion of the *X. americanum*-group, *Systematic Parasitol.,* 16, 35, 1990.

37. **Lamberti, F. and Carone, M.,** A dichotomous key for the identification of species of *Xiphinema* (Nematoda : Dorylaimida) within the *X. americanum* - group, *Nematol. Mediterr.,* 19, 341, 1991.

38. **Rey, J. M., Andres, M. F., and Arias, M.,** A computer method for identifying nematode species. I. Genus *Longidorus* (Nematoda : Longidoridae), *Rev. Nematol.,* 11, 129, 1988.

39. **Decraemer, W.,** Identification of Trichodorids, in *Nematode Identification and Expert System Technology,* Fortuner, R., Ed., Plenum Press, New York, 1988, 157.

40. **Hooper, D. J.,** Handling, fixing, staining and mounting nematodes, in *Laboratory Methods for Work with Plant and Soil Nematodes,* Southey, J. F., Ed., MAFF Reference Book 402, HMSO, London, 1985, 87.

41. **Raski, D. J. and Hewitt, W. B.,** Nematode transmission, in *Methods in Virology,* Vol. 1, Maramorosch, K. and Koprowski, H., Eds., Academic Press, New York, 1967, 309.

42. **von Fritzsche, R.,** Methoden der ubertragung pflanzenpathogener viren durch nematoden, *Biol. Zentralb.,* 86, 753, 1967.

43. **Taylor, C. E. and Murant, A. F.,** The use of quintozene (PCNB) as a nematicide, *Proc. 3rd Br. Insect. Fungicide Conf.,* 1965, 514.

44. **Trudgill, D. L., Brown, D. J. F., and McNamara, D. G.,** Methods and criteria for assessing the transmission of plant viruses by longidorid nematodes, *Rev. Nematol.,* 6, 133, 1983.

45. **Brown, D. J. F. and Trudgill, D. L.,** The spread of carnation ringspot virus in soil with or without nematodes, *Nematologica,* 30, 102, 1985.

46. **Trudgill, D. L. and Brown, D. J. F.,** Effect of bait plant on transmission of viruses by *Longidorus* and *Xiphinema* spp., *Annu. Rep. Scott. Hortic. Res. Inst. for 1979,* 1980, p. 120.

47. **Valdez, R. B.,** Transmission of raspberry ringspot virus by *Longidorus caespiticola, L. leptocephalus* and *Xiphinema diversicaudatum* and of arabis mosaic virus by *L. caespiticola* and *X. diversicaudatum, Ann. Appl. Biol.,* 71, 229, 1972.

48. **McNamara, D. G.,** Studies on the Ability of the Nematode *Xiphinema diversicaudatum* (Micol.) to Transmit Raspberry Ringspot Virus and to Survive in Plant-Free Soil, Ph.D. Thesis, University of Reading, Reading, England, 1978.

49. **Jones, A. T. and Mitchell, M. J.,** Oxidising activity in root extracts from plants inoculated with virus or buffer that interferes with ELISA when using the substrate 3,3',5,5'-tetramethylbenzidine, *Ann. Appl. Biol.,* 111, 359, 1987.

50. **Roberts, I. M. and Brown, D. J. F.,** Detection of six nepoviruses in their nematode vectors by immunosorbent electron microscopy, *Ann. Appl. Biol.,* 96, 187, 1980.

51. **Taylor, C. E. and Robertson, W. M.,** The location of raspberry ringspot and tomato black ring viruses in the nematode vector, *Longidorus elongatus* (de Man), *Ann. Appl. Biol.,* 64, 233, 1969.

52. **Taylor, C. E. and Robertson, W. M.,** Sites of virus retention in the alimentary tract of the nematode vectors *Xiphinema diversicaudatum* (Micol.) and *X. index* (Thorne and Allen), *Ann. Appl. Biol.,* 66, 375, 1970.

53. **Campbell, R. N., Grogan, R. G., and Purcifull, D. E.,** Graft transmission of big vein of lettuce, *Virology,* 15, 82, 1961.

54. **Tomlinson, J. A., Smith, B. R., and Garnett, R. G.,** Graft transmission of lettuce big vein, *Nature,* 193, 599, 1962.

55. **Teakle, D. S.,** Association of *Olpidium brassicae* and tobacco necrosis virus, *Nature,* 188, 431, 1960.

56. **Thottappilly, G.,** Untersuchungen uber die Beziehungen zwischen dem Erbsenblattrollvirus und seinen Vektonen sowie uber ein neuss Pilzund Blattlausubertragbanes Virus der Erbse, Ph.D. Thesis, Justus Liebig University, Giessen, 1968.

57. **Nienhaus, F. and Stille, B.,** Ubertragung des Kartoffel-X-Virus durch zoosporen von *Synchytrium endobioticum, Phytopathol. Z.,* 54, 335, 1965.

58. **Lange, L. and Insuza, V.,** Root-inhabiting *Olpidium* species: the *O. radicale* complex, *Trans. Br. Mycol. Soc.,* 69, 377, 1977.

59. **Barr, D. J. S.,** Zoosporic plant parasites as fungal vectors of viruses: taxonomy and life cycle of species involved, in *Developments in Applied Biology 2: Viruses with Fungal Vectors,* Cooper, J. I. and Asher, M. J. C., Eds., Association of Applied Biologists, Wellesbourne, 1988, 123.

60. **Thouvanel, J.-C. and Fauquet, C.,** Further properties of peanut clump virus and studies on its natural transmission, *Ann. Appl. Biol.,* 97, 99, 1981.

61. **Weber, I.,** Cucumber leaf spot virus, in *AAB Descriptions of Plant Viruses,* No. 319, Association of Applied Biologists, Wellesbourne, 1986, 4.

62. **Temmink, J. H. M., Campbell, R. N., and Smith, P. R.,** Specificity and site of *in vitro* acquisition of tobacco necrosis virus by zoospores of *Olpidium brassicae, J. Gen. Virol.,* 9, 201, 1970.

63. **Dias, H. F.,** Transmission of cucumber necrosis virus by *Olpidium cucurbitacearum* Barr & Dias, *Virology,* 40, 828, 1970.

64. **Campbell, R. N., Lecoq, H., Wipf-Scheibel, C., and Sim, S. T.,** Transmission of cucumber leaf spot virus by *Olpidium radicale, J. Gen. Virol.,* 72, 3115, 1991.

65. **Teakle, D. S. and Hiruki, C.,** Vector specificity in *Olpidium, Virology,* 24, 539, 1964.

66. **Kassanis, B. and MacFarlane, I.,** Interaction of virus strain, fungus isolate. and host species in the transmission of tobacco necrosis virus, *Virology,* 26, 603, 1965.

67. **Mowat, W. P.,** *Olpidium brassicae:* electrophoretic mobility of zoospores associated with their ability to transmit tobacco necrosis virus, *Virology,* 34, 565, 1968.

68. **Tomlinson, J. A.,** Chemical control of *Spongospora* and *Olpidium* in hydroponic systems and soil, in *Developments in Applied Biology 2: Viruses with Fungal Vectors,* Cooper, J. I. and Asher, M. J. C., Eds., Association of Applied Biologists, Wellesbourne, 1988, 293.

69. **Van Dorst, H. J. M. and Peters, D.,** Experiences with the freesia leaf necrosis agent and its presumed vector, *Olpidium brassicae,* in *Developments in Applied Biology 2: Viruses with Fungal Vectors,* Cooper, J. I. and Asher, M. J. C., Eds., Association of Applied Biologists, Wellesbourne, 1988, 315.

70. **Mowat, W. P.,** Augusta disease in tulip — a reassessment, *Ann. Appl. Biol.,* 66, 17, 1970.

71. **Lange, L.,** Augustasyge los tulipaner Markundersogelser of tobak nekrose virus (TNV) og dets vektor, *Olpidium brassicae, Tidsskr. Planteavl,* 80, 153, 1976.

72. **Lin, M. C. and Ruan, Y. L.,** On the wheat yellow mosaic virus (WYMV), *Acta Phytopathol. Sinica,* 16, 73, 1986.

73. **Adams, M. J., Swaby, A. G., and Jones, P.,** Confirmation of the transmission of barley yellow mosaic virus (BaYMV) by the fungus *Polymyxa graminis, Ann. Appl. Biol.,* 112, 133, 1988.

74. **Brakke, M. K. and Estes, A. P.,** Some factors affecting vector transmission of soil-borne wheat mosaic virus from root washings and soil debris, *Phytopathology,* 57, 905, 1967.

75. **Jones, R. A. C. and Harrison, B. D.,** The behaviour of potato mop-top virus in soil, and evidence for its transmission by *Spongospora subterranea* (Wallr.) Lagerh., *Ann. Appl. Biol.,* 63, 1, 1969.

76. **Slykhuis, J. T.,** Seasonal transmission of wheat spindle steak mosaic virus, *Phytopathology,* 65, 1133, 1975.

77. **Campbell, R. N.,** Cultural characteristics and manipulative methods, in *Developments in Applied Biology 2: Viruses with Fungal Vectors,* Cooper, J. I. and Asher, M. J. C., Eds., Association of Applied Biologists, Wellesbourne, 1988, 153.

78. **Braselton, J. P.,** Karyology and systematics of Plasmodiophoromycetes, in *Developments in Applied Biology 2: Viruses with Fungal Vectors,* Cooper, J. I. and Asher, M. J. C., Eds., Association of Applied Biologists, Wellesbourne, 1988, 139.

79. **Ijdenberg, P., Kummert, J., and Lepoire, P.,** Complex populations of minor pathogens associated with roots from barley plants infected with barley yellow mosaic virus, *Parasitica,* 42, 137, 1986.

80. **Campbell, R. N. and Grogan, R. G.,** Acquisition and transmission of lettuce big-vein virus by *Olpidium brassicae, Phytopathology,* 54, 681, 1964.

81. **Tomlinson, J. A. and Garrett, R. G.,** Studies on the lettuce big-vein virus and its vector *Olpidium brassicae* (Wor.) Dang., *Ann. Appl. Biol.,* 54, 45, 1964.

82. **Asher, M. J. and Blunt, S. J.,** The ecological requirements of *Polymyxa betae, Proceedings of the 50th Winter Congress of the International Institute of Sugar Beet Research, Brussels,* 45, 1987.

83. **Teakle, D. S.,** Transmission of plant viruses by fungi, in *Principles and Techniques in Plant Virology,* Kado, C. E. and Agrawal, H. O., Eds., Van Nostrand Reinhold, New York, 1972, 248.

84. **Sahtiyanci, S.,** Studien uber einige wurzelparasitare Olpidiaceen, *Arch. Mikrobiol.,* 41, 187, 1962.

85. **Westerlund, F. V., Campbell, R. N., Grogan, R. G., and Duniway, J. M.,** Soil factors affecting the reproduction and survival of *Olpidium brassicae* and its transmission of big vein agent to lettuce, *Phytopathology,* 68, 927, 1978.

86. **Adams, M. J., Swaby, A. G., and MacFarlane, I.,** The susceptibility of barley cultivars to barley yellow mosaic virus (BaYMV) and its fungal vector, *Polymyxa graminis, Ann. Appl. Biol.,* 109, 561, 1986.

87. **Campbell, R. N. and Fry, P. R.,** The nature of the associations between *Olpidium brassicae* and lettuce big-vein and tobacco necrosis viruses, *Virology,* 29, 222, 1966.

88. **Campbell, R. N.,** Effects of benomyl and ribavirin on the lettuce big vein agent and its transmission, *Phytopathology,* 70, 1190, 1980.

89. **Kassanis, B. and MacFarlane, I.,** Transmission of tobacco necrosis virus by zoospores of *Olpidium brassicae, J. Gen. Microbiol.,* 36, 79, 1964.

90. **Campbell, R. N.,** Relationship between the lettuce big-vein virus and its vector, *Olpidium brassicae, Nature (London),* 195, 675, 1962.

91. **Adams, M. J., Jones, P., and Swaby, A. G.,** The effect of cultivar used as host for *Polymyxa graminis* on the multiplication and transmission of barley yellow mosaic virus (BaYMV), *Ann. Appl. Biol.,* 110, 321, 1987.

92. **Canova, A.,** Researches on virus diseases of Gramineae. III. *Polymyxa graminis* vector of wheat mosaic virus, *Phytopathol. Mediterr.,* 5, 53, 1966.

93. **Abe, H. and Tamada, T.,** Association of beet necrotic yellow vein virus with isolates of *Polymyxa betae* Keskin, *Ann. Phytopathol. Soc. Jpn.,* 52, 235, 1986.

94. **Campbell, R. N.,** Longevity of *Olpidium brassicae* in air-dry soil and the persistence of the lettuce big-vein agent, *Can. J. Bot.,* 63, 2288, 1985.

95. **Hiruki, C.,** Recovery and identification of tobacco stunt virus from air-dried resting spores of *Olpidium brassicae, Plant Pathol.,* 36, 224, 1987.

96. **Adams, M. J. and Swaby, A. G.,** Factors affecting the production and motility of zoospores of *Polymyxa graminis* and their transmission of barley yellow mosaic virus (BaYMV), *Ann. Appl. Biol.,* 112, 69, 1988.

97. **Kendall, T. L. and Lommel, S. A.,** Fungus-vectored viruses of wheat in Kansas, in *Developments in Applied Biology 2: Viruses with Fungal Vectors,* Cooper, J. I. and Asher, M. J. C., Eds., Association of Applied Biologists, Wellesbourne, 1988, 37.

98. **Kuwata, S. and Kubo, S.,** Tobacco stunt virus, in *AAB Descriptions of Plant Viruses,* No. 313, Association of Applied Biologists, Wellesbourne, 1986, 4.

99. **Huijberts, N., Blystad, D. R., and Bos, L.,** Lettuce big-vein virus: mechanical transmission and relationships to tobacco stunt virus, *Ann. Appl. Biol.,* 116, 463, 1990.

100. **Teakle, D. S. and Morris, T. J.,** Transmission of southern bean mosaic virus from soil to bean seeds, *Plant Dis.,* 65, 599, 1981.

101. **Shukla, D. D., Shanks, G. J., Teakle, D. S., and Behncken, G. M.,** Mechanical transmission of galinsoga mosaic virus in soil, *Aust. J. Biol. Sci.,* 32, 267, 1979.

102. **Roberts, F. M.,** The infection of plants by viruses through roots, *Ann. Appl. Biol.,* 37, 385, 1950.

103. **Van Dorst, H. J. M.,** Virus disease of cucumbers, *Annu. Rep. Glasshouse Crops Res. and Exp. Statn., Naaldwijk, Netherlands,* 1969, p. 75.

104. **Broadbent, L.,** The epidemiology of tomato mosaic. VIII. Virus infection through tomato roots, *Ann. Appl. Biol.,* 55, 57, 1965.

105. **Rana, G. L. and Roca, F.,** Nematode transmission of artichoke Italian latent virus, 2nd Int. Meet. Globe Artichoke, Bari, Italy, November 21–24, 1973, p. 139.

106. **Lamberti, F. and Bleve-Zacheo, T.,** Two near species of *Longidorus* (Nematoda: Longidoridae) from Italy, *Nematol. Mediterr.,* 5, 73, 1977.

107. **Van Dorst, H. J. M.,** Evidence for the soilborne nature of freesia leaf necrosis, *Neth. J. Plant Pathol.,* 81, 45, 1975.

108. **Fletcher, J. T., Wallis, W. A., and Davenport, F.,** Pepper yellow vein, a new disease of sweet peppers, *Plant Pathol.,* 36, 180, 1987.

109. **Ivanovic, M., Macfarlane, I., and Woods, R. D.,** Viruses transmitted by fungi: viruses of sugarbeet associated with *Polymyxa betae, Rep. Rothamsted Exp. Statn. for 1982,* 1983, p. 189.

110. **Huth, W. and Adams, M. J.,** Barley yellow mosaic (BaYMV) and BaYMV-M: two different viruses, *Intervirology,* 31, 38, 1990.

111. **Reddy, D. V. R., Nolt, B. L., Hobbs, H. A., Reddy, A. S., Rajeshwari, R., Rao, A. S., Reddy, D. D. R., and MacDonald, D.,** Clump virus in India: isolates, host range, transmission and management, in *Developments in Applied Biology 2: Viruses with Fungal Vectors,* Cooper, J. I. and Asher, M. J. C., Eds., Association of Applied Biologists, Wellesbourne, 1988, 239.

112. **Adams, M. J., Jones, P., and Swaby, A. G.,** Purification and some properties of oat golden stripe virus, *Ann. Appl. Biol.,* 112, 285, 1988.

113. **Fauquet, C., Thouvenel, J.-C., Fargette, D., and Fishpool, L. D. C.,** Rice stripe necrosis virus: a soil-borne rod-shaped virus, in *Developments in Applied Biology 2: Viruses with Fungal Vectors,* Cooper, J. I. and Asher, M. J. C., Eds., Association of Applied Biologists, Wellesbourne, 1988, 71.

114. **Usugi, T.,** Epidemiology and management in Japan of soil-borne cereal mosaic viruses with filamentous particles, in *Developments in Applied Biology 2: Viruses with Fungal Vectors,* Cooper, J. I. and Asher, M. J. C., Eds., Association of Applied Biologists, Wellesbourne, 1988, 213.

115. **Tomlinson, J. A. and Hunt, J.,** Studies on watercress chlorotic leaf spot and on the control of the fungus vector (*Spongospora subterranea*) with zinc, *Ann. Appl. Biol.,* 110, 75, 1987.

116. **Koenig, R., Lesemann, D.-E., Huth, W. and Makkouk, K. M.,** Comparison of a new soilborne virus from cucumber with Tombus-, Diantho-, and other viruses, *Phytopathology,* 73, 515, 1983.

117. **Lin, M. T., Kitajima, E. W., and Munoz, J. O.,** Isolation and properties of squash necrosis virus, a possible member of "tobacco necrosis virus" group, *Fitopatol. Brasileira,* 8, 622, 1983.

118. **Brown, D. J. F.,** personal communication.

Chapter 5

INCLUSIONS IN DIAGNOSING PLANT VIRUS DISEASES

J. R. Edwardson, R. G. Christie, D. E. Purcifull, and M. A. Petersen

TABLE OF CONTENTS

0-8493-4284-8/93/$0.00 + $.50

I. INTRODUCTION

A number of more or less effective measures have been used to control virus diseases. In order for controls to be effectively applied, the virus(es) causing the damage must be recognized. One rapid, reliable, and relatively inexpensive method of diagnosing field infections is cytology of inclusions.

Diagnosis of plant virus-induced diseases can be difficult because many of the large number (500 to 600)[1,2] of viruses induce similar symptoms in the same host. A workable classification system, evolved over the past 20 years, has made the diagnostician's job easier by placing most of the named plant viruses into 35 groups.[3,4] Viruses within groups have similar properties, many of which are not shared by viruses in other groups. Obtaining information about certain virus properties permits assignment of the virus to a group, and when this is accomplished the diagnostician can then predict a number of additional properties of the virus.

Virus-induced inclusions have long been used in diagnosing animal virus infections with light microscopy.[5] Although not as extensively used as animal virus inclusions, the value of inclusions for diagnosing plant virus infections with light microscopy has also been recognized for a long time.[6] Kassanis[7] described the nuclear inclusions of tobacco etch potyvirus as a valuable diagnostic character. Diagnosis of important virus infections in the Brassicaceae via cytoplasmic inclusion differences was reported for cauliflower mosaic *Caulimovirus,* turnip crinkle *Tombusvirus,* turnip mosaic *Potyvirus,* turnip yellow mosaic *Tymovirus,*[8] and radish mosaic *Comovirus.*[9] The usefulness of nuclear inclusions for diagnosing *Gomphrena* rhabdovirus infections was pointed out by Kitajima and Costa.[10] Cytoplasmic and nuclear inclusions were applied to diagnosis of infections by viruses in seven groups.[11] Recently, Brunt and co-workers[12] have listed inclusions as being of diagnostic value for 18 groups of viruses.

A large body of literature is devoted to describing the composition, substructure, and morphology of inclusions in thin sections (for reviews see Martelli and Russo;[13,14] Francki and co-workers[15,16]). It is remarkable that only a small proportion of these investigations has been related to disease diagnosis. Application of electron microscopy to diagnosis with inclusions can be found, for example, in *Potyvirus,*[17] *Tymovirus,*[18] and *Tombusvirus* infections.[19]

Most plant virus infections can be diagnosed at the group level and some at the specific or strain levels with cytological techniques. Infections at the group level can be diagnosed with cytological techniques applied to viruses in 19 groups, and additional viruses in 2 groups which are not separable from each other cytologically (Table 1). Most viruses in both the *Furovirus* and *Tobamovirus* groups induce similar cytoplasmic inclusions: monolayers of particles, stacked plates, paracrystals, and fibrous masses, all of which are composed of particles, and X-bodies which contain a variety of components.

Tobamovirus inclusions of virus particles do not stain in Azure A unless heated.[20] The type member of the furoviruses, soil-borne wheat mosaic, induces inclusions which do not require heat to stain in Azure A.[21] If the other furoviruses also exhibit this characteristic, the two groups will be separable from each other on the basis of cytology of their inclusions.

The large numbers of plant viruses confronting the diagnostician may appear to be overwhelming. However, not all viruses and their vectors are present in all growing areas. And the effects of many viruses do not place them in the category of catastrophic plant diseases:[22] tobacco necrosis *Necrovirus* is usually not considered to cause economically important losses;[23] cryptoviruses appear to cause little or no damage;[24] Murant[25] has pointed out that some nepoviruses (9 viruses) can be devastating to crops but most nepoviruses (19 viruses) are of little economic importance; Lane[26] has observed that bromoviruses have not caused economically significant diseases; Francki and co-workers[27] have observed that only a few (6 of 77) rhabdoviruses are known to cause diseases of economic importance; most carlaviruses (46 of 50 viruses) do not produce serious diseases;[28] only 2 of 20 carmoviruses have been reported to be economically important pathogens.[29] However, there are numerous viruses in different groups which induce losses in crops that can be described as ranging from economically important to catastrophic. Infections with most of these viruses can be diagnosed cytologically at the group level.

II. CYTOLOGY OF INCLUSIONS

A. INTRODUCTION
There are three aspects to the cytology of virus-induced inclusions.

1. *Types of inclusions* — Virus-induced inclusions may consist of (1) aggregated virus particles (all groups) (Plate 1A to C)*, or coat protein shells (tymoviruses); (2) aggregated noncapsid proteins (such as tenuiviruses, potyviruses, see Section III); (3) altered cell constituents (peroxisomes in *Tombusvirus* infections, chloroplasts in *Tymovirus* infections) (Plate 1F); or (4) combinations of some of the preceding types.
2. *Location within cells* — Certain types of inclusions occur only in the cytoplasm (such as *Potyvirus* cylindrical inclusions, see Section III.B, and potato virus X laminate inclusion components [Plate 1D]), others only in nuclei (such as those of the geminiviruses, see Section III.A), while some may occur in both cytoplasm and nuclei (crystalline inclusions induced by bean yellow mosaic *Potyvirus*) (Plate 1E), and others in vacuoles and cytoplasm (crystalline inclusions induced by cucumoviruses).

* Plates 1 to 9 follow page 104.

Table 1. Use of Inclusions for Diagnosis of Infections at the Group or Family Level by Light Microscopy (LM) and Electron Microscopy (EM)

Virus group	Inclusions	LM	EM
Alfalfa mosaic	Hexagonally packed layers of particles in rafts or whorls in the cytoplasm and vacuoles.	−[a]	+[b]
Bromovirus	Cytoplasmic crystalline inclusions of virus particles. Particles in cytoplasm and nuclei.	−	−
Capillovirus	Cytoplasmic particle aggregates in phloem containing proliferated and dilated ER. Particle aggregates in nuclei. The small membrane-bound bodies (characteristic of *Closterovirus* infections) are absent.	−	+
Carlavirus	Cytoplasmic banded-bodies and paracrystals of virus. Irregularly shaped inclusions of ER interspersed with particles and ribosomes. ER in inclusions is accentuated by increased electron-opacity.	+	+
Carmovirus	Cytoplasmic and vacuolar aggregates of particles. Mitochondria are altered into superficially resembling multivesiculate bodies of the Tombusviruses in Galinsoga mosaic and cucumber leaf-spot infections. Crystalline inclusions of particles in xylem of blackgram mottle virus-infected cells.	−	−
Caulimovirus	Cytoplasmic lacunose inclusions of granular and fibrillar proteinaceous material containing aggregated and scattered virus particles. Some members reported to induce nuclear inclusions. Inclusion matrices stain red-magenta; particle aggregates stain blue in Azure A.	+	+
Closterovirus	Cytoplasmic banded-bodies, paracrystals, and fibrous masses of particles. Small membrane-bound bodies containing fibrils. Inclusions predominantly in the phloem.	+	+
Commelina yellow mottle virus group	Cytoplasmic inclusions of aggregated particles sometimes in paracrystalline arrays. Fibrillar and granular viroplasmic inclusions may also be present; however, only a few group members have been studied cytologically.	−	−
Comoviruses	Cytoplasmic inclusions of aggregated particles often in crystalline arrays. Particles aggregate in xylem. Vacuolate-vesiculate cytoplasmic inclusions. Inclusions are similar to *Sobemovirus* inclusions.	+	+
Cryptovirus	Particles in viroplasms in phloem parenchyma cytoplasm. Small aggregates of particles in xylem and vesicles containing fibrils in phloem parenchyma cytoplasm (radish yellow edge virus). Particle aggregates in phloem parenchyma cytoplasm (ryegrass cryptic virus).	−	−
Cucumovirus	Cytoplasmic and vacuolar aggregates of particles often in crystalline arrays. Inclusions are predominantly in the mesophyll.	−	−
Dianthovirus	Cytoplasmic aggregates of particles; also electron-opaque material occurs in the cytoplasm.	−	−

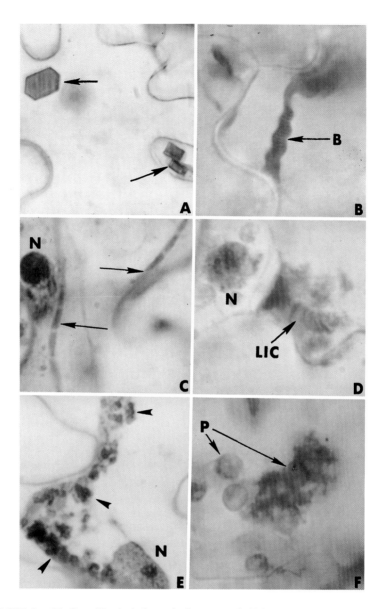

PLATE 1. (A) Crystalline inclusions of tobacco mosaic *Tobamovirus* particles in Turkish tobacco. (Magnification × 970). (B) Banded-body inclusion (**B**) induced by papaya mosaic *Potexvirus* in *Nicotiana benthamiana*. (C) Banded-body inclusions (arrows) induced by pea streak *Carlavirus* in *Pisum sativum* var. Alaska, nucleus (**N**). (D) Laminate inclusion components (**LIC**) induced by potato virus X *Potexvirus*, nucleus (**N**) in NN Turkish tobacco. (E) Crystalline inclusions (arrow heads) induced by bean yellow mosaic *Potyvirus* in *Pisum sativum* var. Alaska. (F) Turnip yellow mosaic *Tymovirus*-induced inclusion of aggregated particles and plastids (**P**) in *Brassica perviridis* var. Tendergreen Mustard. (B to F, magnification × 1940). A, C, E stained in O-G; B, D, F stained in Azure A.

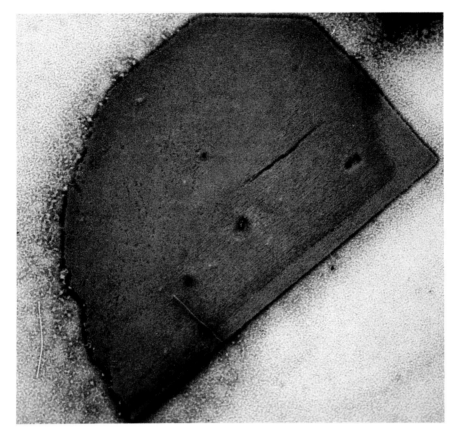

PLATE 2. Laminated aggregate from a cylindrical inclusion induced by *Bidens* mottle *Potyvirus* in *Bidens pilosa*. Leaf dip stained in uranyl acetate. (Magnification × 51,500).

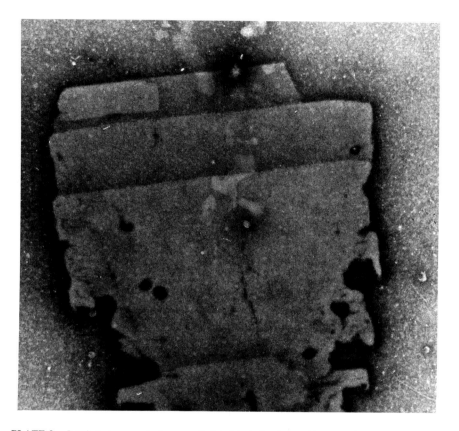

PLATE 3. Laminated aggregate from a cylindrical inclusion induced by pokeweed mosaic *Potyvirus* in *Phytolacca americana*. Leap dip stained in uranyl acetate. (Magnification × 51,500).

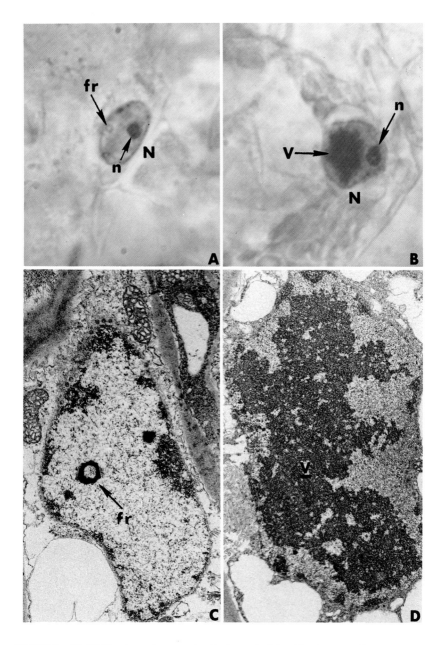

PLATE 4. (A, B) *Nicotiana tabacum* stained in Azure A. (Magnification × 1940). (A) Nucleus
(**N**) containing a nucleolus (**n**) and Euphorbia mosaic *Geminivirus*-induced fibrous ring inclusion
(**fr**). (B) Nucleus (**N**) containing a nucleolus (**n**) and Euphorbia mosaic *Geminivirus*-induced
aggregated virus inclusion (**V**). (C, D) Thin sections of *Lycopersicon esculentum* infected with an
uncharacterized *Geminivirus*. (Magnification × 8400). (C) *Geminivirus*-induced fibrous ring nuclear inclusion (**fr**). (D) *Geminivirus*-induced virus aggregate nuclear inclusion (**V**).

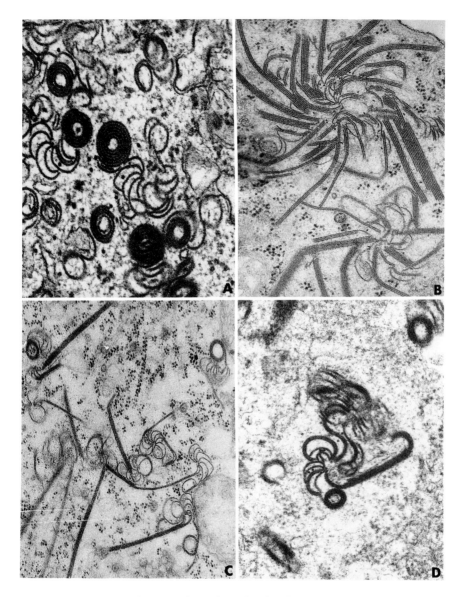

PLATE 5. Thin sections of *Potyvirus*-induced cytoplasmic cylindrical inclusions. (A) Type 1 cylindrical inclusions (pinwheels and scrolls) induced by an uncharacterized *Potyvirus* in *Cucurbita pepo* var. Early Prolific Straight Neck. (Magnification × 63,000). (B) Type 2 cylindrical inclusions (pinwheels and laminated aggregates) induced by tobacco etch *Potyvirus* in *Nicotiana tabacum* var. Havana 425. (Magnification × 50,000). (C) Type 3 cylindrical inclusions (pinwheels, scrolls, and laminated aggregates) induced by turnip mosaic virus *Potyvirus* in *Brassica perviridis* var. Tendergreen Mustard. (Magnification × 20,000). (D) Type 4 cylindrical inclusions (pinwheels, scrolls, and short, curved laminated aggregate) induced by zucchini yellow mosaic *Potyvirus* in *Cucurbita pepo* var. Small Sugar Pumpkin. (Magnification × 52,500).

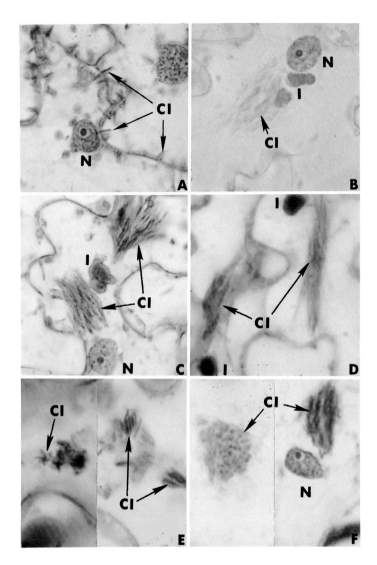

PLATE 6. Cylindrical inclusions induced by potyviruses.(A) Pepper mottle *Potyvirus*-induced cylindrical inclusions (**CI**) on cell walls, early infection, nucleus (**N**) in *Nicotiana tabacum* var. Turkish. (Magnification × 970). (B) Pepper mottle *Potyvirus*-induced cylindrical inclusions (**CI**), longer infection, irregular inclusions (**I**). (Magnification × 970). (C) Pepper mottle *Potyvirus*-induced cylindrical inclusions (**CI**), infection longer than in A or B, irregular inclusions (**I**). (Magnification × 970). (D) Papaya ringspot *Potyvirus* (W strain)-induced cylindrical inclusions (**CI**) and irregular inclusions (**I**) in *Cucurbita pepo* var. Small Sugar Pumpkin. (Magnification × 1940). (E) Tobacco etch *Potyvirus-induced cylindrical inclusions* (**CI**) end view (left arrow), side views (right arrows). (Magnification × 1940). (F) Blackeye cowpea mosaic *Potyvirus*-induced cylindrical inclusions (**CI**) end views (left arrow), side views (right arrow) in *Vigna unguiculata* var. Knuckle Purple Hull. (Magnification × 1940). A, C–F stained in O-G; B stained in Azure A.

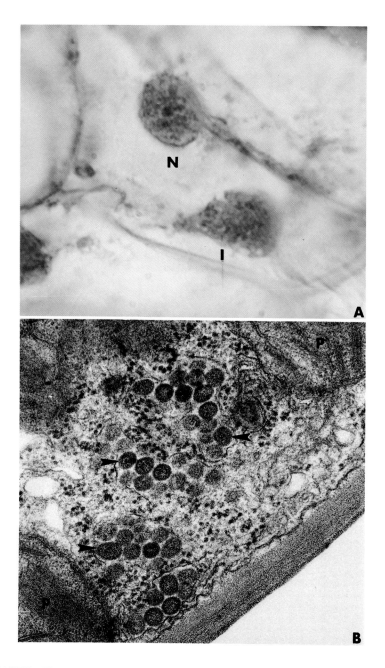

PLATE 7. Tomato spotted wilt virus-infected tissues. (A) Cytoplasmic inclusion of aggregated virus particles (**I**) below nucleus (**N**) in *Nicotiana tabacum*. Stained in O-G. (Magnification ×1940). (B) Cytoplasmic inclusion of aggregated virus particles (arrow heads) between plastids (**P**) in *Nicotiana benthamiana*. (Magnification × 52,500).

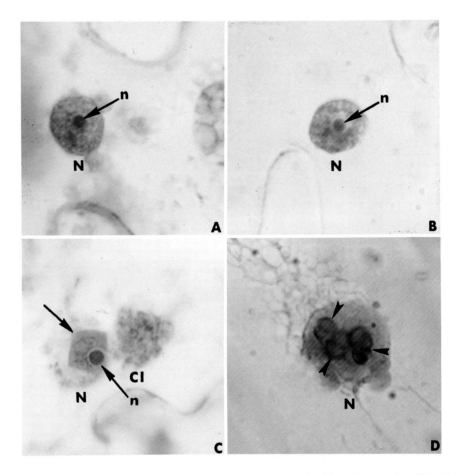

PLATE 8. *Nicotiana tabacum* cells. (Magnification × 1940). (A) Healthy cell with nucleus (**N**) and nucleolus (**n**) stained in O-G. (B) Healthy cell with nucleus (**N**) and nucleolus (**n**) stained in Azure A. (C) Tobacco etch *Potyvirus* (severe strain)-induced cytoplasmic cylindrical inclusions (**CI**) and an end view of a crystalline inclusion (arrow) covering the nucleolus (**n**) in the nucleus (**N**). Stained in O-G. (D) Tobacco etch *Potyvirus* (mild strain)-induced bipyramidal inclusions (arrow heads) in nucleus (**N**). Stained in bromophenol blue.

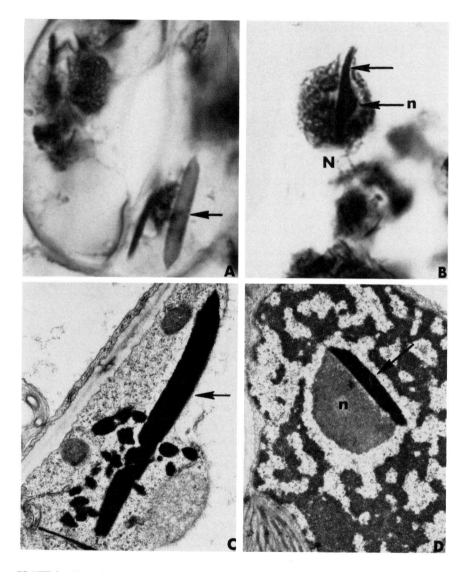

PLATE 9. *Vicia faba* infected with the *Vicia* isolate of bean yellow mosaic *Potyvirus;* A, B stained in O-G. (Magnification × 1970.) C, D, thin sections. (A) Virus-induced large crystalline boat-shaped cytoplasmic inclusion (arrow). (B) Virus-induced crystalline nuclear inclusion (arrow) closely associated with a nucleolus (**n**); nucleus (**N**). (C) Virus-induced boat-shaped cytoplasmic crystalline inclusion (arrow). (Magnification × 15,000). (D) Virus-induced crystalline nuclear inclusion (arrow) closely associated with the nucleolus (**n**). (Magnification × 11,000).

Table 1. Use of Inclusions for Diagnosis of Infections at the Group or Family Level by Light Microscopy (LM) and Electron Microscopy (EM) (continued)

Virus group	Inclusions	LM	EM
Fabavirus	Cytoplasmic crystalline inclusions of particles, also circular tubules and rectangular tubules of particles. Cytoplasmic inclusions of proliferated ER associated with granular material.	−	−
Furovirus	Viruses in this group induce inclusions of the same morphology as those of the *Tobamovirus* group. The type-member (soil-borne wheat mosaic virus) induces inclusions which stain in Azure A in the absence of heat. Other group members require testing for this property. At present, furo- and tobamoviruses cannot be separated cytologically, but they can be differentiated from viruses in other groups.	(+)[c]	(+)
Geminivirus	Nuclear inclusions of aggregated particles and fibrous ring-shaped inclusions, predominantly in the phloem. Particle aggregates stain blue and rings stain blue-green.	+	+
Hordeivirus	Cytoplasmic and nuclear aggregates of particles.	−	−
Ilarvirus	Cytoplasmic and nuclear aggregates of particles. However, only two group members have been studied cytologically.	−	−
Luteovirus	Cytoplasmic aggregates of particles predominantly in the phloem.	+	+
Maize chlorotic dwarf virus group	Cytoplasmic inclusions of dense granular material containing virus particles predominantly in phloem.	−	−
Marafavirus	Cytoplasmic crystalline inclusions of particles, particles in single rows within tubules, particles scattered or aggregated in epidermal and phloem parenchyma vacuoles.	−	−
Necrovirus	Cytoplasmic crystalline inclusions of particles.	−	−
Nepovirus	Cytoplasmic vesiculate inclusions, particle aggregates, single rows of particles within tubules.	−	−
Parsnip yellow fleck virus group	Cytoplasmic inclusions of ribosome-studded vesicles and straight tubules about 30 nm diameter. Tubules about 45 nm diameter usually contain single rows of particles. Those associated with plasmodesmata are sheathed by cell wall material. Dandelion yellow mosaic virus induced a crystalline inclusion of particles in cytoplasm.	−	−
Pea enation mosaic virus group	Cytoplasmic crystalline inclusions of particles. Nuclear inclusions of virus aggregates. Double membrane-bound vesicles containing DNA originate in nuclei and are extended into the cytoplasm. Predominantly in phloem.	+	+
Potexvirus	Cytoplasmic aggregates of virus in the form of banded bodies. Virus aggregates also in fibrous masses.	+	+
Potyvirus	Cytoplasmic cylindrical inclusions of proteinaceous sheets forming pinwheels in cross section, bundles in longitudinal section. Nuclear inclusions induced by several members.	+	+

Table 1. Use of Inclusions for Diagnosis of Infections at the Group or Family Level by Light Microscopy (LM) and Electron Microscopy (EM) (continued)

Virus group	Inclusions	LM	EM
Reoviridae	Cytoplasmic irregularly shaped inclusions containing fibrils and crystals of particles. Single rows of particles in tubules. Predominantly in phloem. Inclusions in the two genera are indistinguishable but can be used to distinguish viruses in those genera from viruses in other groups. Rice ragged stunt inclusions, same as above, except particles in tubules have not been reported in plant cells.	+	+
Rhabdoviridae	Aggregates of virus particles in nuclei and cytoplasm. Viroplasms in nuclei and cytoplasm. Particle aggregates stain light red in Azure A.	+	+
Sobemovirus	Cytoplasmic crystalline inclusions of particles. Particle aggregates and crystals in nuclei. Xylem is blocked by virus particles. Inclusions are similar to those induced by comoviruses.	+	+
Tenuivirus	Crystalline and paracrystalline inclusions of closely associated noncapsid protein fibers are induced in cytoplasm, nuclei, and vacuoles. Inclusions are unstained in Azure A.	+	+
Tobamovirus	Cytoplasmic and vacuolar inclusions of particles in crystalline array in stacked-plate or monolayer forms. Fibrous masses of particles. Paracrystals and angled-layer aggregate inclusions of particles have been observed induced by several members. X-bodies are induced by most members. At present tobamo- and furoviruses cannot be separated cytologically, but they can be differentiated from viruses in other groups. Staining of all inclusions in Azure A except X-bodies requires heat (60°C, 1 min).	(+)	(+)
Tobravirus	Cytoplasmic particle aggregates in the form of tiered crystals. Aggregated mitochondria, particles with short axes aligned in mitochondria.	−	−
Tospovirus	Cytoplasmic aggregates of particles occur in cisternae of ER. One commonly occuring strain of tomato spotted wilt virus does not produce intact particles but induces striated inclusions containing RNA.	+	+
Tombusvirus	Cytoplasmic multivesiculate bodies arise from enlarged vesiculated peroxisomes. In vacuoles, particle aggregates occur in spherical membrane-bound inclusions.	+	+
Tymovirus	Chloroplast peripheries exhibit flask-shaped vesicles containing fibrils. Chloroplasts clump in most virus-host combinations. Empty coat protein shells accumulate in large masses in nuclei. Nuclear inclusions are unstained in Azure A.	+	+

[a] − Indicates that inclusions are not diagnostic for infections.

[b] + Indicates that inclusions are diagnostic for infections.

[c] (+) Indicates that inclusions induced by furoviruses and tobamoviruses cannot be used to differentiate viruses in one of these groups from viruses in the other group. However, inclusions induced by furo- and tobamoviruses can be used to distinguish these viruses from those in other groups.

3. *Tissue specificity* — Some inclusions may be restricted predominantly to a single cell type, such as crystalline and particle aggregate inclusions induced in mesophyll cells by cucumoviruses, or the fibrous and banded-body inclusions of the closteroviruses in phloem tissues.

The types of inclusion, their locations within cells, and their tissue restrictions are all used by the diagnostician to identify viruses inducing the inclusions.

B. LIGHT MICROSCOPY
1. Advantages
The techniques involved in light microscopy of virus-induced inclusions are simple, relatively inexpensive, and rapid. A research light microscope is cheaper to acquire and easier to maintain than an electron microscope. Much larger volumes of tissue can be examined in the light microscope than in the electron microscope. Light microscopy has demonstrated that inclusions are neither uniformly distributed nor necessarily correlated with symptom expression. Difficulties reported in finding certain inclusions in thin sections of tissues infected with viruses such as cowpea chlorotic mottle *Bromovirus* (CCMV),[14] broad bean stain (BBSV),[30] and true broad bean mosaic (TBBMV) comoviruses[30] can be explained by the fact that these studies using electron microscopy failed to take advantage of the extensive sampling capacity inherent in light microscopic techniques which can readily detect inclusions induced by these viruses.[20,21] Most cytological studies with light and electron microscopy devoted to diagnosis have attained identification at the group level. Diagnosis at the species and strain level with virus-induced inclusions is summarized in Table 2. It is also possible to achieve identification at the species or strain level by combining cytological with serological techniques.

2. Stains for Diagnosis
With the application of appropriate stains, inclusions of different types are easily differentiated from each other and from normal cell constituents in the light microscope. Several stains have been used in studies of inclusions. We have found the Azure A and the combination of Calcomine Orange-Luxol Brilliant Green[11,20] (O-G stain) suitable for differentiating inclusions induced by viruses in most of the groups defined by the International Committee on Taxonomy of Viruses (ICTV) (Table 1). When these stains are applied (see

Table 2. Inclusions Diagnostic at the Species and Strain Levels

Virus	Inclusions
Alfalfa mosaic virus (monotypic group)	See Table 1
Cowpea mosaic *Comovirus*	Cytoplasmic, vesiculated inclusions, fibrous inclusions. Single rows of particles in membranes. Nuclear, crystals; nuclear membranaceous
Pea enation mosaic virus (monotypic group)	See Table 1
Papaya mosaic *Potexvirus*	Cytoplasmic, wavy banded-bodies. Nuclear, crystalline inclusions near nucleoli
Potato X *Potexvirus*	Cytoplasmic, laminate inclusions, banded-bodies
Bean yellow mosaic *Potyvirus* (*Vicia* isolate)	Cytoplasmic, type 2 cylindrical and crystalline, some of which are boat-shaped. Nuclear, crystalline, some of which are elongated boat-shaped
Beet mosaic *Potyvirus*	Cytoplasmic, type 2 cylindrical. Nuclear, globular lacunose containing fibers
Blackeye cowpea mosaic *Potyvirus*	Cytoplasmic, type 1 cylindrical. Nuclear, proteinaceous fibrous and elongated
Celery mosaic *Potyvirus*	Cytoplasmic, type 1 cylindrical, and proteinaceous paracrystalline. Nuclear, proteinaceous fibrous
Clover yellow vein *Potyvirus* (pea necrosis strain)	Cytoplasmic, type 2 cylindrical, and long crystalline needles. Nuclear long crystalline needles
Datura shoestring *Potyvirus*	Cytoplasmic, type 3 cylindrical, and aggregates of flattened tubules. Nuclear, aggregated tubules
Henbane mosaic *Potyvirus*	Cytoplasmic, type 3 cylindrical, and paracrystalline
Pepper mottle *Potyvirus*	Cytoplasmic, type 4 cylindrical which are extremely long; also has irregularly shaped (amorphous) inclusions
Plum pox *Potyvirus*	Cytoplasmic, type 2 cylindrical, and elongated proteinaceous crystalline and paracrystalline. Nuclear, elongated crystalline and paracrystalline
Potato A *Potyvirus* (Gugerli isolate)	Cytoplasmic, type 4 cylindrical. Nucleolar, lacunose, globular
Sweet potato russet crack *Potyvirus*	Cytoplasmic, type 4 cylindrical, crystalline
Tobacco etch *Potyvirus* (mild etch strain)	Cytoplasmic, type 2 cylindrical. Nuclear, bipyramidal
Tobacco etch *Potyvirus* (severe etch strain)	Cytoplasmic, type 2 cylindrical. Nuclear, truncated 4-sided pyramid
Tobacco etch *Potyvirus* (Madison isolate)	Cytoplasmic, type 2 cylindrical. Nuclear, octahedral
Tobacco vein mottling *Potyvirus*	Cytoplasmic, type 1 cylindrical, proteinaceous crystalline, fibrous. Nuclear, fibrous elongated
Watermelon mosaic virus-2 *Potyvirus* (original Florida isolate)	Cytoplasmic, type 3 cylindrical. Nuclear, thin flat proteinaceous plate
Zucchini yellow fleck *Potyvirus*	Cytoplasmic, type 1 cylindrical, and fimbriate. Nuclear, fimbriate
Tomato spotted wilt *Tospovirus*	See Table 1

Appendix I at the end of this chapter) to epidermal strips, free-hand sections, cryostat sections, or tissues obtained by abrasion (either the lower or upper epidermis can be removed by gentle abrasion with 600 mesh emery cloth or sandpaper)[11] the following reactions occur: (1) in Azure A, the nucleoplasm is colorless, chromation is blue, nucleoli are red to magenta, while plastids, mitochondria, and peroxisomes are unstained, as are inclusions consisting of protein. Inclusions containing ribonucleoprotein stain red to magenta and those containing deoxyribonucleoprotein stain blue. (2) In the Orange-Green combination, the nucleoplasm stains pale orange, chromatin green, nucleoli olive-green, plastids, mitochondria, and peroxisomes are green, while inclusions containing any type of protein also stain green.[11,20] Small inclusions may be obscured by plastids. Plastids can be dissociated by floating tissues on a 5% solution of Triton X-100 for 5 min prior to staining.[20]

3. Immunofluorescence Microscopy

Immunofluorescence microscopy involves purifying immunoglobulins from antisera,[31] conjugating the globulins with a fluorescent dye such as fluorescein isothiocyanate (FITC) or tetramethylrhodamine isothiocyanate (TRITC),[32] and staining epidermal strips or free-hand or cryostat sections of infected tissues. The poor penetration of the conjugate into plant cells has been improved by the addition of dimethyl sulfoxide (DMSO),[33] which also enhances staining. Penetration of cell walls by antibodies has also been enhanced by the use of enzymes.[34] If the appropriate antiserum has been used, the conjugate will combine with inclusions which will fluoresce in the presence of ultraviolet light. The virus can be identified at the species or strain level with this technique. Since protein A labeled with either FITC or TRITC is now available (LKB Instruments Ltd.), it is no longer necessary to conjugate fluorescent compounds with specific antisera.[32] The immunofluorescence techniques in current use require long times devoted to incubation of tissues with antibodies (about 1.5 h) and saline (about 1 h). They also involve difficulties in penetrating cell walls with dye conjugates, and require relatively expensive equipment.

4. Protein A-Gold Conjugates

Some of the disadvantages have been overcome by using light microscopy with colloidal gold particles (about 15 nm in diameter)[35] in place of fluorescence dyes.[36] Here, gold particles are conjugated with protein A, epidermal strips of infected tissues are treated with pectinase and cellulase and then incubated with antiserum. After several rinses to eliminate nonspecific reactions, protein A-gold is added to the incubated tissues. The protein-gold binds to the antibody which is located on antigenic sites within cells. A blue filter enhances the contrast of the red gold particles with the remainder of the cell constituents, which are blue in transmitted light.[36]

C. ELECTRON MICROSCOPY

Examination of infected tissues in thin sections with the electron microscope has provided much information on the constituents and substructure of inclusions. Most of this information is not attainable through light microscopy. Although many virus infections can be diagnosed at the group level with information on inclusions obtained by light microscopy, some can only be diagnosed via electron microscopy (Table 1). The subject of thin-sectioning of plant materials infected with viruses has been reviewed recently by Martelli and Russo,[14] Francki and co-workers,[15,16] and by Lesemann[37] for filamentous viruses. Diagnosis of virus infections by examining inclusions in thin sections is not a rapid procedure. Small pieces of tissue (about 1×2 or 1×3 mm) are fixed, dehydrated, and embedded in plastic. Portions of the embedded tissue are sectioned with an ultra-microtome and the sections are transferred to grids and stained. Inclusions are then searched for in these small samples which represent very small portions of tissues. Such searches are often not carried out in conjunction with light microscope examination of infected tissues from the same source. In this situation, few, if any, inclusions may be encountered in an apparently large amount of thin-sectioned tissue. Inclusions are not uniformly distributed whether or not they are confined to specific cell types.[11] We consider that terms in the literature such as ''rare'', ''inconsistent'', ''erratic'', or ''labile'', when applied to inclusions in thin sections, indicate that the investigators lacked an appreciation of sampling problems.

Much information that is useful for diagnosis has been obtained through electron microscopy of plant extracts.[38,39] Fragments of diagnostic cylindrical inclusions exhibiting striation periodicity of about 5 nm in negative stain indicate the presence of a *Potyvirus*. The shapes of some of the laminated aggregate components of cylindrical inclusions are distinctive and can be used to diagnose infections at the species level[40] (Plates 2 and 3). It should be possible to extend diagnosis of *Potyvirus* infections by studying differences in inclusion morphology in negative stains. It may be possible to use negatively stained inclusions in other groups for diagnosis, such as the laminate inclusion components induced by potato virus X *Potexvirus,* and the tubes and scrolls induced by maize rough dwarf and oat sterile dwarf fijiviruses (Reoviridae).

III. DIAGNOSIS OF INFECTIONS AT THE VIRUS GROUP LEVEL

Inclusions that are diagnostic for infections at the group level are described in Table 1. The majority of viruses represented in the 35 groups in Table 1 can be diagnosed at the group level by light microscopy of epidermal strips or by electron microscopy of thin sections. Diagnosis of infections at the species level has also been attained through light or electron microscopy (Table 2). Some of the descriptions of inclusions in Tables 1 and 2 are based on both light and electron microscope studies of virus-infected cells, and utilize staining reactions, location of inclusions in cells and tissues, as well as morphology of the inclusions. Other descriptions are based only on electron microscopy of thin sections.

Descriptions including micrographs of some diagnostic inclusions should give the reader an idea of the capacity of cytological studies for rapid and reliable diagnosis. We have selected for more detailed comments three groups of plant viruses, members of which are widely distributed, have wide host ranges, and infect most of the world's economically valuable species.

A. GEMINIVIRUSES

The *Geminivirus* group contains 35 definite members.[4] Several of these cause economically important losses in such crops as beans, cassava, corn, potatoes, and wheat in the temperate and tropical areas. Geminiviruses have been separated into subgroup I: viruses with host ranges limited primarily to the Gramineae, transmitted by leafhoppers but not mechanically, and containing a monopartite DNA genome; subgroup II: viruses with somewhat wider host ranges among dicotyledonous plants, transmitted by leafhoppers and mechanically with difficulty, and containing a monopartite DNA genome; subgroup III: viruses with narrow host ranges among dicotyledonous plants, transmitted by whiteflies and experimentally by mechanical inoculations, containing a bipartite DNA genome.[4] Tomato pseudo curly top, a subgroup II *Geminivirus* which is transmitted by treehoppers, induces the same types of nuclear inclusions as do members of subgroups I to III.[52] Inclusions induced by viruses in all subgroups are confined mainly to phloem tissues. When stained with Azure A, nuclear inclusions appear as large, blue-staining bodies (Plate 4B) which are found to consist of aggregated virus particles when viewed in thin sections (Plate 4D). In addition to the inclusions of aggregated virus particles, small rings also occur in the nuclei (Plate 4A and C). Both types of inclusion contain DNA. The ring-shaped inclusions are faintly stained blue-green in Azure A and are difficult to detect by light microscopy. The inclusions of aggregated virus that are detected prior to and during early symptom development become increasingly difficult to find in tissues with long-standing infections.

Kim and Carr[53] proposed that the nuclear inclusions induced by whitefly-transmitted geminiviruses were diagnostic, based on thin section studies. In thin sections the large nuclear inclusions have been found to be made up of aggregated virus particles, which sometimes occur in crystalline arrays, while the ring-shaped inclusions are fibrous in nature. The leafhopper- and treehopper-transmitted geminiviruses that have been studied cytologically also induce aggregated virus particle inclusions and fibrous rings in nuclei. Tymoviruses also induce nuclear inclusions.[18] These inclusions consist of aggregated empty capsids which stain green in the O-G combination and are unstained in Azure A. Tenuiviruses induce fibrous inclusions in cytoplasm, vacuoles, and nuclei. With the exception of rice stripe virus,[54] which induces nuclear inclusions of aggregated virus (staining red-magenta in Azure A), *Tenuivirus*-induced nuclear inclusions consist of fibrous noncapsid protein (unstained in Azure A). Some viruses in other groups induce nuclear inclusions, such as zucchini yellow fleck *Potyvirus*,[55] blackeye cowpea mosaic *Potyvirus*,[56] and papaya mosaic *Potexvirus*.[20] None of the viruses outside the *Geminivirus* group induce nuclear inclusions which are similar to those of the geminiviruses.

In a recent investigation of cytoplasmic inclusions induced by strains of maize streak *Geminivirus* and isolates serologically related to it, four types of inclusions were reported: crystalline, noncrystalline, sheet-like, and open lattice.[41] These observations suggest the possibility of using differences in inclusions to diagnose some *Geminivirus* infections at the specific and strain levels.

B. POTYVIRUSES

The *Potyvirus* group is the largest and economically the most important of the plant virus groups; 158 viruses are listed as members or possible and probable members of the group,[4] and there are probably many more as yet inadequately described potyviruses. Nineteen potyviruses have been reported to be seed-transmitted.[57] Most potyviruses are transmitted by aphids in a nonpersistent manner but there are no vectors reported for 18 members of the group. One member is whitefly-transmitted, five members are fungal-borne, and five are mite-transmitted.[1] Serological relationships have been established between aphid- and the whitefly-transmitted potyvirus,[60] aphid- and mite-transmitted,[60,61] and aphid- and a fungus-transmitted potyvirus.[62]

Potyviruses induce unique cytoplasmic cylindrical inclusions. Cylindrical inclusions have been reported to possess helicase activity.[63] Portions of tobacco vein mottling *Potyvirus*-induced cylindrical inclusions have been observed in thin sections subjected to immunogold labeling as early as 10 h after inoculation of protoplasts.[64] These inclusions are recognized as a main characteristic of the group[4,58] and are diagnostic for infections by potyviruses.[17,57] The large number of viruses in the group adds to the usual difficulties encountered in attaining diagnoses at the species level. In an effort to reduce the number of comparisons required for diagnosis of *Potyvirus* infections at the species level, the group has been separated into four subdivisions[56,59] based on the differences in types of cylindrical inclusions: subdivision I contains 35 viruses inducing type 1 cylindrical inclusions (pinwheels and scrolls) (Plate 5A); subdivision II contains 44 viruses inducing type 2 inclusions (pinwheels and laminated aggregates) (Plate 5B); subdivision III contains 14 viruses inducing type 3 inclusions (pinwheels, scrolls, and laminated aggregates) (Plate 5C); subdivision IV contains 18 viruses inducing type 4 inclusions (pinwheels, scrolls, and short, usually curved laminated aggregates) (Plate 5D); the types of cylindrical inclusions induced by 28 potyviruses remain to be determined.[57] Assignment of a virus to one of these subdivisions can reduce comparisons required to move from diagnosis at the group level to diagnosis at the specific level. Each strain of a potyvirus induces only one type of cylindrical inclusion and variation in types of cylindrical inclusions between strains and isolates of most individual potyviruses has not been observed (examples are tobacco etch, type 2; bean common mosaic, type 1; dasheen mosaic, type 3; pepper mottle, type 4). However, some strains of the same *Potyvirus* do induce different types of cylindrical inclusions, such as an Australian isolate of potato virus Y with type 1 inclusions and a Canadian isolate of potato virus Y with type 4 inclusions;[59] an Iris strain of turnip mosaic virus with type 4 inclusions[65] and a Florida strain inducing type 3 inclusions;[66] and peanut mottle virus Kuhn isolate with type 3 inclusions[20] and the Venezuelan isolate inducing type 2 inclusions.[67] The existence of such strains complicates the use of the subdivisions for virus diagnosis. Although subdivisions are useful for diagnosis they have no apparent taxonomic usefulness. They were not intended for application to classification.[56,57,59] They are not the equivalent of subgroups.

The cylindrical cytoplasmic inclusions are composed of proteinaceous sheets with a striation periodicity of about 5 nm. These distinctive inclusions are not likely to be confused with inclusions induced by viruses in other groups. They appear first at the periphery of the cell (Plate 6A). Eventually they congregate in massive inclusions which are found in the central portions of the cell, often near nuclei (Plate 6B to F). These inclusions are made up of numerous cylindrical inclusions, crystal-containing microbodies, endoplasmic reticulum (ER), vesicles, ribosomes, and scattered virus particles.

Although definition of the original types of cylindrical inclusions was based on electron microscopy of thin sections, it has been demonstrated that the different types of cylindrical inclusions can be distinguished by light microscopy.[20,57] Since these inclusions are proteinaceous they do not stain in Azure A (Plate 6B). In the Orange-Green combination the individual cylindrical inclusions stain green, as do the massive cytoplasmic inclusions, of which they are a major component. The thin plates which comprise the central portion of the cylindrical inclusion are not resolved in the light microscope. However, the scrolls appear as dots when viewed end-on and remain as dots through considerable change in the focal plane (Plate 6F), while in side view they appear as lines which are rapidly lost from view when the focal plane is changed (Plate 6F). Laminated aggregates appear as lines and remain as lines when the focal plane is changed in the light microscope (Plate 6E).

C. TOMATO SPOTTED WILT VIRUS

Tomato spotted wilt virus (TSWV) strains constitute a monotypic group of plant viruses.[3] A comparison of TSWV properties and those of viruses in the family Bunyaviridae (viruses infecting warm- and cold-blooded vertebrates, and arthropods) led Milne and Francki[68] to conclude that TSWV should be considered as a possible member of this family. It is now placed as the only virus in the genus *Tospovirus* in the Bunyaviridae.[4] The virus has a worldwide distribution and causes major losses in many crop species. It is transmitted by several species of thrips[69] and has been reported to be seed-transmitted in tomato and florists Cineraria.[70]

Serological relationships between TSWV strains have been reported.[71,119,120] Until recently, however, there have been few reports of serological detection of TSWV. It has been pointed out that the reason for this situation was the difficulty in preparing sufficient purified virus for use as an immunogen.[72] Brunt and co-workers,[12] in their recent survey of viruses of tropical plants, do not mention serology as a diagnostic tool for TSWV infections. In 1981 Francki and Hatta[72] stated that the most reliable method for TSWV identification was by observing its characteristic particles in thin sections. While particle diameter measurements range from about 55 to 90 to 120 nm, most measurements of TSWV spherical particles fall between 80 and 85 nm.[69] Since 1981, serological techniques have been applied more often to TSWV detection through enzyme-linked immunosorbent assay (ELISA)[73] and direct tissue blotting.[74]

In thin-section studies the characteristic TSWV particles are reported to aggregate in the cisternae of ER (Plate 7B). This type of inclusion has been reported associated with TSWV in a variety of infected tissues beginning with Ie's 1964[75] studies and continuing with those of Bertaccini and Bellardi (1990)[76] and Khurana and co-workers (1990),[77] and is considered unique for TSWV infections. Aggregation of virus particles between ER membranes is not a property that is unique to TSWV. The comoviruses broad bean true mosaic and radish mosaic exhibit virus particles sandwiched between membranes in the cytoplasm of infected cells;[16] many potyviruses such as pokeweed mosaic[78] and beet mosaic[79] induce formation of bridles (single layers of particles between membranes which extend into vacuoles). Rhabdoviruses such as Sonchus yellow net[80] and northern cereal mosaic[81] induce particle aggregates enclosed in ER. However, the sizes and shapes of the particles of these viruses differ strikingly from those of TSWV.[116,117] Particle aggregates have been described as occurring only in the cytoplasm of TSWV-infected cells by all investigators except Kitajima,[82] who reported individual as well as aggregated particles enclosed in ER and also aggregates of particles between membranes of the nuclear envelope in tobacco root cells infected with the Vira-Cabeca strain.

Milne,[86] Ie,[87] and Francki and Hatta[72] described electron-opaque proteinaceous material in the cytoplasm of TSWV-infected cells. This material was interpreted to be viroplasm. Ie[87] described spherical particles of about 33 to 38 nm in diameter in the viroplasm material and Ie[87] and Francki and Hatta[72] reported the viroplasms to be striated with a periodicity of about 4 to 5 nm. However, Ie[88] investigated two TSWV isolates, obtained after multiple sap-transmissions, in which no particles were detected, but which did contain diffuse masses of "amorphous" material. Ie[88] concluded that the "amorphous" material represented a defective form of TSWV that was present in some, if not all, isolates of the virus. Mixtures of strains must be common in TSWV infections. Whether viroplasms exist in TSWV-infected tissues seems questionable.

TSWV-induced inclusions have also been studied in the light microscope (Plate 7A). Living cells of infected tobacco were observed by Bald[83] to contain loose aggregates which increased in size and fused into large inclusions. Kobatake and co-workers[84] reported induction of a cytoplasmic inclusion larger than nuclei in infected *Datura stramonium* epidermis stained with Phloxine-Methylene Blue.[85] Vacuolated cytoplasmic inclusions in infected tobacco stained magenta in Azure A and green in the Orange-Green combination.[21]

IV. DIAGNOSIS OF INFECTIONS AT THE SPECIES AND STRAIN LEVELS

Most cytological studies devoted to diagnosis have identified viruses at the group level by recognizing inclusions common to the group. Diagnosis has also been achieved with some viruses and strains by combining information on group-specific inclusions with knowledge of other types of inclusions (Table 2). Following are summaries of some cytological investigations which have led to virus identification at the specific level and for some viruses at the strain level.

A. POTYVIRUSES

Cylindrical inclusions and nuclear inclusions of potyviruses stain green in the Orange-Green combination (Plate 6A and C to F) and are unstained in Azure A (Plate 6B). Tobacco etch virus (TEV) induces type 2 cylindrical inclusions (pinwheels and laminated aggregates). The laminated aggregates in the TEV cylindrical inclusions appear as lines and remain as lines when the focus of the light microscope is changed (Plate 6E). TEV also induces proteinaceous crystalline nuclear inclusions with shapes unlike nuclear inclusions induced by other potyviruses. Kassanis[7] stated that TEV-induced nuclear inclusions were diagnostic for infections by the virus. The nuclear inclusions are occasionally observed in the cytoplasm.[93] Several different strains and isolates of TEV induce nuclear inclusions of different shapes.[20] The inclusions of the mild etch strain are bipyramidal (Plate 8D) those of the severe etch strain are thin, truncated, four-sided pyramids (Plate 8C) and the Madison isolate induces octahedral inclusions.[118] Other TEV strains and isolates also exhibit consistent induction of nuclear inclusions with morphologies different from those described above; some of them have been distinguished cytologically in mixed infections.[94] Pepper mottle virus (PepMoV) induces type 4 cylindrical inclusions (pinwheels, scrolls, and short, usually curved, laminated aggregates). The cylindrical inclusions of PepMoV are much longer than those induced by other potyviruses and exhibit pronounced sharp tips (Plate 6C). PepMoV also induces cytoplasmic, irregularly shaped inclusions which stain green in O-G (Plate 6C) and red to magenta in Azure A (Plate 6B). These inclusions vary in size from those that are smaller than plastids to those that are larger than nuclei. PepMoV has not been reported to induce nuclear inclusions.

Blackeye cowpea mosaic virus (B1CMV) induces type 1 cylindrical inclusions in the cytoplasm. In the light microscope the scrolls of the cylindrical inclusions appear as dots in end-on views (Plate 6F). In side view the scrolls appear as lines which rapidly disappear when the focus is changed. In *Crotalaria spectabilis*, B1CMV induces elongated fibrous nuclear inclusions. Bean yellow mosaic virus (BYMV) induces type 2 cylindrical inclusions and numerous crystalline inclusions in nuclei and cytoplasm of infected cells. The crystalline inclusions stain green in O-G (Plate 1E) and do not stain in Azure A. The nuclear inclusions are closely associated with nucleoli.

Fluorescence microscopy was used to detect capsid, cylindrical, and nuclear inclusion proteins in the plasmalemma of pea (*Pisum sativum*) leaves 48 h after inoculation with BYMV.[95] Small crystalline inclusions appeared in nuclei 72 h after inoculation. About 7 days after inoculation, most nuclei contained many crystalline inclusions; some of the nuclear membranes ruptured and nuclear inclusions were extruded into the cytoplasm. The cytoplasmic crystalline inclusions immunofluoresced specifically with nuclear protein antiserum.[95] The *Vicia* isolate of BYMV induces boat-shaped crystalline inclusions in addition to cuboidal crystalline inclusions in nuclei and cytoplasm.[56,96] The boat-shaped crystalline inclusions are distinctive and permit the diagnostician to separate this strain of BYMV from the others (Plate 9). These viruses can be distinguished from each other by the differences in their inclusions. They can also be distinguished from other potyviruses within or outside their respective subdivisions (Table 2).

B. POTEXVIRUSES

The diagnostic, complex, cytoplasmic banded-body inclusions as well as the fibrous aggregates of virus particles induced by potexviruses stain green in the Orange-Green combination and magenta in Azure A. The banded-body inclusions contain virus particles aggregated side by side and form uniform bands which are separated by very narrow, unstained bands. Sometimes bands that are approximately the width of the *Potexvirus* particles become so closely appressed that they form wide bands. The bands surround a central region of cytoplasm which is vesiculated and contains ribosomes and scattered virus particles. *Potexvirus* banded-bodies may be dissociated by mechanical injury involved in producing epidermal strips, by alcohol, aqueous stains, water, or buffers. The inclusions are adequately preserved for light microscopy by fixation with glutaraldehyde. Banded-bodies are dissociated into fibrous masses in fixatives commonly employed for electron microscopy. The inclusions are well preserved in dilute osmium tetroxide.[20] Diagnostic banded-body cytoplasmic inclusions are also induced by carlaviruses and closteroviruses. Carlavirus banded-bodies are narrower and simpler in construction (Plate 1C) than those of the potexviruses or closteroviruses. *Closterovirus* banded-bodies occur predominantly in the phloem while those of potexviruses and carlaviruses do not. The banded-body inclusions reported in some infections by bean yellow mosaic and Agropyron mosaic potyviruses are small and apparently they do not surround central cores of cytoplasmic materials. They are not diagnostic inclusions for potyvirus infections.

Potato virus X (PVX) is the type member of the *Potexvirus* group. In addition to inducing cytoplasmic banded-body and fibrous incusions of aggregated particles, PVX induces cytoplasmic laminate inclusion components. These inclusions have been proposed as diagnostic for PVX infections and the possibility of using them to distinguish PVX strains has been suggested by Milne.[89] The laminate inclusion components (LIC) consist of smooth or beaded proteinaceous sheets arranged in bundles or scrolls. The sheets are about 3 to 4 nm thick and the beads on the sheets are ribosome-like bodies about 11 to 14 nm in diameter. The beaded sheets are destroyed by potassium permanganate and digested by subtilisin. Large numbers of virus particles aggregate between the sheets.[90,91] In the light microscope the LIC stain green in the Orange-Green combination and magenta in Azure A (Plate 1D). The proteinaceous sheets do not stain in Azure A but the virus particles associated with them do, and the ribosome-like beads may also stain in Azure A. LIC are not induced by other potexviruses or by viruses in other groups.[92] The only kinds of inclusions which might be confused with LIC are the potyvirus-induced cylindrical inclusions, but the proteinaceous sheets making up the central portions of cylindrical inclusions have different configurations (pinwheels) than those of the potexvirus LIC, and they do not contain ribosome-like bodies. Cylindrical inclusions do not stain in Azure A, whereas LIC stain magenta.

Papaya mosaic *Potexvirus* (PapMV) induces banded-body inclusions with the same staining capacities as those induced by other potexviruses. However, the shape of the PapMV banded-bodies distinguishes them from banded-body inclusions in other members of the *Potexvirus* group and from those in other groups (Plate 1B). PapMV banded-bodies have a wavy outline and in some orientations they appear to be conical.[20] PapMV also induces proteinaceous, fibrous inclusions closely associated with nucleoli. These inclusions stain green in the Orange-Green combination and are unstained in Azure A.[20]

C. COMOVIRUSES

Cytoplasmic vacuolate-vesiculate inclusions and virus particles in crystalline arrays were suggested as main characteristics of the *Comovirus* group.[92] The presence of these inclusions and the blockage of xylem elements by aggregates of *Comovirus* particles have been proposed as diagnostic for *Com-*

ovirus infections.[21] In the O-G combination the vacuolate-vesiculate inclusions stain green and in Azure A they stain red to magenta. The crystalline aggregates of virus particles stain green in O-G and magenta to violet in Azure A. Vacuolate-vesiculate cytoplasmic inclusions have been reported in cowpea mosaic virus (CPMV)-infected cells,[20] as have inclusions of aggregated virus,[42] of virus in crystalline arrays,[43] and of aggregated particles blocking xylem elements.[20,44] Elongated inclusions of aggregated proteinaceous fibrils have been reported in vacuoles.[44,45] Two types of nuclear inclusions have been found in CPMV-infected tissues, crystalline inclusions of virus[46] and aggregates of membranes.[47] The inclusions induced by CPMV in the cytoplasm and vacuoles are closely similar to those induced by other comoviruses. However, the nuclear inclusions distinguish CPMV from other comoviruses.

V. DIAGNOSIS OF FIELD INFECTIONS

There have been 395 plant viruses sufficiently characterized for taxonomists to place them as members of defined groups,[4] with 320 as probable or possible members. However, viruses with properties not fitting any existing group continue to be reported.[48,49] The diagnostician does not have to contemplate dealing with all these pathogens. His interest in distinguishing viruses is usually confined to those infecting one or, at most, a few crop species. Although some viruses infect a crop in the field, their effects may be of minor or no economic significance.

Natural infections of beans (*Phaseolus vulgaris*) have been reported for 44 viruses from 17 groups and for 3 ungrouped viruses (Table 3). Ten of these viruses have been described as economically important. Infections involving each of these ten viruses can be diagnosed cytologically at the group level and infections induced by four of them can be diagnosed cytologically at the species level. Bean common mosaic *Potyvirus* induces type 1 cylindrical inclusions and no nuclear inclusions. Bean yellow mosaic and clover yellow vein potyviruses induce type 2 cylindrical inclusions, but their nuclear inclusions differ from each other and from peanut mottle virus. Some strains of peanut mottle induce type 2, and others induce type 3 cylindrical inclusions. One isolate inducing type 2 cylindrical inclusions also induced thin crystalline nuclear inclusions.[115] Southern bean mosaic *Sobemovirus* induces inclusions of aggregated particles in the xylem which distinguish it from viruses in all other groups except the comoviruses. Nuclear inclusions of aggregated particles, sometimes in crystalline arrays, separate southern bean mosaic virus from comoviruses.

Peppers (*Capsicum annuum*) have been reported to be infected in the field with 37 viruses in 17 groups (Table 4). Ten of these viruses have been reported to induce economically significant reductions in yield and quality of peppers. Infections involving each of these ten viruses can be diagnosed cytologically at the group level and infections induced by two of them can

Table 3. Bean (*Phaseolus vulgaris*) Natural Infections

Virus	Diagnostic inclusions	
	Virus group	Specific virus
Abutilon mosaic *Geminivirus*[a]	+	−
Bean common mosaic *Potyvirus*[b,50]	+	+
Bean distortion dwarf *Geminivirus*	+	−
Bean golden mosaic *Geminivirus*[51]	+	−
Bean leaf roll *Luteovirus*	+	−
Bean line pattern mosaic (unclassified)	−	−
Bean mild mosaic *Carmovirus*	−	−
Bean pod mottle *Comovirus*	+	−
Bean rugose mosaic *Comovirus*	+	−
Bean yellow mosaic *Potyvirus*[97]	+	+
Bean yellow vein-banding (unclassified)	−	−
Beet curly top *Geminivirus*[98]	+	−
Beet western yellows *Luteovirus*	+	−
Broad bean true mosaic *Comovirus*	+	−
Broad bean wilt *Fabavirus*	−	−
Clover yellow mosaic *Potexvirus*	+	−
Clover yellow vein *Potyvirus*[99]	+	+
Cotton leaf crumple *Geminivirus*	+	−
Cowpea chlorotic mottle *Bromovirus*	−	−
Cowpea mild mottle *Carlavirus*	+	+
Cowpea mosaic *Comovirus*	+	+
Cowpea severe mosaic *Comovirus*	+	−
Cucumber mosaic *Cucumovirus*[100]	−	−
Euphorbia mosaic *Geminivirus*	+	−
Horsegram yellow mosaic *Geminivirus*	+	−
Milk vetch dwarf *Luteovirus*	+	−
Passionfruit woodiness *Potyvirus*	+	−
Pea early browning *Tobravirus*	−	−
Pea enation mosaic virus	+	+
Peanut mottle *Potyvirus*[101]	+	+
Peanut stunt *Cucumovirus*[102]	−	−
Quail pea mosaic *Comovirus*	+	−
Rhynchosia mosaic *Geminivirus*	+	−
Southern bean mosaic *Sobemovirus*[103]	+	+
Soybean dwarf *Luteovirus*	+	−
Soybean mosaic *Potyvirus*	+	−
Squash leaf curl *Geminivirus*	+	−
Subterranean clover stunt *Luteovirus*	+	−
Sunn hemp mosaic *Tobamovirus*	+	−
Tobacco necrosis *Necrovirus*	−	−
Tobacco ringspot *Nepovirus*	−	−
Tobacco streak *Ilarvirus*	−	−
Tobacco yellow dwarf *Geminivirus*[104]	+	−
Tomato aspermy *Cucumovirus*	−	−
Tomato black ring *Nepovirus*	−	−
Tomato ringspot *Nepovirus*	−	−
Tomato spotted wilt *Tospovirus*	+	+

[a] Information on all viruses without citation numbers is from Reference 21.
[b] Literature citations are for viruses inducing economically important effects on beans.

Table 4. Pepper (*Capsicum annuum*) Natural Infections

Virus	Diagnostic inclusions	
	Virus group	Specific virus
Alfalfa mosaic virus[a]	+	+
Beet curly top *Geminivirus*[b][111]	+	−
Beet western yellows *Luteovirus*	+	+
Belladonna mottle *Tymovirus*	+	+
Bell pepper mottle *Tobamovirus*	+	−
Broad bean wilt *Fabavirus*	−	−
Chino del tomate *Geminivirus*	+	−
Cucumber mosaic *Cucumovirus*[109]	−	−
Maladie du Anaheim *Rhabdovirus*	+	−
Moroccan pepper *Tombusvirus*	+	−
Pepper mild mosaic *Potyvirus*	+	−
Pepper mild mottle *Tobamovirus*[112]	+	−
Pepper mild tigre *Geminivirus*	+	−
Pepper mottle *Potyvirus*	+	+
Pepper ringspot *Tobravirus*	−	−
Pepper severe mosaic *Potyvirus*	+	−
Pepper veinal mottle *Potyvirus*[107]	+	−
Peru tomato virus *Potyvirus*[108]	+	−
Potato leaf roll *Luteovirus*	+	−
Potato virus M *Carlavirus*	+	−
Potato virus S *Carlavirus*	+	−
Potato virus X *Potexvirus*[105]	+	+
Potato virus Y *Potyvirus*[106]	+	−
Tobacco etch *Potyvirus*[106]	+	+
Tobacco leaf curl *Geminivirus*	+	−
Tobacco mosaic *Tobamovirus*	+	−
Tobamovirus-Ob *Tobamovirus*[110]	+	−
Tobacco necrosis *Necrovirus*	−	−
Tobacco rattle *Tobravirus*	−	−
Tobacco streak *Ilarvirus*	−	−
Tomato aspermy *Cucumovirus*	−	−
Tomato black ring *Nepovirus*	−	−
Tomato bushy stunt *Tombusvirus*	+	−
Tomato mosaic *Tobamovirus*[113]	+	−
Tomato ringspot *Nepovirus*	−	−
Tomato spotted wilt *Tospovirus*	+	+
U2-tobacco mosaic *Tobamovirus*	+	−

[a] Information on all viruses without citation numbers is from Reference 114.
[b] Citation numbers are for viruses inducing economically significant yield and quality reductions in peppers.

be diagnosed at the species level. Potato X *Potexvirus* is distinguished from the other viruses infecting peppers by its unique cytoplasmic inclusions. These are the large virus particle aggregate inclusions, some in the form of banded-bodies (group-specific inclusions), and the laminate inclusion components (virus-specific inclusions). Tobacco etch *Potyvirus* is distinguished from other viruses infecting peppers by its cytoplasmic type 2 cylindrical inclusions (group specific) and by its crystalline nuclear inclusions (virus specific).

Most of the viruses in Tables 3 and 4 do not have worldwide distribution. It seems unlikely that diagnosticians would need to consider every virus that has been found naturally infecting the crop(s) of his concern.

In field plantings, infections with more than one virus are often encountered. Techniques that are designed to identify viruses at the species level, such as ELISA, are usually not designed to detect mixed infections in the field. Cytological studies can detect mixed infections involving viruses in different groups and can often detect mixed infections of viruses in the same group.

VI. CONCLUDING REMARKS

Cytological studies of inclusions can provide answers to some of the problems diagnosticians face, quickly and relatively inexpensively. The light microscope stains discussed here provide good differentiation between normal cell constituents and virus-induced inclusions, bacteria, fungi, and mycoplasmas. Examination of tissues for inclusions can be used to indicate whether a virus is the causal agent of a disease; in most infections the group to which the virus belongs; and in some, even the particular virus or strain which is present. Mixed infections, often encountered in the field, are not readily detected with other techniques but usually they are readily detected with light microscopy.

Cytological information on inclusions is also used indirectly in diagnosis. No laboratory or culture collection center possesses antisera to even a majority of plant viruses. The ability to assign most viruses to groups via light microscopy provides the diagnostician-serologist information which assists his search for the appropriate antisera to test.

Cytological techniques are rather easily mastered. At least, that is our judgment as the result of several workshops which we have conducted on the subject. Mastery of the techniques must be followed by experience in finding, observing, and recording the properties of inclusions. That experience may have to be very broad for workers in disease diagnostic laboratories, or may be confined to viruses infecting a single species of interest to a plant breeder. Whatever the required breadth of experience in cytology of virus infections may be, its value will certainly be enhanced by updating information about inclusions with additions to microscope and projection slides, drawings, and micrographs of inclusions and the collection of reprints on inclusion cytology.

REFERENCES

1. **Matthews, R. E. F.,** Classification and nomenclature of viruses, *Interviology,* 17, 4, 1982.
2. **Brown, F.,** Minutes of the 17th Meeting of the Executive Committee of the International Committee on Taxonomy of Viruses, Edmonton, August 8–9, 1987, p. 1.
3. **Wildy, P.,** Classification and nomenclature of viruses, First Report of the International Committee on Nomenclature of Viruses, *Monogr. Virol.,* 5, 1, 1971.

4. **Francki, R. I. B., Fauquet, C., Knudson, D. L., and Brown, F.,** Classification and nomenclature of viruses. Fifth Report of the International Committee on Taxonomy of Viruses, *Arch. Virol.,* (Suppl. 2), 1991, 450.

5. **Malherbe, H. H. and Strickland-Cholmley, M.,** *Viral Cytopathology,* CRC Press, Boca Raton, FL, 1980, 78.

6. **McWhorter, F. P.,** Plant virus inclusions, *Annu. Rev. Phytopathol.,* 3, 287, 1965.

7. **Kassanis, B.,** Intranuclear inclusions in virus infected plants, *Ann. Appl. Biol.,* 26, 705, 1939.

8. **Milicic, D., Stefanac, Z., and Mamula, D.,** Intracellular changes induced by crucifer viruses, in *Plant Virology, Proc. 6th Conf. Czech. Plant Virol. Olomouc 1967,* Blatny, C., Ed., Academia, Prague, 1969, 54.

9. **Stefanac, Z. and Ljubesic, N.,** Inclusion bodies in cells infected with radish mosaic, *J. Gen. Virol.,* 13, 51, 1971.

10. **Kitajima, E. W. and Costa, A. S.,** Morphology and developmental stages of *Gomphrena* virus, *Virology,* 29, 523, 1966.

11. **Christie, R. G. and Edwardson, J. R.,** Light microscopic techniques for detection of plant virus inclusions, *Plant Dis.,* 70, 273, 1986.

12. **Brunt, A., Crabtree, K., and Gibbs, A.,** *Viruses of Tropical Plants,* C.A.B. International, Wallingford, 1990, 707.

13. **Martelli, G. P. and Russo, M.,** Plant virus inclusion bodies, *Adv. Virus Res.,* 21, 175, 1977.

14. **Martelli, G. P. and Russo, M.,** Use of thin sectioning for visualization and identification of plant viruses, *Methods Virol.,* 8, 143, 1984.

15. **Francki, R. I. B., Milne, R. G., and Hatta, T.,** *Atlas of Plant Viruses,* Vol. 1, CRC Press, Boca Raton, FL, 1985, 222.

16. **Francki, R. I. B., Milne, R. G., and Hatta, T.,** *Atlas of Plant Viruses,* Vol. 2, CRC Press, Boca Raton, FL, 1985, 284.

17. **Edwardson, J. R.,** Electron microscopy of cytoplasmic inclusions in cells infected with rod shaped viruses, *Am. J. Bot.,* 53, 359, 1966.

18. **Lesemann, D. E.,** Virus group-specific and virus-specific cytological alterations induced by members of the Tymovirus Group, *Phytopathol. Z.,* 90, 315, 1977.

19. **Russo, M., DiFranco, A., and Martelli, G. P.,** Cytopathology in the identification and classification of tombusviruses, *Intervirology,* 28, 134, 1987.

20. **Christie, R. G. and Edwardson, J. R.,** Light and electron microscopy of plant virus inclusions, *Fla. Agric. Exp. Stn. Monogr.,* 9, 150, 1977.

21. **Edwardson, J. R. and Christie, R. G.,** *Handbook of Viruses Infecting Legumes,* CRC Press, Boca Raton, FL, 1991, 504.

22. **Klinkowski, M.,** Catastrophic plant diseases, *Annu. Rev. Phytopathol.,* 8, 37, 1970.

23. **Uyemoto, J. K.,** Tobacco necrosis and satellite viruses, in *Handbook of Plant Virus Infections and Comparative Diagnosis,* Kurstak, E., Ed., Elsevier/North Holland, New York, 1981, 123.

24. **Boccardo, G., Milne, R. G., Luisoni, E., Lisa, V., and Accotto, G. P.,** Three seed-borne cryptic viruses containing double stranded RNA isolated from white clover, *Virology,* 147, 29, 1985.

25. **Murant, A. F.,** Nepoviruses, in *Handbook of Plant Virus Infections and Comparative Diagnosis,* Kurstak, E., Ed., Elsevier/North Holland, New York, 1981, 197.

26. **Lane, L. C.,** Bromoviruses, in *Handbook of Plant Virus Infections and Comparative Diagnosis,* Kurstak, E., Ed., Elsevier/North Holland, New York, 1981, 333.

27. **Francki, R. I. B., Kitajima, E. W., and Peters, D.,** Rhabdoviruses, in *Handbook of Plant Virus Infections and Comparative Diagnosis,* Kurstak, E., Ed., Elsevier/North Holland, New York, 1981, 455.

28. **Wetter, C. and Milne, R. G.,** Carlaviruses, in *Handbook of Plant Virus Infections and Comparative Diagnosis,* Kurstak, E., Ed., Elsevier/North Holland, New York, 1981, 695.

29. **Morris, T. J. and Carrington, J. C.,** Carnation mottle virus and viruses with similar properties, in *The Plant Viruses,* Vol. 3, Koenig, R., Ed., Plenum Press, New York, 1988, 73.
30. **Russo, M., Castellano, M. A., and Martelli, G. P.,** The ultrastructure of broad bean stain and broad bean true mosaic virus infections, *J. Submicrosc. Cytol.,* 14, 149, 1982.
31. **Miller, T. J. and Stone, H. D .,** The rapid isolation of ribonuclease-free immunoglobulin G by protein A-sepharose affinity chromatography, *J. Immunol. Methods,* 24, 111, 1978.
32. **Hiebert, E., Purcifull, D. E., and Christie, R. G.,** Purification and immunological analyses of plant viral inclusion bodies, *Methods Virol.,* 8, 225, 1984.
33. **Hebert, D. C., Weaker, F. J., and Sheridan, P. J.,** Evaluation of the ability of Tween 20, dimethylsulphoxide and Triton-X 100 to enhance immunochemical staining of autoradiograms, *Histochem. J.,* 14, 161, 1982.
34. **Nishiguchi, M., Motogoshi, F., and Oshima, N.,** Further investigation of a temperature-sensitive strain of tobacco mosaic virus: its behavior in tomato leaf epidermis, *J. Gen. Virol.,* 46, 497, 1980.
35. **Horisberger, M. and Rosset, J.,** Colloidal gold, a useful marker for transmission and scanning electron microscopy, *J. Histochem. Cytochem.,* 25, 295, 1977.
36. **Ko, N. J.,** A new approach to plant virus identification by light microscopy, *Plant Prot. Bull.,* 30, 1, 1988.
37. **Lesemann, D. E.,** Cytopathology, in *The Plant Viruses,* Vol. 4, Milne, R. G., Ed., Plenum Press, New York, 1988, 179.
38. **Brandes, J. and Wetter, C.,** Classification of elongated plant viruses on the basis of particle morphology, *Virology,* 8, 99, 1959.
39. **Brandes, J.,** Identifizierung von gestreckten pflanzenpathogenen Viren auf morphologischer Grundlage, *Mitt. Biol. Bundesanst. Land Forstwirts. (Berlin-Dahlem),* 110, 1, 1964.
40. **Christie, S. R.,** unpublished.
41. **Pinner, M. S., Plaskitt, D. A., Medina, V., and Markham, P. G.,** Morphology of viral inclusions in monocotyledons infected with geminiviruses, *VIIIth Int. Congr. Virol.* Aug. 26–31, Berlin, Abstr. P79-029, 1990.
42. **Van der Scheer, C. and Groenewegen, J.,** Structure in cells of *Vigna unguiculata* infected with cowpea mosaic virus, *Virology,* 46, 493, 1971.
43. **Edwardson, J. R. and Christie, R. G.,** Comoviruses, in *Viruses Infecting Forage Legumes,* I, Florida Agricultural Experiment Station Monogr. 14, 1986, 109.
44. **Kitajima, E. W.,** Citopatologia e localizacao de virus de milho e de leguminosas alimenticias nas plantas infetadas e nos vectores, *Fitopatol. Brasileira,* 4, 24, 1979.
45. **Lin, N. S.,** Gold-IgG complexes improve the detection and identification of viruses in leaf-dip preparations, *J. Virol. Methods,* 8, 181, 1984.
46. **Pares, R. D. and Whitecross, M. I.,** Gold-labelled antibody decoration (GLAD) in the diagnosis of plant viruses by immuno-electron microscopy, *J. Immunol. Methods,* 9, 107, 1982.
47. **Patterson, S. and Verduin, B. J. M.,** Applications of immunogold labelling in animal and plant virology, *Arch. Virol.,* 97, 1, 1987.
48. **Milne, R. G.,** Quantitative use of the electron microscope decoration technique for plant virus diagnosis, *Acta Hortic.,* 234, 321, 1988.
49. **Bock, K. R., Guthrie, E. J., and Pearson, M. N.,** Notes on East African plant virus diseases. IX. Cucumber mosaic virus, *E. Afr. Agric. Forest. J.,* 41, 81, 1975.
50. **Francki, R. I. B., Randles, J. W., Chambers, T. C., and Wilson, S. B.,** Some properties of purified cucumber mosaic virus (Q strain), *Virology,* 28, 729, 1966.
51. **Francki, R. I. B. and Habili, N.,** Stabilization of capsid structure and enhancement of immunogenicity of cucumber mosaic virus (Q strain) by formaldehyde, *Virology,* 48, 309, 1972.

52. **Christie, R. G., Ko, N. J., Falk, B. W., Hiebert, E., Lastra, R., Bird, J., and Kim, K. S.,** Light microscopy of geminivirus-induced nuclear inclusion bodies, *Phytopathology,* 76, 124, 1986.

53. **Kim, K. S. and Carr, R.,** Characteristic ultrastructure and cytochemistry of plant cells infected with whitefly-transmitted geminiviruses, in *Workshop Plant Virus Detection, Agric. Exp. Stn. Univ. Puerto Rico, Rio Piedras,* 1982, 25, 1982.

54. **Yamashita, S., Doi, Y., and Yora, K.,** Intracellular appearance of rice stripe virus, *Ann. Phytopathol. Sco. Jpn.,* 51, 637, 1985.

55. **Martelli, G. P., Russo, M., and Vovlas, C.,** Ultrastructure of zucchini yellow fleck virus infections, *Phytopathol. Mediterr.,* 20, 193, 1981.

56. **Edwardson, J. R.,** Some Properties of the Potato Virus Y-Group, Florida Agricultural Experiment Station Monogr. No. 4, 1974, 398.

57. **Edwardson, J. R. and Christie, R. G.,** The Potyviruses, Florida Agricultural Experiment Station Monogr. No. 16, 1991, 1244.

58. **Fenner, F. F.,** Classification and nomenclature of viruses, Second Report of the International Committee on Taxonomy of Viruses, *Intervirology,* 7, 4, 1976.

59. **Edwardson, J. R., Christie, R. G., and Ko, N. J.,** Potyvirus cylindrical inclusions — subdivision IV, *Phytopathology,* 74, 1111, 1984.

60. **Shukla, D. D., Ford, R. E., Tosic, M., Jilka, F., and Ward, C. W.,** Possible members of the potyvirus group transmitted by mites or whiteflies share epitopes with aphid-transmitted definitive members of the group, *Arch. Virol.,* 105, 143, 1989.

61. **Lesemann, D. E. and Vetten, H. J.,** The occurrence of tobacco rattle and turnip mosaic viruses in *Orchis* ssp., and of an unidentified potyvirus in *Cypripedium calceolus, Acta Hortic.,* 164, 45, 1985.

62. **Stanarius, A., Proeseler, G., and Richter, J.,** Immunelektronenmikroskopische Untersuchungen zur serologischen Verwandtschaft des Gerstenegelbmosaik-Virus (barley yellow mosaic) und des Milden Gerstenmosaik-Virus (barley mild mosaic virus) mit anderen gestreckten Viren, *Arch. Phytopathol. Pflanzenschutz,* 25, 303, 1989.

63. **Lain, S., Riechmann, J. L., Martin, M. T., and Garcia, J. A.,** Homologous potyvirus and flavivirus proteins belonging to a superfamily of helicase-like proteins, *Gene,* 82, 357, 1989.

64. **Murphy, J. F., Jarlfors, U., and Shaw, J. G.,** Development of cylindrical inclusions in potyvirus-infected protoplasts, *Phytopathology,* 81, 371, 1991.

65. **Inouye, N. and Mitsuhata, K.,** Turnip mosaic virus isolated from Iris, *Nogaku Kenkyu,* 57, 1, 1978.

66. **Edwardson, J. R. and Purcifull, D. E.,** Turnip mosaic virus induced inclusions, *Phytopathology,* 60, 85, 1970.

67. **Herold, F. and Munz, K.,** Peanut mottle virus, *Phytopathology,* 59, 663, 1969.

68. **Milne, R. G. and Francki, R. I. B.,** Should tomato spotted wilt virus be considered as a possible member of the family Bunyaviridae?, *Intervirology,* 22,72, 1984.

69. **Edwardson, J. R. and Christie, R. G.,** Tomato spotted wilt virus group, in Viruses Infecting Forage Legumes III, Florida Agricultural Experiment Station Monogr. No. 14, 1986, 563.

70. **Crowley, N. C.,** The effect of developing embryos on plant viruses, *Aust. J. Biol. Sci.,* 10, 443, 1957.

71. **Tsakiridis, J. P. and Gooding, G. V.,** Tomato spotted wilt virus in Greece, *Phytopathol. Mediterr.,* 11, 42, 1972.

72. **Francki, R. I. B. and Hatta, T.,** Tomato spotted wilt virus, in *Handbook of Plant Virus Infections and Comparative Diagnosis,* Kurstak, E., Ed., Elsevier/North Holland, New York, 1981, 492.

73. **Cho, J. J., Mau, R. F. L., Gonsalves, D., and Mitchell, W. C.,** Reservoir weed hosts of tomato spotted wilt virus, *Plant Dis.,* 70, 1014, 1986.

74. **Hsu, H. T. and Lawson, R. H.**, Direct blotting for detection of tomato spotted wilt virus in *Impatiens, Plant Dis.*, 75, 292, 1991.
75. **Ie, T. S.**, An electronic microscope study of tomato spotted wilt virus in the plant cell, *Neth. J. Plant Plathol.*, 72, 114, 1964.
76. **Bertaccini, A. and Bellardi, M. G.**, Tomato spotted wilt virus infecting *Gloxinia* in Italy, *Phytopathol. Mediterr.*, 29, 205, 1990.
77. **Khurana, S. M. P., Garg, I. D., Behl, M. K., and Singh, M. N.**, Potential reservoirs of tomato spotted wilt virus in the North West Himalayas, *Natl. Acad. Sci. Lett.*, 13, 297, 1990.
78. **Kim, K. S. and Fulton, J. P.**, Electron microscopy of pokeweed mosaic virus, *Virology*, 37, 297, 1969.
79. **Russo, M. and Martelli, G. P.**, Cytology of *Gomphrena globosa* L. plants infected by beet mosaic virus (BMV), *Phytopathol. Mediterr.*, 8, 65, 1969.
80. **Van Beek, N. A. M., Lohuis, D., Dijkstra, J., and Peters, D.**, Morphogenesis of Sonchus yellow net virus in cowpea protoplasts, *J. Ultrastruct. Res.*, 90, 294, 1985.
81. **Toriyama, S.**, Electron microscopy of developmental stages of northern cereal mosaic virus in wheat plant cells, *Ann. Phytopathol. Soc. Jpn.*, 42, 563, 1976.
82. **Kitajima, E. S.**, Electron microscopy of Vira-Cabeca virus (Brazilian tomato spotted wilt virus) within the host cell, *Virology*, 26, 89, 1965.
83. **Bald, J. G.**, A strain-host combination for studying tomato spotted wilt virus inclusions, *Phytopathology*, 52, 723, 1962.
84. **Kobatake, H., Osaki, T., Yoshioka, A., and Inouye, T.**, Tomato spotted wilt virus, *Ann. Phytopathol. Soc. Jpn.*, 42, 287, 1976.
85. **Christie, R. G.**, Rapid staining procedures for differentiating plant virus inclusions in epidermal strips, *Virology*, 31, 268, 1967.
86. **Milne, R. G.**, An electron microscope study of tomato spotted wilt virus in sections of infected cells and in negative stain preparations, *J. Gen. Virol.*, 6, 267, 1970.
87. **Ie, T. S.**, Electron microscopy of developmental stages of tomato spotted wilt virus in plant cells, *Virology*, 43, 468, 1971.
88. **Ie, T. S.**, A sap-transmissible defective form of tomato spotted wilt virus, *J. Gen. Virol.*, 59, 387, 1982.
89. **Milne, R. G.**, Potato virus X, *Rothamsted Exp. Stn. Rep.*, 1968, 126, 1969.
90. **Shalla, T. A. and Shepard, J. F.**, The form and composition of laminate inclusion components in potato virus X-infected cells, *Phytopathology*, 62, 788, 1972.
91. **Shalla, T. A. and Shepard, J. F.**, The structure and antigenic analysis of amorphous inclusion bodies induced by potato virus X, *Virology*, 49, 654, 1972.
92. **Edwardson, J. R. and Christie, R. G.**, Use of virus-induced inclusions in classification and diagnosis, *Annu. Rev. Phytopathol.*, 16, 31, 1978.
93. **Sheffield, F. M. L.**, The cytoplasmic and nuclear inclusions associated with severe etch virus, *J. R. Microscop. Soc.*, 61, 30, 1941.
94. **Christie, R. G.**, unpublished.
95. **Chang, C. A., Purcifull, D. E., Hiebert, E., and Edwardson, J. R.**, Immunofluorescence evidence for the origin of nuclear inclusions and cytoplasmic crystals induced by bean yellow mosaic virus, *Phytopathology*, 76, 1061, 1986.
96. **Randles, J. W., Davies, C., Gibbs, A. J., and Hatta, T.**, Amino acid composition of capsid protein as a taxonomic criterion for classifying the atypical S strain of bean yellow mosaic virus, *Aust. J. Biol. Sci.*, 33, 245, 1980.
97. **Tapio, E.**, Virus diseases of legumes in Finland and in the Scandinavian Countries, *Ann. Agric. Fenniae*, 9, 1, 1970.
98. **Bennett, C. W.**, The curly top disease of sugarbeet and other plants, *Am. Phytopathol. Soc. Monogr.*, 7, 20, 1971.
99. **Provvidenti, R.**, Inheritance of resistance to clover yellow vein virus in *Pisum sativum*, *J. Hered.*, 78, 126, 1987.

100. **Horvath, J.,** The role of some plants in the ecology of cucumber mosaic virus with special regard to bean, *Acta Phytopathol. Acad. Sci. Hung.,* 18, 217, 1983.
101. **Provvidenti, R. and Chirko, E. M.,** Inheritance of resistance to peanut mottle virus in *Phaseolus vulgaris, J. Hered.,* 78, 402, 1987.
102. **Echandi, E. and Hebert, T. T.,** Stunt of beans incited by peanut stunt virus, *Phytopathology,* 61, 328, 1971.
103. **Tremaine, J. H. and Hamilton, R. I.,** Southern bean mosaic virus, *CMI/AAB Descriptions of Plant Viruses,* No. 274, Association of Applied Biologists, Wellesbourne, 1983.
104. **Thomas, J. E. and Bowyer, J. W.,** Tobacco yellow dwarf virus, *CMI/AAB Descriptions of Plant Viruses,* No. 278, Association of Applied Biologists, Wellesbourne, 1984.
105. **Rao, K. N., Appa Rao, A., and Reddy, D. V. R.,** A ringspot strain of potato virus X on chilli *(Capsicum annuum), Ind. Phytopathol.,* 23, 69, 1970.
106. **Zitter, T. A.,** Naturally occurring pepper virus strains in south Florida, *Plant Dis. Rep.,* 56, 586, 1972.
107. **Lana, A. F. and Adegbola, M. O. K.,** Important virus diseases in West African crops, *Rev. Plant Pathol.,* 56, 849, 1977.
108. **Fribourg, C. E. and Fernandez-Northcote, E. N.,** Peru tomato virus, *CMI/AAB Descriptions of Plant Viruses,* No. 255, Association of Applied Biologists, Wellesbourne, 1982.
109. **Sharma, O. P. and Singh, J.,** Reaction of different genotypes of pepper to cucumber mosaic and tobacco mosaic viruses, *Capsicum Newsl.,* 4, 47, 1985.
110. **Csillery, G. and Rusko, J.,** The control of a new tobamo virus strain by a resistance linked to anthocyanin deficiency in pepper, Capsicum annuum, Eucarpia Work. Group 4th Meet., 1980, p. 40.
111. **Thomas, P. E. and Mink, G. I.,** Beet curly top virus, *CMI/AAB Descriptions of Plant Viruses,* No. 210, Association of Applied Biologists, Wellesbourne, 1979.
112. **Pares, R. D.,** A tobamovirus infecting Capsicum in Australia, *Ann. Appl. Biol.,* 106, 469, 1985.
113. **Pares, R. D.,** Serological comparison of an Australian isolate of Capsicum mosaic virus with Capsicum tobamovirus isolates from Europe and America, *Ann. Appl. Biol.,* 112, 609, 1988.
114. **Edwardson, J. R.,** unpublished.
115. **Xiong, Z.,** Purification and Partial Characterization of Peanut Mottle Virus and Detection of Peanut Stripe Virus in Peanut Seeds, M.Sc. Thesis, University of Florida, Gainesville, 1984.
116. **Ie, T. S.,** A sap-transmissible, defective form of tomato spotted wilt virus, *J. Gen. Virol.,* 59, 387, 1982.
117. **Francki, R. I. B., Milne, R. G., and Hatta, T.,** Tomato spotted wilt virus group, in *Atlas of Plant Viruses,* Vol. 1, CRC Press, Boca Raton, FL, 1985, 101.
118. **Edwardson, J. R. and Christie, R. G.,** Light microscopy of inclusions induced by viruses infecting peppers *(Capsicum* spp.), *Fitopatol. Brasileira,* 4, 341, 1979.
119. **Wang, M. and Gonsalves, D.,** ELISA detection of various tomato spotted wilt virus isolates using specific antisera to structural proteins of the virus, *Plant Dis.,* 74, 154, 1990.
120. **DeAvila, A. C., Huguenot, C., Resende, R. O., Kitajima, E. W., Goldbach, R. W., and Peters, D.,** Serological differentiation of 20 isolates of tomato spotted wilt virus, *J. Gen. Virol.,* 71, 2801, 1990.

APPENDIX I

I. Preparation of the orange-green (O-G) protein stain. Stain powders should be prepared separately as follows:
    A. Add 1 g of Calcomine Orange RS[a] to 100 ml of 2-methoxyethanol, stir thoroughly and filter.
    B. Add 1 g of Luxol Brilliant Green[a] BL to 100 ml of 2-methoxyethanol, stir thoroughly and filter. The stains should be stored in brown bottles and will keep indefinitely if tightly capped. Prepare the final staining solution by mixing one part distilled water, one part of the orange dye, and eight parts of the green dye. This solution is stable and can be used as needed.

II. Preparation of the Azure A nucleoprotein stain:
    A. Azure A powder[a] should be stirred into 100 ml of 2-methoxyethanol to achieve a 0.1% dye content (Azure A powders vary in dye content). This stain will keep indefinitely if capped and stored in a brown bottle.
    B. Prepare a 0.2 M solution of dibasic sodium phosphate. It is important to use a hydrated (not anhydrous) form. Prepare the final staining solution by adding one part of the phosphate solution to nine parts of Azure A. This solution must be prepared fresh with each staining sequence. Do *not* reuse it.

III. Staining procedure for both the O-G stain and the Azure A stain:
    A. Place epidermal peels, free-hand sections, or abraded pieces into one of the stains.
    B. Stain for 10 min at room temperature or for 10 s at full power in a microwave oven.
    C. Destain for 30 s to 1 min in 95% ethanol (ETOH). *Caution,* prolonged periods in the ethanol will remove the stains. If 2-methoxyethyl acetate (2-MEA) is available, a mixture of 70 parts 2-MEA to 30 parts 95% ETOH is preferred because this mixture more effectively controls destaining.
    D. Place in 2-MEA for 1 to 2 min. This step stops the destaining reaction and samples can remain in it for an extended period of time. This step can be omitted if 2-MEA is not available.
    E. Mount the O-G stained material in Euparal Vert[b] (green color). Mount the Azure A stained material in standard Euparal[b] (straw color).

IV. Triton X-100 treatment: When the O-G combination is used, stained plastids often obscure small inclusions. The plastids can be dissolved by treating tissue pieces with a 2% solution of Triton X-100 (Rohm and Haas Co., Philadelphia, PA 19105) for 5 min *before* staining. This treatment is especially useful for detecting the cylindrical inclusions of potyviruses, particularly during their early stages of development when these small inclusions are located at the cell periphery. Both Triton-treated and untreated controls should be included, since Triton dissociates certain inclusions.

[a] We have obtained good results with the dyes obtained from Aldrich Chemical Company, P.O. Box 355, Milwaukee, WI 53201.
[b] Euparal may be obtained from Carolina Biological Supply, Burlington, NC 27215.

Chapter 6

VIRUS PURIFICATION IN RELATION TO DIAGNOSIS

Richard Stace-Smith and Robert R. Martin

TABLE OF CONTENTS

0-8493-4284-8/93/$0.00 + $.50

I. INTRODUCTION

A century has passed since the publication of Iwanowski's[1] classic paper reporting that the causal agent of tobacco mosaic disease could pass through a filter impenetrable to bacteria. In the intervening years hundreds of virus or virus-like diseases of plants have been described. Francki et al.[2] list 394 viruses that have been assigned to 35 Families or Groups. In addition they list 320 viruses as probable or possible members of these groups. However, there are many other virus-like diseases that have not been characterized in sufficient detail to be able to allocate them even tentatively to one of the established virus Families or Groups. Many of these diseases have been assigned a name but a name is of little value unless there is a firm basis for detection and identification, so that meaningful comparisons can be made between a virus occurring in a particular crop in one part of the world and a similar virus occurring in the same or a different crop in another part of the world.

Accurate diagnosis often depends on a comparison of many distinct properties of the agent. Some of these properties can be determined using crude sap, but in general such properties are capable of providing only a tentative diagnosis, or of assigning the causal virus to a virus Family or Group. In many circumstances attempts must be made to study as many properties as possible in order to achieve accurate diagnosis. For this reason, considerable attention is usually directed towards virus purification in the course of identifying a virus. However, even with the application of the best tools available in a well-equipped virology laboratory, many of the plant viruses have defied attempts at purification. This is particularly true for those viruses that occur in woody hosts and for which no herbaceous host is known.

Much information that is useful in identifying a virus, or deciding that it may possibly be a new virus can be obtained by using crude virus preparations or by studying the virus in situ. Most studies involving determination of host range, symptom expression and mode of transmission are carried out using crdue virus obtained from an infected plant. Cytopathology using either light or electron microscopy is studied with fixed tissue. Perhaps the most significant technical achievement in diagnostic plant virology during the 1960s was the development of quick methods for the preparation of plant sap extracts for examination in the electron microscope. Application of the negative stain techniques to determine particle morphology can be made adequately with crude sap, and indeed it is preferable to use crude sap rather than purified preparations when determining the modal length of virus particles. By making full use of the above tests, one may either make a positive identification of an unknown virus or obtain sufficient information to make a tentative diagnosis. This tentative diagnosis may be confirmed using serological tests, providing of course that the required antiserum is available. However, if the properties determined on crude preparations are not sufficiently diagnostic, an investigator has little choice but to proceed with attempts at vrius purification.

Virus purification is undertaken with the objectives of obtaining virus preparations of both high quality and good yield. It is not always possible to achieve both of these objectives with an unknown virus but, if only one of the two objectives is achieved, much can be done with partially purified virus or with a relatively small quantity of highly purified virus.

II. THE TAXONOMIC APPROACH

A. STEPWISE DETERMINATION

In identifying an unknown virus, the usual process involves the stepwise determination of as many characters as may be required to match the information about an unknown virus with that of a known, previously-described virus. After experience with a particular crop in a particular location, identification may be no more complicated than field observations of the symptoms induced, or field observations followed by sap transmissions to a few diagnostic host plants. This represents the simplest scenario that is commonly followed in many diagnostic laboratories. However, when the diagnosis involves an unfamiliar crop, or a familiar crop being grown in a geographical region that has not been previously sampled, the identification process may be more complicated and an increasing number of character determinations may be necessary in order to achieve the desired match. In such situations, a stepwise process aimed at providing information on as many characters as required is undertaken. Brunt et al.[3] applied the terms "pragmatic" and "logical" to the two phases of identification, recognizing that, in practice, a mixture of the two is commonly used.

The "pragmatic" approach to identification is based on the fact that each virus has a definitive host range that may be confined to one or a few plant families and that plants are usually easier to identify than viruses. When the natural host or hosts of the virus are determined, a check of records will establish which viruses have previously been isolated from that species. Suspects can then be checked by serological or nucleic acid hybridization tests. If this pragmatic process yields negative or inconclusive results, the identification strategy next moves to the "logical" phase.

The logical, or taxonomic approach to identification is aimed at establishing the taxonomic group to which the virus belongs by determining some of its "group-specific" characters, such as the shape and size of the particles. These characters may place the unknown virus in a taxonomic group and, supplemented with host range information, they may reveal whether the unknown virus is a described member of the likely group.

B. PLANT VIRUS SYSTEMATICS

Plant virologists recognize that there is no universally applicable set of characters for defining virus species. In practice, a virus species is a collection of virus isolates whose known properties are so similar that there is no reason to distinguish between them and give them separate names.[4] Many of the viruses that have been given separate names share a set of characters and may be considered to belong to the same virus group. Relatively few characters are actually needed to identify viruses as members of a particular group, but it is not common for a single characteristic to uniquely define a group. This principle was used to establish 16 virus groups, some monotypic, in 1971.[5] Additional groups have been added as the International Committee on Taxonomy of Viruses approved proposed new groups. There are now 35 well defined groups[2] and some 10 to 15 others are in the process of being defined.[3]

The "logical" or "taxonomic" way to identify an unknown virus is to determine some of the properties that are useful in making an assignment to a virus group. The key characteristics that may indicate to which group an unknown virus belongs are: (a) the appearance and dimensions of the virus particle as determined by examining negatively stained sap from an infected plant in an electron microscope; (b) an estimate of the number of sedimenting components into which the particles separate, and the sedimentation coefficient of those components, by centrifuging purified preparations; and (c) determining the number and size(s) of nucleic acid and protein species in its particles. These would then be followed by determination of "species-specific" characters such as natural and experimental host range, symptoms in particular host species and vector species. Researchers without access to an electron microscope and an ultracentrifuge will of necessity have to rely on a greater number of less discriminating tests that can be carried out on non-purified preparations. In practice, several tests will be carried out concurrently and, when the possible identity is determined, confirmatory serological or nucleic acid hybridizations tests will be performed.

When the unknown virus does not match with any known virus, it becomes important to collect and store information on as many characters as possible to aid in future comparative studies.

C. THE TAXONOMIC APPROACH TO PURIFICATION

The purification of plant viruses is more of an art than a science. this is because there is no predetermined procedure that an investigator can follow when undertaking the purification of an unknown virus. The usual process in deciding on a protocol is to use crude preparations to determine some of the characters of the unknown and, with this information, attempt to assign the unknown virus as a possible member of one of the known virus groups. If this step is successful, one can then make the assumption that the tentative member of the group shares many of the properties of the known members of the group, including stability when subjected to various purification protocols. This may be termed the taxonomic approach to virus purification. It is by no means infallible because members of a group frequently react in different ways to a particular purification procedure. Even strains of the same virus may require different protocols for effective purification. However, the approach at least provides a starting point upon which improvements can be made.

In this chapter, we plan to couple the taxonomic approach to plant virus identification with the taxonomic approach to plant virus purification. We will do this by summarizing the current literature on purification of viruses in the recognized virus groups. The groups are listed in alphabetical order, using virus names, group names and acronyms included in Appendix 2 of Matthews.[6]

1. Alfalfa Mosaic Virus Group

Alfalfa mosaic virus (AMV) has bacilliform shaped particles of varying lengths. The virus is sap-transmitted, seed-transmitted in some hosts and aphid-transmitted in a nonpersistent manner. This virus which has a very broad host range[7] and has many properties in common with the ilarviruses. Both have tripartite genomes with four species of RNA encapsidated in the virus particles. The three larger RNAs of both groups are infectious with either the addition of RNA 4 or coat protein. Also, the coat proteins of some of the ilarviruses can be used to activate the three large RNAs of AMV and vice versa.[8]

The virus is rather easy to purify. Virus can be extracted in phosphate buffer from 0.05 to 0.5 M at a pH near neutrality. Addition of reducing reagents ascorbic acid or mercaptoethanol are commonly used.[7] EDTA added to the buffers during purification helps stabilize the virus. After clarification the virus can be precipitated with polyethylene glycol (PEG) and NaCl. This is often followed by differential centrifugation and sucrose density centrifugation. Some strains of this virus reach very high titers yielding greater than 1 mg/kg of tissue.[9]

2. Bromovirus Group

The bromoviruses are particularly suited to studies of virus structure and plant virus replication because they are stable, easily isolated in gram quantities and can be reassembled from protein and RNA.

The bromoviruses are best purified from young plants that have been infected for 1 to 2 weeks. Plants that have been infected for a longer time still contain ample virus but there is a loss of specific infectivity with increasing age of the infection. Potential complications arise during purification because of the limited pH range of stability of the virus particles in the absence of magnesium ions (pH 3 to pH 6) and the propensity of the particles to precipitate with polyions at low ionic strength.[10] The bromoviruses can be readily purified with or without an ultracentrifuge. The purification procedure that is used for the type member, brome mosaic virus, works equally well with broad bean mottle and cowpea chlorotic mottle, two other members of the group. Infected barley tissue is homogenized with an equal weight of 0.5 M sodium acetate buffer, pH 4.5, and then emulsified with a small amount of chloroform. After low-speed centrifugation, the virus is precipitated by adding PEG to 6% (w/v) and stirring for 15 min. Following low-speed centrifugation, the pellet is suspended in distilled water to about 1/10 the original volume and emulsified with a few ml of chloroform. The virus is reprecipitated with PEG. The virus is relatively pure at this point, but further purification can be achieved by differential centrifugation.[11]

3. Capillovirus Group

This newly designated group of plant viruses[12] somewhat resembles the closteroviruses. The two members, apple stem grooving virus (ASGV) and potato virus T (PVT) are closely related serologically. The particles have a flexuous, open structure superficially like that of a closterovirus, but are shorter, thicker and a little less flexuous, and the helix has a somewhat shorter pitch.[13]

Bentonite treatment is effective for the clarification of both ASGV[14] and PVT.[15] Systemically infected Chenopodium quinoa leaves are triturated in 0.06 M phosphate buffer, pH 7.0, filtered, and the filtrate shaken with 3 to 4% bentonite suspension. The mixture is clarified by low-speed centrifugation, and the treatment repeated with small additions of bentonite until the extract is straw yellow in color. This extract is then precipitated with 5 to 6% PEG and the resuspended pellets purified further by differential centrifugation and sucrose gradient centrifugation. Yield is 3 to 10 mg virus per kg leaf tissue.

4. Carlavirus Group

Some carlaviruses occur in relatively high concentrations in their natural or experimental host and have been readily purified. Others, particularly those that infect woody host plants and for which no herbaceous host is known, are more difficult to purify. When purification is done directly from a woody host, the inclusion of substances such as the alkaloid nicotine and polyvinylpyrrolidone (PVP) in the extraction buffer is helpful.[16] There is a tendency for carlavirus particles to aggregate during purification. This aggregation is minimized by extracting the virus using a buffer at a high pH (pH 8.0 to 8.5) and at a moderately high molarity (e.g., 0.1 to 0.2 M).

Ellis et al.[17] obtained high yields of purified elderberry carlavirus by homogenizing infected elderberry leaves in a 0.1 M sodium borate, 0.1 M EDTA buffer containing 2% PVP, 0,5% nicotine alkaloid and 0.02 M mercaptoethanol, adjusted to pH 8.2. After homogenizing, 2% (v/v) Triton X-100 was added and the extract stirred overnight at room temperature. The extract was then squeezed through nylon tricot and the virus precipitated by the addition of 1% sodium chloride and 8% PEG. Following differential centrifugation, the virus was further purified by equilibrium gradient centrifugation in 40% (w/v) cesium chloride.

5. Carmovirus Group

The viruses in this group are relatively stable and occur in high concentrations in infected tissue of suitable propagative hosts. Purification has been achieved by using a variety of protocols but the one that is preferred by most workers involves homogenizing infected tissue in sodium acetate buffer, pH 5.0, as a first step.[18,19] The advantage of this extraction procedure is that it not only releases the virus but also clarification is simultaneously achieved. The acidification step should be used with some caution because some of the viruses in this groups are isolectric in acidic conditions and may precipitate during clarification.[20] After expressing through cheesecloth and a low speed centrifugation, the virus can be further purified by precipitation with PEG followed by differential centrifugation, or by differential centrifugation alone. Further purification is accomplished by centrifugation on sucrose density gradients.

6. Caulimovirus Group

Caulimoviruses were the first plant viruses found to have a double-stranded DNA genome. They are transmitted by aphids in nature and each virus has a restricted host range. They are assembled in the cytoplasm of their host cells, where the virus particles accumulate in inclusion bodies. The group has eleven established members and six possible.[2]

Most members of this group infect herbaceous hosts and virus can be purified from one or more herbaceous hosts. One member, strawberry vein banding virus, has a narrow host range in the genus Fragaria.[21] Attempts to purify virus from *Fragaria* spp. has been largely unsuccessful due to difficulties in eliminating dense, viscous host contaminants. Partial success was achieved by Morris et al.[22] by centrifuging partially purified virus on a step gradient consisting of a layer of 1.5 g/cm³ CsCl in 30% sucrose (3 ml) and a layer of 30% sucrose (7 ml). The strawberry extract was layered above the sucrose and centrifuged in a swinging bucket rotor.

The type member of the group, cauliflower mosaic virus, has been purified by several methods, three of which are outlined in detail by Shepherd.[23] A recent purification protocol developed by Gong et al.[24] yielded highly purified preparations that could be crystallized. They propagated the virus in turnips (Brassica rapa, cv Just Right), harvested infected leaves 2 to 4 weeks later and used them immediately for purification. Leaves were homogenized in a Tris-HCl, 0.1 M, pH 7.5 buffer containing 40 mM EDTA and 1.5 M urea, 1 g leaf tissue/3 ml buffer. The extract was then mixed with Triton X-100 and mercaptoethanol and the solution stirred at least 30 min at 4°C. After low-speed centrifugation, the supernatant was filtered through miracloth. The centrifugation and filtration was repeated twice and the clarified solution was ultracentrifuged. The pellets were suspended in a buffer (pH 7.5, 0.1 M Tris-HCl containing 10 mM EDTA and 1.25 M urea) and stirred overnight at 4°C. After low-speed centrifugation, the virus was centrifuged through a 10 to 40% (w/v) sucrose density gradient. The virus bands were recovered, diluted at least 5-fold in double distilled water, and concentrated by ultracentrifugation. The virus was then centrifuged through a density gradient using either 10 to 40% sucrose, 10 to 40% cesium chloride, 10 to 40% cesium sulfate, or 30 to 60% Nycodenz to improve the purity of the virus preparation.

7. Closterovirus Group

Closteroviruses present special problems in relation to purification. Their long, flexuous rods are subject to shearing, and become entangled with each other and absorb to host plant constituents during purification. The first problem is how to extract the most virus particles yet retain integrity of the virus particle. Once extracted, the next problem is to purify and concentrate the intact particles without aggregation and shearing. Lee et al.[25] devised an improved purification procedure for increased yield and recovery of intact particles of citrus tristreza virus involving the use of PEG, p-isooctylphenyl ether in the extraction buffer, two PEG precipitation steps, and centrifugation of a preformed step isopycnic cesium sulfate gradient. Yields from this procedure were 20 to 30 mg/kg of bark tissue of *Citrus excelsa*.

Purification of other closteroviruses has been achieved by extraction with a relatively high ratio of buffer:leaf weight (e.g. 5:1) and release of virus by grinding in a mortar and pestle or a leaf press.[26,27] Clarification is a key step in purification, and mild methods, particularly the use or bentonite, have generally been most successful.[28] Gradual addition of bentonite to extracts and sequential removal of the resulting coagulum until the solution is clear has proven to be both gentle and effective in the purification of several closteroviruses.[29] Following clarification, virus particles can be concentrated by PEG precipitation and further purified by rate zonal or isopycnic gradient centrifugation in cesium sulfate. Inclusion of Mg^{++} in the extract improves the yield of some closteroviruses. The yield of beet yellows virus was improved by stirring the clarified sap for 30 min with Triton X-100.[30]

Yield of virus is greatly influenced by the plant used, growth conditions, time of harvest and type of tissue. Virus yield may be as high as 200 to 300 mg/kg of leaves. Monette and James[31] reported improved yield of grapevine virus A by using *in vitro* cultures of infected *Nicotiana benthamiana* as the virus source.

8. Commelina Yellow Mottle Virus Group

The individual viruses in this group have restricted host ranges, in some cases limited to several species wihin the same genus. Purification is therefore done from a natural host of the particular virus, which may be difficult material to work with. For example, cacao swollen shoot infected cacao contains mucilaginous and polyphenolic materials which are difficult to overcome during purification. The problem was solved by extracting leaf tissue in ten to twenty times its weight in a phosphate buffer containing added protein such as 1 to 2% w/v egg albumen, bovine serum albumen or hide powder.[32]

Banana streak virus was purified by extracting fresh infected banana leaf tissue in a Tris-citrate buffer containing sodium sulfite, PVP and Triton X-100. The homogenate was clarified by blending in chloroform followed by low-speed centrifugation. Virus in the aqueous supernatant was concentrated by differential centrifugation and further purified by centrifugation in a pre-formed cesium chloride gradient containing 10% sucrose.[33] A similar procedure was used to purify a sugarcane bacilliform virus[34] and commelina yellow mottle virus.[35]

9. Comovirus Group

The Comovirus group consists of 14 member viruses, each of which shares all or nearly all of the main characteristics of cowpea mosaic virus, the type member. The host range of each member virus is confined to a few genera. Most of the member viruses have legumes as natural hosts; others occur in brassicas, cucurbits and solanaceous hosts. Source of virus for purification would be determined by the most appropriate natural host.

The comoviruses occur in moderate to high concentrations in plant sap. Since the viruses are stable, and tolerate clarification by organic solvents, most purification procedures involve the use of either n-butanol or butanol-chloroform as clarification agents. The purification protocol worked out by van Kammen and de Jager[36] for cowpea mosaic virus has, with minor modifications, been used successfully to purify several of the comoviruses. Freshly harvested leaves were homogenized with twice their weight of 0.1 *M* phosphate buffer, pH 7.0. The homogenate was squeezed and given a low-speed centrifugation. The supernatant fluid was stirred for 1 min with 0.7 volume of a 1:1 mixture of chloroform and n-butanol. After low-speed centrifugation, the clear aqueous layer was removed, and the virus precipitated by adding

PEG to 4% and NaCl to 0.2 *M*, and stirring for 60 min at room temperature. The precipitate was pelleted by low-speed centrifugation and resuspended in phosphate buffer. After low-speed centrifugation, the supernatant fluid containing the virus was layered on top of 1 ml of 40% (w/v) sucrose and given a high-speed centrifugation. The clear virus pellet was dissolved in sterile distilled water and given a low-speed centrifugation to remove possible contaminants. This procedure yielded about 2 g/kg leaf tissue of highly purified virus.

10. Cryptovirus Group

Cryptovirus particles are reasonably stable and survive purification procedures well. The main problem is that the viruses in this group have a low initial concentration in the host tissues; hence the yield of purified virus is low. Yield estimates vary from 50 to 150 ug/kg of tissue.[37]

The viruses must be extracted from naturally infected plants, because, unlike those in other groups, they have no natural vectors and are not graft transmissible. However, they are transmitted through seed with high frequency and, where both parents are carriers, 100% of progeny seedlings may be carriers. Tissue is usually extracted in a phosphate buffer, 0.1 to 0.5 *M*, pH 7 to 8. Such antioxidants as thioglycollate, DIECA or sodium sulfite are added. In earlier protocols, the extracted sap was clarified by shaking with chloroform, sometimes mixed with up to an equal volume of n-butanol. However, there is now concern about the effect of the organic solvents on the integrity of the virus particles, and more recent procedures are avoiding their use.

A procedure that has been devised for the purification of white clover cryptic viruses 1 and 2[38] and carnation cryptic virus[39] avoids the use of chloroform for clarification or high molarity salts in the later stages of purification. Tissue is frozen at −80°C, ground in a centrifugal grinding mill at 20,000 rpm, and the resultant paste is suspended in 0.1 *M* citrate buffer, pH 5.6, containing 1 m*M* EDTA, and filtered through a nylon stocking. Thioglycollic acid is then added to the sap to 0.5% (v/v) final concentration, and stirred for 30 min, and Nonidet P40 is added to 1% (v/v), with stirring for a further 30 min. The homogenate is then clarified by low-speed centrifugation and 10% PEG (w/v) and 1% NaCl (w/v) are dissolved in the supernatant. After stirring for 2.5 hr, the mixture is centrifuged and the resultant pellets suspended in 0.05 M potassium phosphate buffer, pH 7.0. The precipitate is clarified by low-speed centrifugation and the supernatant ultracentrifuged through a 20% sucrose cushion. The alternate centrifugation is repeated and the final product subjected to sucrose gradient centrifugation.

11. Cucumovirus Group

The type member of this group is cucumber mosaic virus (CMV). Other definitive members are tomato aspermy virus (TAV) and peanut stunt virus

(PSV). All of these viruses are readily transmitted mechanically to a wide range of host plants belonging to numerous families. There is a wide choice of suitable propagative hosts but most investigators use *Nicotiana* spp. for CMV and TAV and *Vigna sinensis* for PSV. All three viruses have been purified but yields and specific infectivity are very much dependent on the strain, the clarification procedure and the conditions of the suspending medium. TAV is considerably easier to purify that the other two viruses. In general, the protocols that are satisfactory for the purification of CMV will work well with TAV and PSV.

Scott[40] devised a protocol for purifying the Y strain of CMV based on the extraction of virus with a mixture of chloroform and 0.5 M citrate buffer, pH 6.5 (1 g:1 ml:1 ml). Following low-speed centrifugation, the aqueous phase was dialysed against 0.005 M borate buffer, pH 9.0, and the virus concentrated by three cycles of differential centrifugation. While this method was effective with the Y strain, it did not work well with some other strains. Lot et al.[41] used a similar protocol for purifying a tomato isolate of CMV, except that they suspended the pellets in solutions containing low concentrations of citrate buffer with 2% Triton X-100.

The above methods were based on the use of organic solvents for clarification. Mossop et al.[42] worked with a strain of CMV that was partially or totally destroyed when clarified with organic solvents such as chloroform, diethyl ether or carbon tetrachloride. They successfully purified this strain by homogenizing infected tissue in 3 vol (w/v) of 0.1 M sodium phosphate buffer containing 0.1% each of thioglycollic acid and DIECA, pH 8.0. After straining and low-speed centrifugation, Triton X-100 (2%) was added to the supernatant and, after stirring for 15 min, the virus was pelleted by high-speed centrifugation. The pellets were resuspended in the extraction buffer and the particulate material was removed by low-speed centrifugation. Following ultracentrifugation through a 10% sucrose cushion, the virus was further purified by sucrose gradient centrifugation.

12. Dianthovirus Group

This group consists of three members: the type member, carnation ringspot virus (CRSV), red clover necrotic mosaic virus (RCNMV) and sweet clover necrotic mosaic virus (SCNMV). All three viruses are stable and reach high concentrations in infected plants. As a consequence, a variety of purification protocols have been developed, all of which are acceptable. Yields of purified virus range from 150 to 300 mg/kg of infected tissue.

CRSV is usually purified from infected cowpea or *Nicotiana clevelandii* plants. The procedure of Tremaine and Dodds[43] is effective with most strains. It involved extracting infected tissue in 0.2 M sodium acetate buffer (pH 5.0) containing 0.02 M DIECA and 0.1% mercaptoethanol. The extract was adjusted to pH 5.0 and left for 4 hr at 5°C before clarifying by low-speed centrifugation. The virus was precipitated by adding PEG to 8% and, following low-speed centrifugation, the pellets were suspended in 0.1 M sodium acetate buffer (pH 5.0) and given one cycle of differential centrifugation. Further purification was achieved by sucrose density gradient centrifugation.

A similar procedure has been used to purify the other two viruses except that RCNMV was clarified by adding n-butanol to 8.5%[44] and SCNMV by adding an equal volume of butanol-chloroform to the extracted sap.[45]

13. Fabavirus Group

This small group consists of broad bean wilt virus (serotype I and II) and Lamium mild mosaic virus. Purification is difficult because many protocols result in a loss of virus during the early stages, apparently because of particle aggregation and absorption to host materials. Xu et al.[46] devised a protocol which resulted in minimal loss of virus, and yielded from 20 to 80 mg/kg of tissue. The viruses were propagated in either *Chenopodium quinoa* or pea. Infected tissue was ground in a phosphate buffer containing EDTA, thioglycollic acid and Triton X-100. The homogenate was filtered, mixed with chloroform, and given a low-speed centrifugation. The virus in the aqueous phase was concentrated by differential centrifugation and further purified by sucrose gradient centrifugation.[46-48]

14. Furovirus Group

Furoviruses are labile, fungal-transmitted, rod-shaped viruses having particles that superficially resemble tobamoviruses. Unlike the tobamoviruses, particles characteristically occur in low concentrations. These viruses are purified with difficulty, mainly because: (i) they are unstable *in vitro;* (ii) tend to aggregate and/or fragment during purification; and (iii) occur in low concentration in host plants, and/or dissociate only slowly from the intercellular inclusion in which they occur.[49] A procedure developed for the purification of soil-borne wheat mosaic virus[50] has been used successfully to purify other furoviruses. Infected leaf tissue is homogenized in a borate buffer (0.5 M, pH 9.0) and the expressed sap clarified by stirring with an equal volume of one part ethylene dichloride and two parts chloroform. Following differential centrifugation, the aqueous phase is subjected to alternate centrifugation followed by sucrose gradient centrifugation. Similar procedures, usually involving the addition of Triton X-100[51,52] have been used to purify other furoviruses. Some of the furoviruses are sedimented during low speed centrifugation of sap, presumably as a result of aggregation of virus particles to each other or with large cellular components.[53]

15. Geminivirus Group

The viruses belonging to this group are divided into three subgroups (I, II and III) on the basis of their genome organization, insect vector, and host range. Subgroup I and II viruses are vectored by leafhoppers and contain a monopartite genome. Viruses in subgroup III are vectored by whiteflies and contain a bipartite genome. Viruses in subgroup I infect monocots, while those in subgroup II and II infect dicots. Virus concentration in tissue is relatively low. Final yields of purified virus are generally not reported but may be up to 3 mg/kg fresh leaf tissue.[54]

The best characterized member of subgroup I is maize streak virus. It exists as a number of strains, infecting a wide range of grasses and cereals. Several purification schedules have been used. The one developed by Pinner et al.[55] appears to work well with a range of isolates. Infected tissue was harvested and stored at −80°C, ground in liquid nitrogen and homogenized in two volumes of sodium acetate buffer, pH 4.5, and one volume of chloroform. The mixture was subjected to differential centrifugation and the resulting pellets resuspended in either 0.01 M sodium acetate buffer, pH 4.5, or 0.01 M phosphate buffer, pH 7.0. The virus was further purified by sucrose gradient centrifugation.

Viruses belonging to subgroups II and III have been purified using procedures similar to those outlined above for the purification of maize streak virus. Briddon et al.[56] purified African cassava mosaic virus (subgroup III) by grinding 40 g of systemically infected *Beta vulgaris* tissue in liquid nitrogen to a fine powder before suspending it in 40 ml buffer (0.01 M sodium phosphate, pH 7.0, containing 0.001 M EDTA and 1% monothioglycerol). The suspension was digested with 1 g Driselase for 1 hr at 28°C before adding 40 ml chloroform. The semipurified preparation was further purified by pelleting through a sucrose cushion, and the final pellet resuspended in 0.5 ml of 0.001 M sodium phosphate buffer.

Several other viruses belonging to subgroups II and III have been purified in a similar manner, although some protocols concentrate the virus by the addition of PEG[54,57] and centrifugation through cesium sulfate as a final step in purification.[57]

16. Hordeivirus Group

Purification of the hordeiviruses has been achieved, but yield of purified virus depends greatly on the host, environmental factors, length of infection and virus strain.[58] When devising purification procedures, potential problems include the strong negative charge of the particles, resulting in interaction with polycations, particularly at low ionic strengths. The carbohydrate composition of the virus particles may encourage interaction with host lectins. Some detergents which solubilize these viruses result in an increase in yield of particles but a decrease in biological activity.[59] High ionic strength borate buffer (pH 9.0) give high yields but particles tend to aggregate. Freezing the tissue is not recommended, because recovery of virus from tissue stored at −20°C decreases with time of storage. Following extraction, viruses in this group can be clarified by adding chloroform-butanol (1:1) and the virus concentrated by ammonium sulfate precipitation.

17. Ilarvirus Group

Most of the ilarviruses have woody plants as their natural hosts, and are therefore best isolated from experimental herbaceous hosts. These viruses usually have a wide experimental host range[8] but occasionally the host range is narrow.[60] Antioxidants are often required for mechanical transmission of the viruses from woody hosts to herbaceous hosts. Peak titers of some these viruses in herbaceous plants are maintained for only short periods of time[60] while others maintain high titers over longer periods.[61]

Many of these viruses have been purified successfully using the protocol of Fulton.[62] Either systemically infected leaves or heavily infected inoculated leaves are homogenized in 0.02 M phosphate buffer, pH 8.0, containing 0.02 M mercaptoethanol and DIECA at a ratio of 1.5 to 2 ml/gram of leaf tissue. After low speed centrifugation the supernatant is mixed thoroughly with hydrated calcium phosphate and given a low-speed centrifugation. After a high-speed centrifugation and resuspension of the pellets, the material is brought to pH 4.8 to 5.0 with citric acid and given another low-speed centrifugation. The supernatant is adjusted to neutral pH and given a cycle of differential centrifugation. The addition of EDTA to the purification buffers often increases the stability of these viruses, suggesting that divalent cations are not involved in virus stability.

Chloroform[61] also has been used successfully for the initial clarification of several ilarviruses. Also, PEG has been used to concentrate ilarviruses to reduce the number of cycles of differential centrifugation required.[60,61] Some of the ilarviruses are quite unstable to acidification as a method of clarification. Therefore a protocol similar to that described by Fulton[63] should be used for the first attempt at purifying an unknown *Ilarvirus*.

The ilarviruses can be further purified by sucrose density gradient centrifugation. There are usually 2 to 4 virus bands observed in the sucrose density gradients due to the presence of different sized particles. In cesium gradients the ilarviruses often give only a single band since the proportion of nucleic acid and protein is similar for the different sized particles.

20. Luteovirus Group

Luteoviruses have relatively stable, small, isometric particles that occur in low concentrations in their hosts. They are restricted primarily to the phloem tissue which make extraction difficult. The purification of luteoviruses has been improved greatly by enzymatic maceration of host tissues to disrupt phloem cells and release virus particles.[64] All procedures used today for luteovirus purification include an enzymatic degradation step.[65-68] Even with the enzymatic maceration step yields of luteoviruses rarely exceed 1 mg/kg of tissue.

Environmental conditions under which the plants are grown are very important for luteovirus purification, especially the temperature at which plants are grown. Carrot redleaf and beet western yellows viruses gave very poor yields from plants grown at temperatures above 22°C whereas potato leafroll virus did much better at temperatures above 25°C.[66] Time after inoculation can make a large difference in yields when purifying luteoviruses.

Fresh or frozen tissue is usually homogenized in a mechanical blender with 2 to 3 volumes of buffer. The tissue may be powdered in liquid nitrogen before adding it to the buffer.[66] The buffer used is often slightly acidic (e.g. pH 6.0 to 6.5) because the maceration enzymes have an optimal pH below 6.0 but the luteoviruses are most stable above pH 6.0. Thus, the pH of the buffer is a compromise between the optima for enzyme activity and virus stability. Enzymatic maceration at pH 6.0 is suitable for most luteoviruses but potato leafroll virus gives much better yields at pH 7.0.[66] Most luteoviruses can be homogenized in phosphate or citrate buffer. Carrot redleaf virus is the exception in that it does not survive in phosphate buffer during extraction.[69] The extraction buffer often contains mercaptoethanol as a reducing agent, and sodium azide to prevent microbial growth during the enzyme incubation step.

Industrial grade pectinases used to clarify fruit juices or wines are often used as the macerating enzyme. Most enzymes require a 4 to 16 hr incubation step at temperatures from 20 to 37°C. Enzymes used in luteovirus purification include: Celluclast (Novo Laboratories, Wilton, CT, USA); Extractase P20X (Finnsugar Biochemics, Schaumberg, IL, USA); and Ultrazym 100 (Schweizerische Ferment AG, Base, Switzerland).[66] After the enzymatic maceration 1% Triton X-100 is often added and the suspension stirred for 1 to 3 hr. Clarification is by the addition of (1/6 to 2/3 volumes) chloroform-butanol (1:1) and vigourous shaking or stirring for 10 to 30 min. After low-speed centrifugation the virus is precipitated from the aqueous phase by the addition of 8% PEG and 1 to 2% NaCl. After 1 hr incubation the virus is pelleted by low-speed centrifugation and resuspended in a small volume of buffer.

The concentrated virus is further purified by one or two cycles of differential centrifugation with one high-speed centrifugation usually including a sucrose cushion. The final purificationis by rate zonal density centrifugation in sucrose gradients. Most luteoviruses are not stable in cesium chloride and therefore isopycnic gradients using this salt are not recommended.

When working with a new luteovirus or a suspected luteovirus it is advisable to test growing infected plants at 18 to 20°C and at 25 to 28°C since some of those viruses do better either at the lower or higher temperatures. It is also suggested that the pH of the maceration buffer be tested at pH 6.0 and 7.0 as virus yield can differ significantly between the two values.

19. Maize Chlorotic Dwarf Virus Group

Maize chlorotic dwarf virus is the only recognized member of this group. The virus has a narrow host range, limited to members of the Poaceae, and is transmitted in a semi-persistent manner by leafhoppers but not by inoculation with sap.

A purification schedule developed by Louie et al.[70] has been refined somewhat over the years. Hunt et al.[71] incorporated modifications that resulted in a simplified purification and greater virus yields. This simplified procedure has been further refined by Gingery and Nault.[72] Infected leaves are ground in 0.3 *M* potassium phosphate buffer, pH 7.0, containing 0.5% mercapto-ethanol (1 g tissue per 5 ml buffer). The extract is squeezed and clarified by emulsifying with chloroform (1/3 volume). After differential centrifugation, the pellets are suspended in 0.3 *M* potassium phosphate buffer, pH 7.0, clarified with 1/4 volume chloroform, and layered directly onto a 10 to 40% sucrose gradient. Virus recovered from the gradients is pelleted by high speed centrifugation and the pellets resuspended in 0.01 *M* potassium phosphate buffer containing 0.01 M EDTA. Yield of purified virus is about 2 to 3 mg/ kg of tissue.

20. Marafivirus Group

A variety of techniques have been used to purify the three viruses in this group. Maize rayado fino virus, the type member, has been purified from infected maize leaves by homogenizing in a phosphate buffer and clarifying by low speed centrifugation[73] or by the addition of chloroform.[74] The virus is concentrated by either PEG precipitation or differential centrifugation, followed by sucrose gradient centrifugation. Further purification is by iso-pycnic banding in cesium chloride.[73] Extraction with organic solvents yields significantly higher quantities of virus protein shells but leads to an extensive degradation of the RNA.[75]

Bermuda grass etched-line virus was purified by homogenizing infected leaves in sodium citrate, pH 6.5, and clarifying the extract by adjusting the pH to 5.0. Following two cycles of differential centrifugation, the virus was centrifuged through a sucrose density gradient column.[76]

21. Necrovirus Group

The necrovirus group contains two established members, tobacco necrosis virus (TNV), and Chenopodium necrosis virus, with two other possible members.

TNV has a wide natural host range and several strains distinguished by biological and serological criteria. In nature, it is usually confined to root tissue of infected plants although it is readily sap-inoculated to the leaves of many experimental hosts, where the virus induces local lesions but does not invade systemically. The source of virus for purification is the inoculated leaves of cowpea, bean, cucumber, *Nicotiana* spp. or Chenopodium quinoa. Under favorable conditions for plant growth, lesions begin to appear within 24 hr and the virus is purified from such leaves within 3 to 4 days of inoculation. Because of the high physical and chemical stability of the virus, leaves are often stored frozen until the virus is isolated.[77]

An effective purification protocol has recently been reported by Adam et al.[78] Leaves of inoculated Nicotiana benthamiana or Chenopodium quinoa were harvested 3 days post-inoculation and frozen. One part frozen leaves was homogenized in three parts (w/v) 0.1 *M* sodium acetate buffer, pH 6.0, for 1 min. Chloroform (1.5 ml/g of tissue) was then added and the homogenization continued for a further 30 sec. The homogenate was given a low-speed centrifugation and the aqueous phase given a high-speed centrifugation into a sucrose cushion. The resultant pellets were suspended in 0.05 *M* sodium acetate buffer, pH 6.0, and further purified by sucrose gradient centrifugation.

22. Nepovirus Group

Most nepoviruses have wide natural and experimental host ranges. In the field they are often found in perennial species. For this reason, as a first step in purification the virus is usually transmitted to a suitable herbaceous host such as cucumber, tobacco or *Chenopodium quinoa*. Most viruses in the group reach moderately high concentrations in crude sap of herbaceous hosts and purification has been successfully achieved for most members of the group. Yields are low to moderate, ranging from 10 to 50 mg/kg leaf tissue.

Tobacco ringspot virus, the type member of the group, is stable and easy to purify. Squash, cucumber, petunia, cowpea, French bean, tobacco and *Nicotiana clevelandii* have been used as sources for purification.[79] Steere[80] devised a purification procedure involving the addition of two volumes of a 1:1 mixture of n-butanol and chloroform to one volume of crude juice, stirring the resultant suspension for 15 min, and separating the emulsion by low-speed centrifugation. The aqueous phase was removed and stored at 22°C for 12 to 16 hr, during which time most of the non-virus protein was denatured. The virus was then clarified and concentrated by differential centrifugation. A modification of the butanol-chloroform procedure that is equally satisfactory involves the addition of n-butanol to a final concentration of 8.5%.[81]

Tomato ringspot virus and several other nepoviruses do not respond well to clarification with organic solvents and gentler procedures are required. Some can be clarified by the addition of ammonium sulfate, and magnesium-activated bentonite has been used for purifying others.[82] A purification protocol devised initially for the purification of tomato ringspot virus (Stace-Smith, unpublished) has been used for purifying several other nepoviruses. Leaves of infected *Nicotiana clevelandii* are homogenized (2 ml/g of tissue) in 0.03 *M* dibasic sodium phosphate buffer containing 0.02 *M* ascorbic acid and 0.02 *M* mercaptoethanol, pH 8.0. The homogenate is expressed through nylon tricot and the sap clarified by low-speed centrifugation. The supernatant is adjusted to pH 5.0 and left overnight at 4°C. Following low-speed centrifugation, the virus is precipitated by adding 8% PEG plus 1% NaCl and stirring for 1 hr at 4°C. The virus is pelleted by low speed centrifugation and the pellet suspended in at least 1/10 the original volume of 0.05 *M* sodium citrate, pH 7.5. After differential centrifugation, the concentrated virus is centrifuged into a 10 to 40% sucrose gradient made in 0.05 *M* sodium citrate buffer.

23. Parsnip Yellow Fleck Virus Group

Parsnip yellow fleck virus reaches only low concentrations in sap of plants grown under warm, summer conditions in a greenhouse. The virus multiplies best in spinach held at 15°C under growth chamber conditions. About 19 days after inoculation, the inoculated and uninoculated leaves are extracted in a phosphate buffer containing EDTA and thioglycollic acid. The extract is clarified either by adding an equal volume of diethyl ether or by adding 8.5% (w/v) n-butanol. After low speed centrifugation, the virus particles are precipitated from the clarified supernatant fluid by adding PEG. Following differential centrifugation, subsequent purification is by centrifugation in sucrose density gradients.[83]

A similar procedure was used to purify dandelion yellow mosaic virus from infected *Chenopodium quinoa*.[84]

24. Pea Enation Mosaic Virus Group

Pea enation mosaic virus, the only member of the group, has relatively stable small isometric particles that occur in low concentrations in host tissues. This virus has many properties in common with the luteoviruses, notably the facts that some isolates are aphid transmissible in a persistent manner, virus titers are low in host tissues, and the genome organization of RNA-1 is similar.[85] However, the virus is not phloem limited and can be transmitted mechanically, which are properties distinct from those of luteoviruses. Since the virus is not phloem limited, enzyme extraction has not been used for virus purification. Isolates that are maintained by mechanical transmission may lose the ability to be aphid transmissible. The loss of aphid transmissibility is associated with the absence of a second protein detectable in purified virus preparations.[86]

The infected material usually is homogenized in 2 volumes of 0.15 to 0.2 M sodium acetate[87] or 0.1 M potassium phosphate buffer pH 6.0[88] containing an equal volume of chloroform. The addition of $MgCl_2$ to the buffer used during purification has been reported to increase the yield of virus.[87] After expressing through cheesecloth followed by a low speed centrifugation, the virus can be further purified by precipitation with PEG, followed by differential centrifugation or by differential centrifugation alone. Further purification is accomplished by centrifugation on sucrose density gradients made up with sodium acetate buffer.

25. Plant Bunyaviruses

Among the plant viruses, tomato spotted wilt virus, the only member of the plant bunyaviruses, is unique in its particle morphology and genome structure. The virus particles are more or less isometric and enveloped. They contain four different proteins, an internal nucleocapsid protein of M_r 27K, two membrane proteins of 78K and 58K, and a large protein of 200 K. Since the virus has enveloped particles, organic solvents or nonionic detergents cannot be used in purification. For purification, infected tissue is homogenized in a buffer containing sodium sulfite and, after low-speed centrifugation, the virus is purified from either the supernatant or pellet, as both may contain large quantities of virus.[89] Some workers discard the low-speed supernatant and purify from the low-speed pellet only.[90-92] Others discard the low-speed pellet and proceed with purification of virus contained in the supernatant.[93] Either procedure is satisfactory in view of the high yield of virus in infected tissue. Resuspended pellets may then be given a low-speed centrifugation and concentrated by high-speed centrifugation. The resuspended pellets may be clarified by mixing with an antiserum prepared against healthy plant sap before being subjected to sucrose density gradient centrifugation.[89] Final purification may be achieved by equilibrium centrifugation in a cesium chloride gradient.[93]

26. Plant Reoviruses

Reoviruses are large isometric particles about 70 nm in diameter that occur in low concentrations in their host tissues. They are usually found in all parts of the host plant. Yields of purified virus are low, generally 1 to a few mg/kg of tissue. This is the only group of plant viruses that contain dsRNA in the virus particle. Intact particles consist of a double protein shell. The outer shell breaks down readily leaving a particle about 40 to 50 nm in diameter. Particles contain 12 (subgroup 1) or 10 (subgroups 2 and 3) pieces of dsRNA. Electrophoresis of dsRNA extracted from infected host tissues or from purified particles is diagnostic by the number of bands obtained.

Virus is purified from infected leaves or roots. In the case of rice gall dwarf virus and maize rough dwarf virus root tissue of infected plants are the best source of virus for purification. Tissues are homogenized in cold phosphate buffer pH 7.5 (1g/1.5 to 2.0 ml of buffer) containing a reducing agent, 0.01 M DIECA and 0.005 M EDTA and the homogenate is expressed through cheesecloth. The extract is clarified with carbon tetrachloride,[94] an equal volume of Freon 113[95] or chloroform.[96] The virus is concentrated by high speed centrifugation[95] or by precipitation with 6% PEG and 0.3 M NaCl.[94] The virus pellets are resuspended in either phosphate buffer pH 7.0 containing EDTA or histidine buffer, pH 7.0, containing 0.01 M MgCl$_2$. The final purification is by centrifugation through sucrose gradients in the same buffers. To remove traces of plant material the viruses are usually centrifuged through a second sucrose gradient. Yields of virus are often low, about 1 mg/kg of tissue.

In the case of maize rough dwarf virus the purified particles are often 55 to 60 nm in diameter. These are virus particles without the outer protein shell. Boccardo and Milne[97] found that dithiobis(succinimidyl) propionate (DSP) stabilized intact particles. The optimal concentration of DSP for stabilizing the intact particles was 0.001 to 0.01 mg/ml.

27. Plant Rhabdoviruses

Rhabdoviruses infecting plants have been given many different names but it is not clear how many distinct viruses are represented because antisera to only a few of the named viruses have been prepared. Rhabdoviruses consist of an outer envelope consisting of lipid and membrane proteins and a nucleoprotein core. The envelope of some rhabdoviruses are derived from the inner membrane of the nuclear membrane while others have the envelope derived from cytoplasmic membranes, the group is subdivided on the basis of the origin of the membrane. All viruses in this group are very large (S values about 1000) and quite fragile.

Most purification procedures that have been developed involve clarification of initial extract by filtration through celite pads, either before[98-100] or after[101] a discontinuous sucrose gradient centrifugation. Alternative clarification procedures utilize acidification to pH 5.5.[102] The virus may be concentrated by pelleting through a sucrose pad[99] or by PEG precipitation followed by sucrose density gradient centrifugation.[98] Column chromatography on calcium phosphate gel was successfully used with lettuce necrotic yellows virus.[100]

When particles of rhabdoviruses are treated with a nonionic detergent, the surface glycoproteins are solubilized and the nucleocapsid is exposed.[103] The soluble antigens are useful for serological comparisons.[98,99,104]

28. Potexvirus Group

Many of the potexviruses occur in high concentration in their natural or experimental host and this, together with the stability of their particles, has resulted in the development of purification protocols that yield large quantities of highly purified preparations.

While the 18 recognized members of the potexvirus group present few purification problems, there are 21 viruses that have been classified as possible members of the group, and most of these have not been successfully purified. For example, leaf dip preparations of rhododendron leaves showing symptoms of infection with Rhododendron necrotic ringspot virus contain particles that resemble potexvirus particles but purification attempts have failed.[105] Similarly, strawberry plants infected with strawberry mild yellow edge disease have been shown to contain flexuous rods that resemble the potexviruses but the virus has not been purified.[106] In general, those potexviruses that infect woody hosts, and for which no herbaceous host is known, are difficult to purify.

A variety of methods have been devised to purify the potexviruses. Clarification of extracted sap has been achieved by organic solvents, acidification and absorption. Concentration is usually done by PEG precipitation, and partial purification by sucrose gradient centrifugation. Further purification can be achieved by equilibrium gradient centrifugation in cesium chloride. The various purifiction methods that have been used successfully with the potexviruses have been reviewed by Purcifull and Edwardson.[107] The potexviruses are stable and occur in high concentrations, which results in yields of about 250 mg/kg of infected tissue.

29. Potyvirus Group

The potyvirus group is by far the largest of the plant virus groups. It contains over 74 definitive members and 84 probable or possible members.[2] Most of these viruses are transmitted in a non-persistent manner by aphids. Other members are transmitted by fungi, whiteflies or mites. The taxonomy of such a large group is inevitably unsatisfactory and incomplete. Recent molecular evidence indicates that the possible members of the group share many characteristics with the recognized potyviruses.[108] Four genera within a Potyviridae family are being proposed (Chapter 1).

Some of the definitive potyviruses are easily purified and good yields (up to 20 mg/kg of leaf tissue) are readily obtained. Others have proved extremely difficult to purify and often give only low yields. Different strains of potyviruses frequently differ in their ease of handling and may require different purification procedures. A major problem encountered during purification of different potyviruses is a tendency for irreversible aggregation of the virus during extraction, or in later stages of purification, resulting in considerable loss of particles during low speed centrifugation. Urea, usually at $1.0\ M$, and Triton X-100 at 0.5 to 5.0% (v/v) have been incorporated into the extraction buffer to reduce aggregation. Aggregation is also reduced by using extraction buffers of high pH (8.0 to 8.5) and high molarity (0.1 to 0.5).

The choice of source plant is important. Many potyviruses have restricted host ranges, thereby offering little choice for virus propagation. Where the host range permits, potyviruses should be propagated in a species known to support high concentrations of virus, such as *Nicotiana clevelandii, N. benthamiana, Phaseolus vulgaris* or *Pisum sativum*.

Protein precipitants are now widely used to assist in sedimentation of potyviruses from clarified sap. Ammonium sulfate is now rarely used but PEG (4% w/v) in the presence of $0.1\ M$ sodium chloride is usually effective. Some potyviruses have been reported to be irreversibly precipitated by PEG but this may be attributed to inadequate resuspension of the precipitated virus.[109] Final purification is frequently achieved by sucrose density gradient centrifugation or by cesium sulfate step gradient centrifugation[110,111] or sucrose-cesium sulfate centrifugation.[112]

Barley yellow mosaic virus has been proposed as the type species of a genus containing fungal-transmitted potyviruses. (Chapter 1) Each of the viruses in this genus has a narrow natural and experimental host range. Propagation of these viruses is carried out from sap-inoculated seedlings of suitable cereal cultivars, with the inoculated seedlings being placed in a growth chamber controlled at 15°C during day and 13°C at night[113] or cooler.[114] Infected tissue is homogenized in a phosphate buffer and clarified with carbon tetra-chloride, followed by differential centrifugation and sucrose density gradient centrifugation[115] or isopycnic centrifugation in cesium chloride.[116] Final pellets are suspended in 0.1 *M* citrate buffer, pH 7.0.[117]

Wheat streak mosaic virus, transmitted by an eriophyid mite, was purified by grinding infected wheat leaves in a blender with two volumes of 0.01 *M* potassium phosphate buffer and adjusting to pH 6.1 prior to low-speed centrifugation.[118] The supernatant was adjusted to pH 7.5 to 8.0 and brought to 10 m*M* in trisodium citrate and 1% (v/v) in Triton X-100. The extract was centrifuged through a 4 ml pad of 20% sucrose and the pellets suspended in 10 m*M* trisodium citrate (pH 8.0) and clarified by low-speed centrifugation. For final purification, samples were given a sucrose density gradient centrifugation.

30. Sobemovirus Group

The particles of most sobemoviruses are stabilized by pH- and cation-dependent bonds. In designing reliable purification procedures, consideration must by given to the effects of buffers and other chemicals on the structure and stability of the virus particles.[119] Buffers below pH 7.0 should be used, and chelating agents avoided. A method that works satisfactorily for the purification of several viruses in the group involves blending leaves in sodium acetate, pH 5.0, following low-speed centrifugation and precipitation of the virus by the addition of PEG.[120] The preparation may be concentrated by high speed centrifugation through a sucrose cushion, and further purified by isopycnic centrifugation in cesium chloride at pH 5.0.[119]

31. Tenuivirus Group

The viruses comprising this group have flexuous, threadlike particles that assume several configurations, including branched filamentous structures.[121] Although the particles are now relatively easy to purify, early attempts at purification were unsuccessful. All protocols use one or more antioxidants in the extraction buffer. Initial extracts are clarified with chloroform or carbon tetrachloride and the clarified extracts are concentrated by PEG precipitation. Further purification is achieved by sucrose density gradient centrifugation and isopycnic centrifugation in cesium chloride.[121-124] Reported yields of purified virus are as low as 10 mg/kg fresh tissue for rice hoja blanca virus[125] to as high as 300 mg/kg for rice stripe virus.[126]

32. Tobamovirus Group

The tobamoviruses are very stable and present in large quantities in infected plants. Yields can be as high as 10 g/kg of fresh tissue but yields between 1 and 3 g/kg of tissue are more common. For these reasons they were, for many years, the most studied among the plant viruses. Gooding and Hebert[127] devised a simple technique for the purification of TMV based on concentrating the virus by the addition of polyethylene glycol (PEG) rather than ultracentrifugation. Infected tissue is homogenized in a phosphate buffer, clarified by adding 8 ml of n-butanol per 100 ml extract and, after low-speed centrifugation, the virus is concentrated by the addition of 4% PEG and 1% sodium chloride prior to low-speed centrifugation. Further purification is obtained by a second PEG precipitation, followed by equilibrium gradient centrifugation in cesium chloride. This technique, or modifications of it, have been used to purify several tobamoviruses.[128,129]

33. Tobravirus Group

The tobraviruses are characterized as having a genome consisting of two species of ssRNA contained in straight tubular particles of two specific lengths, known as long (L) and short (S) particles. All tobravirus strains can give rise to isolates that do not produce virus rods through loss of the RNA 2 contained in the S particles.[130] These isolates, known as NM-type, lack the gene that codes for the coat protein and therefore do not produce virus particles.

Normal tobraviruses, that contain both L and S particles are readily purified, with systemically infected leaves yielding from 20 to 100 mg of virus per kg of tissue. For purification, infected leaves are homogenized in a buffer, and the extract clarified by blending with a half volume of a 1:1 n-butanol-chloroform mixture.[131,132] The supernatant fluid is frozen overnight[132] or for several days[133] before concentrating by high speed centrifugation[132] or by the addition of PEG. Separation of the S and L particles may be achieved by two cycles of centrifugation in sucrose gradients.[134]

34. Tombusvirus Group

Tombusviruses are present in high concentrations in their hosts and, under the right conditions for purification, can give high yields or purified virus. Extraction buffers at pH values above neutrality and containing chelating agents such as EDTA cause the particles to swell and become destabilized. Purification procedures have therefore been devised to take account of the requirements for preserving particle integrity. Since the virus is unstable at pH values above neutrality and is not precipitated at pH 4.5 to 5.0, the standard purification procedure involves homogenizing infected tissue in sodium acetate buffer, pH 5.0, containing ascorbic acid.[135-138] The extract is reasonably clear but it can be further clarified by adjusting to pH 4.5 followed by a low speed centrifugation.[137] The virus is concentrated by precipitation with PEG followed by high-speed centrifugation. Further purification is achieved by equilibrium centrifugation in cesium chloride.[135,139]

35. Tymovirus Group

The tymoviruses occur in high concentrations in their hosts and, since the viruses are structurally stable, it is relatively easy to obtain in large quantities of highly purified virus. The concentration in leaf sap is often 50 to 500 mg/kg.[140,141] All members of the group have at least two classes of particles constructed from a single species of polypeptide. Whatever method of purification is selected, it leads to a preparation containing virus and empty protein shells. Matthews[142] clarified infected sap with ethanol, precipitated the virus by addition of saturated ammonium sulfate, and further purified by dialysis at pH 4.8 followed by two cycles of differential centrifugation. The viral components in purified preparations were separated on cesium chloride gradients.

REFERENCES

1. **Iwanowski, D.,** Ueber die Mosaikkrankheit der Tabakspflanze, *Bull. Acad. Imp. Sci. St.-Petersbourg,* 3, 65, 1892.
2. **Francki, R. I. B., Fauquet, C. M., Knudson, D. L., and Brown, F.,** Classification and nomenclature of viruses, Fifth Report of the International Committee on Taxonomy of Viruses, *Arch. Virol.,* Supplementum 2, Springer-Verlag, Wein, 1991.
3. **Brunt, A., Crabtree, K., and Gibbs, A., Eds.,** *Viruses of Tropical Plants,* C.A.B. International, Wallingford, 1990.
4. **Gibbs, A. and Harrison, B.,** *Plant Virology, The Principles,* Edward Arnold Ltd., London, 1976.
5. **Harrison, B. D., Finch, J. T., Gibbs, A. J., Hollings, M., Shepherd, R. J., Valenta, V., and Wetter, C.,** Sixteen groups of plant viruses, *Virology,* 45, 356, 1971.
6. **Matthews, R. E. F.,** *Plant Viology,* Academic Press, London, 1991.
7. **Hull, R.,** Alfalfa mosaic virus, *Adv. Virus Res.,* 15, 365, 1969.
8. **Fulton, R. W.,** Ilarviruses, in *Handbook of Plant Virus Infections and Comparative Diagnosis,* Kurstak, E., Ed., Elsevier/North-Holland, Amsterdam, 1981, 377.
9. **Jaspers, E. M. J. and Bos, L.,** Alfalfa mosaic virus, CMI/AAB Descriptions of Plant Viruses, No. 229, 1981.
10. **Lane, L. C.,** The bromoviruses, *Adv. Virus Res.,* 19, 151, 1974.
11. **Lane, L. C.,** Brome mosaic virus, CMI/AAB Descriptions of Plant Viruses, No. 180, 1977.
12. **Brown, F.,** The classification and nomenclature of viruses: summary of results of meeting of the International Committee on Taxonomy of Viruses in Edmonton, Canada 1987, *Intervirology,* 30, 181, 1989.
13. **Milne, R. G.,** Taxonomy of the rod-shaped filamentous viruses, in *The Plant Viruses,* Vol. 4, Milne, R. G., Ed., Plenum Press, New York, 1988, 3.
14. **Yoshikawa, N. and Takahashi, T.,** Properties of RNAs and proteins of apple stem grooving and apple chlorotic leaf spot viruses, *J. Gen. Virol.,* 69, 241, 1988.
15. **Salazar, L. F. and Harrison, B. D.,** Host range, purification and properties of potato virus T, *Ann. Appl. Biol.,* 89, 229, 1978.
16. **Martin, R. R. and Bristow, P. R.,** A carlavirus associated with blueberry scorch disease, *Phytopathology,* 78, 1636, 1988.
17. **Ellis, P. J., Stace-Smith, R., and Converse, R. H.,** Isolation and some properties of a North American carlavirus in *Sambucus racemosa, Acta Hortic.* 303, 113, 1992.
18. **Lommel, S. A., McCain, A. H., and Morris, T. J.,** Evaluation of indirect enzyme-linked immunosorbent assay for the detection of plant viruses, *Phytopathology,* 72, 1018, 1982.

19. **Riviere, C. J., Pot, J., Tremaine, J. H.,and Rochon, D. M.,** Coat protein of melon necrotic spot carmovirus is more similar to those of tombusviruses than those of carmoviruses, *J. Gen. Virol.,* 70, 3033, 1989.

20. **Morris, T. J. and Carrington, J. C.,** Carnation mottle virus and viruses with similar properties, in *The Plant Viruses,* Vol. 3, Koenig, R., Ed., Plenum Press, New York, 1988, 73.

21. **Frazier, N. W. and Converse, R. H.,** Strawberry vein banding virus, CMI/AAB Descriptions of Plant Viruses, No. 219, 1980.

22. **Morris, T. J., Mullen, R. H., Schlegel, D. E., Cole, A., and Alosi, M. C.,** Isolation of a caulimovirus from strawberry tissue infected with strawberry vein banding virus, *Phytopathology,* 70, 156, 1980.

23. **Shepherd, R. J.,** Cauliflower mosaic virus, *CMI/AAB Descriptions of Plant Viruses,* No. 241, 1981.

24. **Gong, Z. X., Wu, H., Cheng, R. H., Hull, R., and Rossman, M. G.,** Crystallization of cauliflower mosaic virus, *Virology,* 179, 941, 1990.

25. **Lee, R. F., Garnsey, S. M., Brlansky, R. H., and Goheen, A. C.,** A purification procedure for enhancement of citrus tristeza virus yields and its application to other phloem-limited viruses, *Phytopathology,* 77, 543, 1987.

26. **Bar-Joseph, M., Gumpf, D. J., Dodds, J. A., Rosner, A., and Ginzberg, I.,** A simple purification method for citrus tristeza virus and estimation of its genome size, *Phytopathology,* 75, 195, 1985.

27. **Reed, R. R. and Falk, B. W.,** Purification and partial characterization of beet yellow stunt virus, *Plant Dis.,* 73, 358, 1989.

28. **Bar-Joseph, M. and Hull, R.,** Purification and partial characterization of sugar beet yellows virus, *Virology,* 62, 552, 1974.

29. **Lister, R. M. and Bar-Joseph, M.,** Closteroviruses, in *Handbook of Plant Virus Infections and Comparative Diagnosis,* Kurstak, E., Ed., Elsevier/North-Holland, Amsterdam, 1981, 809.

30. **Karasev, A. V., Agranovsky, A. A., Rogov, V. V., Miroshnichenko, N. A., Dolja, V. V., and Atabekov, J. G.,** Viron RNA of beet yellows closterovirus: cell free translation and some properties, *J. Gen. Virol.,* 70, 241, 1989.

31. **Monette, P. L. and James, D.,** Use of in vitro cultures of *Nicotiana benthamiana* for the purification of grapevine virus A, *Plant Cell Tissue Organ Cult.,* 23, 131, 1990.

32. **Brunt, A. A.,** Cocao swollen shoot virus, CMI/AAB Descriptions of Plant Viruses, No. 10, 1970.

33. **Lockhart, B. E. L.,** Purification and serology of a bacilliform virus associated with banana streak disease, *Phytopathology,* 76, 995, 1986.

34. **Lockhart, B. E. L. and Autrey, L. J. C.,** Occurrence in sugarcane of a bacilliform virus related serologically to banana streak virus, *Plant Dis.,* 72, 230, 1988.

35. **Lockhart, B. E. L.,** Evidence for a double-stranded circular DNA genome in a second group of plant viruses, *Phytopathology,* 80, 127, 1990.

36. **Van Kammen, A. and de Jager, C. P.,** Cowpea mosaic virus, CMI/AAB Descriptions of Plant Viruses, No. 197, 1978.

37. **Boccardo, G., Lisa, V., Luisoni, E., and Milne, R. G.,** Cryptic plant viruses, *Adv. Virus Res.,* 32, 171, 1987.

38. **Boccardo, G. and Accotto, G. P.,** RNA-dependent RNA polymerase activity in two morphologically different white clover cryptic viruses, *Virology,* 163, 413, 1988.

39. **Marzachi, C., Milne, R. G., and Boccardo, G.,** *In vitro* synthesis of double-stranded RNA by carnation cryptic virus-associated RNA-dependent RNA polymerase, *Virology,* 165, 115, 1988.

40. **Scott, H.,** Purification of cucumber mosaic virus, *Virology,* 20, 103, 1963.

41. **Lot, H., Marrou, J., Quiot, J.-B., and Esvan, C.,** Contribution à l'étude du virus de la mosaique du Concombre (CMV), II. Méthode de purification rapide du virus, *Ann. Phytopathol.,* 4, 25, 1972.

42. **Mossop, D. W., Francki, R. I. B., and Grivell, C. J.,** Comparative studies of tomato aspermy and cucumber mosaic viruses, V. Purification and properties of cucumber mosaic virus inducing severe chlorosis, *Virology,* 74, 544, 1976.

43. **Temaine, J. H. and Dodds, A. J.,** Carnation ringspot virus, CMI/AAB descriptions of Plant Viruses, No. 308, 1985.

44. **Hollings, M. and Stone, O. M.,** Red clover necrotic mosaic virus, CMI/AAB descriptions of Plant Viruses, No. 181, 1985.

45. **Hiruki, C.,** The dianthoviruses: a distinct group of isometric plant viruses with bipartite genome, *Adv. Virus Res.,* 33, 257, 1987.

46. **Xu, Z. G., Cockbain, A. J., Woods, R. D., and Govier, D. A.,** The serological relationships and some other properties of isolates of broad bean wilt virus from faba bean and pea in China, *Ann. Appl. Biol.,* 113, 287, 1988.

47. **Lisa, V., Luisoni, E., Boccardo, G., Milne, R. G., and Lovisolo, O.,** Lamium mild mosaic virus: a virus distantly related to broad bean wilt, *Ann. Appl. Biol.,* 100, 467, 1982.

48. **Makkouk, K. M., Kumari, S. G., and Bos, L.,** Broad bean wilt virus: host range, purification, serology, transmission characteristics, and occurence in faba bean in West Asia and North Africa, *Neth. J. Plant Pathol.,* 96, 291, 1990.

49. **Brunt, A. A. and Shikata, E.,** Fungus-transmitted and similar labile rod-shaped viruses, in *The Plant Viruses,* Vol. 2, Van Regenmortel, M. H. V. and Fraenkel-Conrat, H., Eds., Plenum Press, New York, 1986, 305.

50. **Gumpf, D. J.,** Purification and properties of soil-borne wheat mosaic virus, *Virology,* 43, 588, 1971.

51. **Tamada, T., Shirako, Y., Abe, H., Saito, M., Kiguchi, T., and Harada, T.,** Production and pathogenicity of isolates of beet necrotic yellow vein virus with different numbers of RNA components, *J. Gen. Virol.,* 70, 3399, 1989.

52. **Kurppa, A.,** Potato mop-top virus: purification, preparation of antisera and detection by means of ELISA, *Ann. Agric. Finn.,* 29, 9, 1990.

53. **Brunt, A. A. and Richards, K. E.,** Biology and molecular biology of furoviruses, *Adv. Virus Res.,* 36, 1, 1989.

54. **Ikegami, M. and Shimizu, S. I.,** Serological studies on mung bean yellow mosaic virus, *J. Phytopathol.,* 122, 108, 1988.

55. **Pinner, M. S., Markham, P. G., Markham, R. H., and Dekker, E. L.,** Characterization of maize streak virus: description of strains; symptoms, *Plant Pathol.,* 37, 74, 1988.

56. **Briddon, R. W., Pinner, M. S., Stanley, J., and Markham, P. G.,** Geminivirus coat protein gene replacement alters insect specificity, *Virology,* 177, 85, 1990.

57. **Morales, F., Niessen, A., Ramirez, B., and Castano, M.,** Isolation and partial characterization of a geminivirus causing bean dwarf mosaic, *Phytopathology,* 80, 96, 1990.

58. **Jackson, A. O. and Lane, L. C.,** Hordeiviruses, in *Handbook of Plant Virus Infections and Comparative Diagnosis,* Kurstak, E., Ed., Elsevier/North-Holland, Amsterdam, 1981, 565.

59. **Brakke, M. K.,** Dispersion of aggregated barley stripe mosaic virus by detergents, *Virology,* 9, 506, 1959.

60. **MacDonald, S. G., Martin, R. R., and Bristow, P. R.,** Characterization of an ilarvirus associated with a necrotic shock reaction in blueberry, *Phytopathology,* 81, 210, 1991.

61. **Adams, A. N., Clark, M. F., and Barbara, D. J.,** Host range, purification and some properties of a new ilarvirus from *Humulus japonicus, Ann. Appl. Biol.,* 114, 497, 1989.

62. **Fulton, R. W.,** Purification and some properties of tobacco streak and tulare apple mosaic viruses, *Virology,* 32, 153, 1967.

63. **Fulton, R. W.,** Serology of viruses causing cherry necrotic ringspot, plum line pattern, rose mosaic, and apple mosaic, *Phytopathology,* 58, 635, 1968.

64. **Takanami, Y. and Kubo, S.**, Enzyme-assisted purification of two phloem-limited plant viruses: tobacco necrotic dwarf and potato leafroll, *J. Gen. Virol.*, 44, 153, 1979.

65. **Ashby, J. W. and Huttinga, H.**, Purification and some properties of pea leafroll virus, *Neth. J. Plant Pathol.*, 85, 113, 1979.

66. **D'Arcy, C. J., Martin, R. R., and Spiegel, S.**, A comparative study of luteovirus purification methods, *Can. J. Plant Pathol.*, 11, 251, 1989.

67. **Casper, R.**, Luteoviruses, in *The Plant Viruses*, Vol. 3, Koenig, R., Ed., Plenum Press, New York, 1988, 235.

68. **Waterhouse, P. W., Gildow, F. E., and Johnstone, G. R.**, Luteoviruses, CMI/AAB Descriptions of Plant Viruses, No. 339, 1985.

69. **Waterhouse, P. M. and Murant, A. F.**, Purification of carrot red leaf virus and evidence from four serolgoical tests for its relationship to luteoviruses, *Ann. Appl. Biol.*, 97, 191, 1981.

70. **Louie, R., Knoke, J. K., and Gordon, D. T.**, Epiphytotics of maize dwarf mosaic and maize chlorotic dwarf diseases in Ohio, *Phytopathology*, 64, 1455, 1974.

71. **Hunt, R. E., Nault, L. R., and Gingery, R. E.**, Evidence for infectivity of maize chlorotic dwarf virus and for a helper component in its leafhopper transmission, *Phytopathology*, 78, 499, 1988.

72. **Gingery, R. E. and Nault, L. R.**, Severe mosaic chlorotic dwarf disease caused by double infection with mild virus strains, *Phytopathology*, 80, 687, 1990.

73. **Gamez, R.**, Maize rayado fino virus, CMI/AAB Descriptions of Plant Viruses, No. 220, 1980.

74. **Gingery, R. E., Gordon, D. T., and Nault, L. R.**, Purification and properties of an isolate of maize rayado fino virus from the United States, *Phytopathology*, 72, 1313, 1982.

75. **Gamez, R. and Leon, P.**, Maize rayado fino and related viruses, in *The Plant Viruses*, Vol. 3, Koenig, R., Ed., Plenum Press, New York, 1988, 213.

76. **Lockhart, B. E. L., Khaless, N., Lennon, A. M., and El Maatauoi, M.**, Properties of Bermuda grass etched-line virus, a new leafhopper transmitted virus related to maize rayado fino and oat blue dwarf viruses, *Phytopathology*, 75, 1258, 1985.

77. **Uyemoto, J. K.**, Tobacco necrosis and satellite viruses, in *Handbook of Plant Virus Infections and Comparative Diagnosis*, Kurstak, E., Ed., Elsevier/North-Holland, Amsterdam, 1981, 123.

78. **Adam, G., Winter, S., and Lesemann, D.-E.**, Characterization of a new strain of tobacco necrosis virus isolated from nutrient feeding solution, *Ann. Appl. Biol.*, 116, 523, 1990.

79. **Stace-Smith, R.**, Tobacco ringspot virus, CMI/AAB Descriptions of Plant Viruses, No. 309, 1985.

80. **Steere, R. L.**, Purification and properties of tobacco ringspot virus, *Phytopathology*, 46, 60, 1956.

81. **Tomlinson, J. A., Shepherd, R. J., and Walker, J. C.**, Purification, properties, and serology of cucumber mosaic virus, *Phytopatholgoy*, 49, 293, 1959.

82. **Murant, A. F.**, Nepoviruses, in *Handbook of Plant Virus Infections and Comparative Diagnosis*, Kurstak, E., Eds., Elsevier/North-Holland, Amsterdam, 1981, 197.

83. **Hemida, S. K. and Murant, A. F.**, Particle properties of parsnip yellow fleck virus, *Ann. Appl. Biol.*, 114, 87, 1989.

84. **Bos, L., Huijberts, N., Huttinga, H., and Maat, D. Z.**, Further charcterization of dandelion yellow mosaic virus from lettuce and dandelion, *Neth. J. Plant Pathol.*, 89, 207, 1983.

85. **Demler, S. A. and DeZoeten, G. A.**, The nucleotide sequence and luteovirus-like nature of RNA 1 of an aphid-transmissible strain of pea enation mosaic virus, *J. Gen. Virol.*, 72, 1819, 1991.

86. **Hull, R.,** Particle differences related to aphid-transmissibility of a plant virus, *J. Gen. Virol.,* 34, 183, 1977.

87. **Mahmood, K. and Peters, D.,** Purification of pea enation mosaic virus and the infectivity of its components, *Neth. J. Plant Pathol.,* 79, 138, 1973.

88. **Thottappilly, G., Bath, J. E., and French, J. V.,** Aphid transmission characteristics of pea enation mosaic virus acquired from a membrane-feeding system, *Virology,* 50, 681, 1972.

89. **Tas, P. W. L., Boerjan, M. L., and Peters, D.,** Purification and serological analysis of tomato spotted wilt virus, *Neth. J. Plant Pathol.,* 83, 61, 1977.

90. **Mohamed, N. A., Randles, J. W., and Francki, R. I. B.,** Protein composition of tomato spotted wilt virus, *Virology,* 56, 12, 1973.

91. **Gonsalves, D. and Trujill, E. E.,** Tomato spotted wilt virus in papaya and detection of the virus by ELISA, *Plant Dis.,* 70, 501, 1986.

92. **deAvila, A. C., Huguenot, C., Resende, R., de O., Kitajima, E. W., Goldbach, R. W., and Peters, D.,** Serological differentiation of 20 isolates of tomato spotted wilt virus, *J. Gen. Virol.,* 71, 2801, 1990.

93. **Law, M. D. and Moyer, J. W.,** A tomato spotted wilt-like virus with a serologically distinct N protein, *J. Gen. Virol.,* 71, 933, 1990.

94. **Omura, T. and Inuoe, H.,** Rice gall dwarf virus, CMI/AAB Descriptions of Plant Viruses, No. 296 1985.

95. **Milne, R. G. and Ling, K. C.,** Rice ragged stunt virus, CMI/AAB Descriptions of Plant Viruses, No. 248, 1982.

96. **Lovisolo, O.,** Maize rough dwarf virus, CMI/AAB Descriptions of Plant Viruses, No. 72, 1971.

97. **Boccardo, G. and Milne, R. G.,** Enhancement of the immunogenicity of the maize rough dwarf virus outer shell with the cross-linking reagent dithoibis(succinimidyl) propionate, *J. Virol. Methods,* 3, 109, 1981.

98. **Milne, R. G. and Conti, M.,** Barley yellow striate mosaic, CMI/AAB Descriptions of Plant Viruses, No. 312, 1986.

99. **Gerber, R. S.,** Discrimination by gel-diffusion serology of digitaria striate, maize sterile stunt and other rhabdoviruses of Poaceae, *Ann. Appl. Biol.,* 116, 259, 1990.

100. **Dietzgen, R. G. and Francki, R. I. B.,** Analysis of lettuce necrotic yellows virus structural proteins with monoclonal antibodies and concanavalin A, *Virology,* 166, 486, 1988.

101. **Falk, B. W. and Weathers, L. G.,** Comparison of potato yellow dwarf virus serotypes, *Phytopathology,* 73, 81, 1983.

102. **Lockhart, B. E. L.,** Occurrence of cereal chlorotic mottle virus in Northern Africa, *Plant Dis.,* 70, 912, 1986.

103. **Dale, J. L. and Peters, D.,** Protein composition of the virions of five plant rhabdoviruses, *Intervirology,* 16, 86, 1981.

104. **Lundsgaard, T.,** Comparison of *Festuca* leaf streak virus antigens with those of three other rhabdoviruses infecting the gramineae, *Intervirology,* 22, 50, 1984.

105. **Coyier, D. L., Stace-Smith, R., Allen, T. C., and Leung, E.,** Viruslike particles associated with a Rhododendron necrotic ringspot disease, *Phytopathology,* 67, 1090, 1977.

106. **Jelkmann, W., Martin, R. R., Lesemann, D.-E., Vetten, H. J., and Skelton, F.,** A new potexvirus associated with strawberry mild yellow edge disease, *J. Gen. Virol.,* 71, 1251, 1990.

107. **Purcifull, D. E. and Edwardson, J. R.,** Potexviruses, in *Handbook of Plant Virus Infections and Comparative Diagnosis,* Kurstak, E., Ed., Elsevier/North-Holland, Amsterdam, 1981, 627.

108. **Ward, C. W. and Shukla, D. D.,** Taxonomy of potyviruses: current problems and some solutions, *Intervirology,* 32, 269, 1991.

109. **Brunt, A. A.**, Purification of filamentous viruses and virus-induced noncapsic proteins, in *The Plant Viruses*, Vol. 4, Milne, R. G., Ed., Plenum Press, New York, 1988, 85.

110. **Zagula, K. R., Barbara, D. J., Fulbright, D. W., and Lister, R. M.**, Evaluation of three ELISA methods as alternatives to ISEM for detection of wheat spindle streak mosaic strains of wheat yellows mosaic virus, *Plant Dis.*, 74, 974, 1990.

111. **Reddick, B. B. and Barnett, O. W.**, A comparison of three potyviruses by direct hybridization analysis, *Phytopathology*, 73, 1506, 1983.

112. **Shukla, D. D., Jilka, J., Tosic, M., and Ford, R. E.**, A novel approach to the serology of potyviruses involving affinity-purified polyclonal antibodies directed towards virus-specific N termini of coat proteins, *J. Gen. Virol.*, 70, 13, 1989.

113. **Usugi, T., Kashiwazaki, S., Omura, T., and Tsuchizaki, T.**, Some properties of nucleic acids and coat proteins of soil-borne filamentous viruses, *Ann. Phytopathol. Soc. Jpn.*, 55, 26, 1989.

114. **Jianping, C. and Adams, M. J.**, Serological relationships between five fungally transmitted cereal viruses and other elongated viruses, *Plant Pathol.*, 40, 226, 1991.

115. **Usugi, T. and Saito, Y.**, [Purification and serological properties of barley yellow mosaic virus and wheat yellow mosaic virus], *Ann. Phytopathol. Soc. Jpn.*, 42, 12, 1976.

116. **Huth, W. and Adams, M. J.**, Barley yellow mosaic virus (BaYMV) and BaYMV-M: two different viruses, *Intervirology*, 31, 38, 1990.

117. **Kashiwazaki, S., Ogawa, K., Usugi, T., Omura, T., and Tsuchizaki, T.**, Characterization of several strains of barley yellow mosaic virus, *Ann. Phytopathol. Soc. Jpn.*, 55, 16, 1989.

118. **Brakke, M. K., Skopp, R. N., and Lane, L. C.**, Degradation of wheat streak mosaic virus capsid protein during leaf senescence, *Phytopathology*, 80, 1401, 1990.

119. **Hull, R.**, The sobemovirus group, in *The Plant Viruses*, Vol. 3, Koenig, R., Ed., Plenum Press, New York, 1988, 113.

120. **Tremaine, J. H. T. and Hamilton, R. I.**, Southern bean mosaic virus, CMI/AAB Descriptions of Plant Viruses, No. 274, 1983.

121. **Gingery, R. E.**, The rice stripe virus group, in *The Plant Viruses*, Vol. 4, Milne, R. G., Ed., Plenum Press, New York, 1988, 297.

122. **Ishikawa, K., Omura, T., and Hibino, H.**, Morophological characteristics of rice stripe virus, *J. Gen. Virol.*, 70, 3465, 1989.

123. **Toriyama, S. and Watanabe, Y.**, Characterization of single- and double-stranded RNAs in particles of rice stripe virus, *J. Gen. Virol.*, 70, 505, 1989.

124. **Hayano, Y., Kakutani, T., Hayashi, T., and Minobe, Y.**, Coding strategy of rice stripe virus: major nonstructural protein is encoded in viral RNA segment 4 and coat protein in RNA complementary to segment 3, *Virology*, 177, 372, 1990.

125. **Morales, F. J. and Niessen, A. I.**, Association of spiral filamentous viruslike particles with rice hoja blanca, *Phytopathology*, 73, 971, 1983.

126. **Toriyama, S.**, Characterization of rice stripe virus: a heavy component carrying infectivity, *J. Gen. Virol.*, 61, 187, 1982.

127. **Gooding, G. V. Jr. and Hebert, T. T.**, A simple technique for purification of tobacco mosaic virus in large quantities, *Phytopathology*, 57, 1285, 1967.

128. **Tobias, I., Rast, A. Th. B., and Matt, D. Z.**, Tobamoviruses of pepper, eggplant and tobacco: comparative host reactions and serological relationships, *Neth. J. Plant Pathol.*, 88, 257, 1982.

129. **Fraile, A. and Garcia-Arenal, F.**, A classification of the tobamoviruses based on comparisons among their 126 K proteins, *J. Gen. Virol.*, 71, 2223, 1990.

130. **Harrison, B. D. and Robertson, D. J.**, Tobraviruses, in *The Plant Viruses*, Vol. 2, Van Regenmortel, M. H. V. and Fraenkel-Conrat, H., Eds., Plenum Press, New York, 1986, 339.

131. **Ghabrial, S. A. and Lister, R. M.**, Coat protein and symptom specification in tobacco rattle virus, *Virology*, 52, 1, 1973.

132. **Russo, M., Gallitelli, D., Vovlas, C., and Savino, V.,** Properties of broad bean yellow band virus, a possible new tobravirus, *Ann. Appl. Biol.,* 105, 223, 1984.
133. **Robinson, D. J. and Harrison, B. D.,** Evidence that broad bean yellow band virus is a new serotype of pea early-browning virus, *J. Gen. Virol.,* 66, 2003, 1985.
134. **Kurrpa, A., Jones, A. T., Harrison, B. D., and Bailiss, K. W.,** Properties of spinach yellow mottle, a distinctive strain of tobacco rattle virus, *Ann. Appl. Biol.,* 98, 243, 1981.
135. **Martelli, G. P., Gallitelli, D., and Russo, M.,** Tombusviruses, in *The Plant Viruses,* Vol. 3, Koenig, R., Ed., Plenum Press, New York, 1988, 13.
136. **Rochon, D. and Tremaine, J. H.,** Cucumber necrosis virus is a member of the tombusvirus group, *J. Gen. Virol.,* 69, 395, 1988.
137. **Gallitelli, D., Hull, R., and Koenig, R.,** Relationships among viruses in the tombusvirus group: nucleic acid hybridization studies, *J. Gen. Virol.,* 66, 1523, 1985.
138. **Hillman, B. I., Morris, T. J., and Schlegel, D. E.,** Effect of low-molecular-weight RNA and temperature on tomato bushy stunt virus symptom expression, *Phytopathology,* 75, 361, 1985.
139. **Rochon, D. M. and Johnston, J. C.,** Infectious transcripts from cloned cucumber necrosis virus cDNA: evidence for a bifunctional subgenomic mRNA, *Virology,* 181, 656, 1991.
140. **Koenig, R. and Lesemann, D.-E.,** Tymoviruses, in *Handbook of Plant Virus Infections and Comparative Diagnosis,* Kurstak, E., Ed., Elsevier/North-Holland, Amsterdam, 1981, 33.
141. **Hirth, L. and Givord, L.,** Tymoviruses, in *The Plant Viruses,* Vol. 3, Koenig, R., Ed., Plenum Press, New York, 1988, 163.
142. **Matthews, R. E. F.,** Some properties of TYMV nucleoproteins isolated in cesium chloride density gradients, *Virology,* 60, 54, 1974.

Chapter 7

SEROLOGICAL PROCEDURES

M. H. V. Van Regenmortel and M.-C. Dubs

TABLE OF CONTENTS

© 1993 by CRC Press, Inc.

I. ANTIGENIC STRUCTURE OF VIRAL PROTEINS

In order to apply serological techniques in plant virology in the most efficient manner, it is useful to have some insight into our current knowledge of protein antigenicity. The viral antigens that are relevant for diagnostic work are proteins and in recent years there has been considerable increase in our understanding of the molecular structure of the antigenic sites of proteins. In this section the structure of viral antigenic determinants will be reviewed and the antigenic properties of two genera of plant viruses, the tobamoviruses and the potyviruses, will be described briefly.

A. TYPES OF EPITOPES

Epitopes or antigenic determinants are those parts of a viral protein that react specifically with antibodies to the protein. Sela[1] divided epitopes in two groups: sequential epitopes, which are defined by a continuous stretch of a polypeptide chain, and conformational epitopes that require the native conformation present in the intact protein in order to be recognized by antibody. Since it is difficult to envisage that an antibody could recognize a sequence of residues independently of its conformation, sequential epitopes should not be equated with conformation-independent epitopes.

Another way of classifying epitopes consists of distinguishing continuous and discontinuous epitopes.[2] A continuous epitope is defined as a linear stretch of contiguous amino acid residues while a discontinuous epitope is defined as a group of spatially adjacent surface residues that are brought together by the folding of the polypeptide chain. Discontinuous epitopes may also arise by the juxtaposition of residues from two separate peptide chains. For instance, a poliovirus epitope was found to be made up of residues from the VP1 and VP2 proteins[3] while an epitope of foot-and-mouth disease virus consisted of residues from the region 141 to 160 of one VP1 molecule together with residues 200 to 213 of a neighboring VP1 molecule.[4]

In practice the label "continuous epitope" is given to any linear peptide fragment of a protein that reacts with antibodies raised against the intact molecule. Frequently, therefore, continuous epitopes correspond to unfolded regions of the intact protein.[5] This has led to the contention that the use of peptides for epitope mapping of proteins tends to give misleading results because it concentrates on the epitopes of denatured molecules.[6] However, it seems unwarranted to conclude that all reported cases of cross-reactivity between proteins and linear peptides are due to the presence of antibodies reactive only with the denatured form of the protein.

There is general agreement today that the majority of protein epitopes are discontinuous, a viewpoint which has gained strength from the structures of the five epitopes that have been elucidated so far by X-ray crystallo-graphy.[7-11] In all examples, the epitopes were discontinuous and comprised residues from a number of surface loops, each epitope totaling as many as 15 to 22 amino acid residues that were in contact with the combining site of the antibody.

There is evidence, however, that the distinction between continuous and discontinuous epitopes is slightly artificial. Replacement studies with peptide analogs of so-called continuous epitopes have established that it is usually possible to replace certain residues of the epitope by any of the 19 other amino acids without affecting the antigenic activity.[12] This means that not every residue in the antigenically active peptide is an interacting residue of the epitope and in close contact with residues of the antibodies. As a result, the continuous epitope is in fact, functionally speaking, discontinuous.

In discussions of viral epitopes, it is useful to distinguish four categories of epitopes, i.e., cryptotopes, neotopes, metatopes, and neutralization epitopes. Cryptotopes are hidden epitopes that become immunologically available only after dissociation, fragmentation, or denaturation of the antigen.[13] The terms neotope and metatope were introduced[14] to characterize some of the epitopes of tobacco mosaic virus (TMV). Neotopes are found on the surface of polymerized viral protein and arise either through conformational changes of the monomers induced by intersubunit bonds or through the juxtaposition of residues from neighboring subunits as in the case of the discontinuous epitopes of poliovirus and foot-and-mouth disease virus mentioned above.

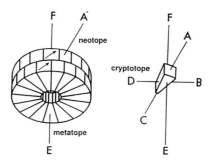

FIGURE 1. Schematic model of the protein subunits of TMV in monomeric form and as a double layer disk showing the location of different types of epitopes. Surface A′ harbors neotopes while surfaces B, C, and D possess cryptotopes. Metatopes are present on the bottom E of the subunit which contains the right radial and left radial α-helices comprising residues 74 to 89 and 114 to 134, respectively.[15]

Metatopes are epitopes that are present in both the dissociated and polymerized forms of the viral protein. The topological distribution of neotopes, metatopes, and cryptotopes on the coat protein of TMV is illustrated in Figure 1.

Neutralization epitopes correspond to epitopes of the virus that are specifically recognized by antibody molecules able to neutralize the infectivity of the virus. Such epitopes can only be identified and measured in a functional assay, the neutralization test, and their properties cannot be analyzed outside of the operational context of neutralizing antibodies and potentially infectable cells.[16] Methods used for localizing viral epitopes have been described in a recent review.[16]

B. ANTIGENIC STRUCTURE OF TOBAMOVIRUSES

The antigenic structure of TMV has been extensively studied by several groups.[17,18] The antigenic valence of TMV, i.e., the number of antibody molecules capable of binding simultaneously to one virus particle, was found to be about 800.[19] The fact that this value is smaller than the number of subunits in the virion (i.e., 2130) is due to steric hindrance caused by the larger size of the antibody molecule (150 kDa) compared to TMV protein (17.5 kDa). Although the antigenic valence of the TMV subunit is 3 to 5,[20] a much larger number of partially overlapping epitopes have been identified in the subunit.

Different approaches have been used to identify the epitopes of TMV and its dissociated protein subunit, such as inhibition studies with peptide fragments, cross-reactivity studies with viral mutants, and binding studies with monoclonal antibodies. The different techniques used for locating continuous epitopes in dissociated TMV protein are listed in Table 1.[21-26] Studies using peptide fragments indicated that virtually the entire sequence of TMV protein was antigenic and that the molecule contained no less than 13 continuous epitopes.

Table 1. Methods Used for Localizing Continuous Epitopes in TMV Protein

Serological technique	Type of probe	Location of epitope (residues)	Ref.
Precipitation inhibition	Tryptic and synthetic peptides	153–158	21
Complement fixation inhibition	Tryptic and synthetic peptides	103–112, 62–68	22,23
ELISA inhibition	Tryptic and synthetic peptides	1–10, 34–39, 55–61, 62–68, 80–90, 108–112, 153–158	24
ELISA	Conjugated synthetic peptides	19–32, 76–88, 90–95, 15–134, 134–146	25
ELISA	Synthetic hexapeptides immobilized on pins	40–50, 100–106	26
ELISA using Mabs	Synthetic hexapeptides immobilized on pins	90–98, 139–145	26

Most continuous epitopes shorter than 10 residues were found to correspond to regions of high segmental mobility in TMV protein.[27] Segmental mobility is mostly found in surface projections of proteins such as loops and turns and the correlation between mobility and antigenicity is therefore linked to the correlation between surface accessibility and antigenicity. In addition, the mobility of a protein segment may improve its complementarity with an antibody combining site through an induced fit phenomenon. However, the correlation with mobility is restricted to fairly short regions of 6 to 8 residues. When longer peptides that have a more rigid conformation, such as α-helices, are used as antigenic probes, regions of TMV protein that are structured and possess low mobility are also found to possess antigenic activity.[25,28]

Information on which amino acid residues of TMV protein are important for antigenicity has also been obtained by comparing the antigenic properties of wild-type virus with that of mutants or related tobamoviruses possessing a small number of amino acid substitutions.[29,30] However, in the 1960s the tertiary structure of the protein was unknown and it was therefore not possible to know if the substitutions affected the antigenicity by altering the conformation of the polypeptide chain or if the mutated residues were directly part of an epitope. An interesting case of substitution that affected the antigenicity was the exchange $Pro_{156} \rightarrow Leu$ which allowed the mutant virion to react with so-called heterospecific antibodies present in all TMV antisera.[14] Such heterospecific antibodies are antibodies that react better with a related antigen (e.g., a mutant) than with the homologous antigen (wild-type TMV) used for immunizing the animal. Such antibodies occur very frequently but are mostly overlooked, simply because the investigator does not test for their presence.[31] When all antibodies capable of reacting with TMV were removed from TMV antisera by prior absorption with the virus, there remained heterospecific antibodies that reacted with the mutant possessing the substitution at position 156.[32,33]

Cross-reactivity studies with mutants and related strains are most informative when carried out with monoclonal antibodies (Mabs). When polyclonal antisera are used in this type of study, those antibodies that recognize an alteration in one epitope may be swamped by the large number of antibodies in the antiserum that continue to react normally with the many unchanged epitopes. In contrast, when Mabs are used, only one epitope is analyzed at a time and the influence of a substitution is then more readily detectable.[34]

Mabs tend to be specific for the three-dimensional structure of the antigen and they are thus particularly suited for analyzing discontinuous epitopes.[32] By means of computer-generated images of the surface residues of the TMV subunit, it was possible to show that some Mabs recognized certain clusters of residues which together formed a number of discontinuous epitopes.[35] Of 18 Mabs raised against TMV, half were specific for the quaternary structure of the virus (i.e., specific for neotopes) and were unable to react with monomeric subunits.

In another study of 30 Mabs raised against TMV protein, more than 80% of the antibodies were unable to bind to any of 18 synthetic peptides representing virtually the entire viral protein.[25] Only three continuous epitopes located in residues 90 to 98, 139 to 145, and 76 to 88 could be identified with this panel of Mabs.[25,26] These epitopes were not accessible to antibodies when the subunits were polymerized and they correspond, therefore, to cryptotopes.

C. ANTIGENIC STRUCTURE OF POTYVIRUSES

The serological properties of potyviruses have been found extremely useful for identifying and classifying the individual members of this very large group of plant viruses. Early work[36] had shown the presence of distant antigenic relationships between different potyviruses and, as more viruses were studied, the pattern of serological relationships among individual potyviruses became increasingly complex. In many cases serological groupings did not correlate with the biological properties of the viruses and there was no clear demarcation line that separated individual virus species as compared to particular virus strains of the same species.[37]

Recent immunochemical and sequence data have clarified many of the earlier puzzling serological observations and have provided explanations for much of the past confusion.[38] The coat protein of individual potyviruses varies considerably in size (263 to 330 residues), mainly because of differences in the length of their N-terminal regions. The N- and C-terminal regions of the coat protein are exposed at the surface of the virus[39-41] and contribute significantly to the antigenic specificity of individual virus species. Mild trypsin treatment of virus particles removes the terminal ends of the polypeptide chains and shortens the coat protein to produce a "core" polypeptide of about 30 kDa. Studies with a number of potyviruses showed that the N-terminal sequences of the core protein of trypsin-treated particles were highly homologous and that enzyme had cleaved the protein at a position equivalent to Asp 68 or Asp 70 in the sequence of Johnson grass mosaic virus (JGMV).[41]

Mild trypsin treatment also removed the C-terminus (from position 286 in JGMV) of the coat protein.

Antisera raised against *Potyvirus* preparations devoid of proteolyzed particles react strongly with intact particles but only weakly with particles containing trypsin-treated protein. These results indicate that the N-terminus of the coat protein which is removed by proteolytic digestion is the immunodominant region of the virus particle.[41] Epitopes located in the N-terminal region of the coat protein generate antibodies specific for individual virus species, whereas epitopes located in the trypsin-resistant core protein region give rise to group-specific antibodies that are able to recognize many different potyviruses. For instance bean yellow mosaic virus (BYMV) and clover yellow vein virus (CYMV) reacted with antiserum to trypsin-resistant core particles of JGMV but not with antiserum to intact JGMV particles.[41]

The immunodominance of the N-terminal region of the coat protein was confirmed by a pepscan analysis[42] of JGMV protein.[43] A total of 296 overlapping synthetic octapeptides immobilized on pins and which covered the entire amino acid sequence of the coat protein of JGMV were tested with JGMV antisera. Antibodies reacting with the central and C-terminal peptide regions appeared in the antisera only after prolonged immunization. Antibodies reactive with octapeptides located in the central region were present in antisera to JGMV core particles and core protein but not in antisera to intact JGMV particles.[43] These results have been confirmed with several other potyviruses and led to the conclusion that earlier contradictory findings on serological relationships among potyviruses were caused by uncontrolled degradation of the N- and C-terminal regions of the virus particles used for immunization.

Both virus- and group-specific Mabs have been produced against several potyviruses. Gugerli and Fries[44] raised Mabs against PVY-N and obtained some antibodies that reacted with a common epitope of 24 different isolates of potato Y *Potyvirus* (PVY) while other antibodies were strictly specific to individual PVY-N isolates.

In an effort to obtain Mabs cross-reacting with a wide range of potyviruses, Jordan and Hammond[45] immunized mice with a mixture of 12 potyviruses in the form of intact particles and viral protein. One Mab, labeled PTY-1, was found to react with at least 135 *Potyvirus* isolates representing 48 different potyviruses. This Mab, PTY-1, appears to recognize a cryptotope located in the conserved, core region of the viral protein.[46]

II. ANTIBODY REAGENTS

A. PRODUCTION OF ANTISERA
1. Immunogens

Virus preparations used for the production of polyclonal antiserum should be free of contaminating plant antigens. Methods for removing plant proteins from virus preparations have been discussed extensively in an earlier review.[33] Even when a preparation appears to be pure by biochemical criteria, it frequently happens that a lengthy immunization schedule leads to the appearance of antibodies to contaminating substances that were present in trace amounts. It is for this reason that the amount of antigenic material administered to animals should be kept to a minimum. Many workers use a dose of 1 to 10 mg of purified virus per injection, although antisera of adequate titer can be obtained by injecting animals with as little as 50 to 100 µg of material.[47] Immunization with large doses of virus does not lead to proportionally higher antibody levels and has the disadvantage that contaminants may then reach a level where they are able to induce an immune response. By adsorbing virus particles to the surface of phenol-killed, acid-treated *Salmonella* bacteria it is possible to immunize rabbits with as little as 20 to 50 µg of virus per injection.[48]

Most plant virus particles are fairly stable at 37°C in the body of the animal and will give rise to antibodies specific for the quaternary structure of virions.[33] Certain unstable viruses, such as the plant reoviruses or the cucumoviruses, which are readily degraded, may give a strong response to dissociated coat protein subunits or even to double stranded RNA.[49] Particle degradation can be prevented by aldehyde treatment, a procedure which can lead to a considerable enhancement of immunogenicity[50,51] and causes only little (if any) alteration of the antigenic properties of the virus.[52,53]

Antisera to viral coat protein can be obtained by immunization with viruses that have been degraded by sodium dodecyl sulfate.[54] Small amounts of protein obtained from bands cut out from gels after electrophoresis have been found also to give rise to antisera of adequate titer.[55] Procedures for preparing antisera to inclusion bodies induced by potyviruses and caulimoviruses have been reviewed by Hiebert et al.[56]

Synthetic peptides corresponding to continuous epitopes of viral coat proteins or of viral-encoded proteins are increasingly used to raise antipeptide antibodies capable of binding to the parent protein. The principles underlying this approach have been explained in several reviews.[16,57,58] The position of continuous epitopes in the protein can be predicted using various algorithms[59] and the corresponding peptides are synthesized by solid-phase synthesis procedures.[60] Rabbits are immunized with unconjugated peptides when they are longer than 15 residues or with peptides conjugated to a carrier protein.[26,57] This approach using antipeptide antibodies has been used to study the appearance of nonstructural proteins in leaves infected with different plant viruses[61-63] and to elucidate the molecular structure of plant virus epitopes.[26,64-66]

2. Immunization Procedures

Rabbits are the most commonly used animal for producing plant virus antisera. A volume of 20 ml of antiserum can be obtained from an immunized rabbit every 2 weeks. The animal is bled by making a small incision in the marginal vein of the ear. Blood can be drawn without discomfort to the animal by placing the ear in a glass container fitted with two outlets and applying a small negative pressure by means of a water vacuum pump.

Chickens are also convenient animals for producing large amounts of antibodies to plant viruses. When laying hens are immunized, antibody can be obtained very easily from the egg yolks by precipitation with polyethylene glycol.[67,68] These animals will produce one egg a day. About 75 mg purified immunoglobulin and 10 mg purified viral antibody can be obtained from one egg.[69] Avian antibodies are particularly valuable in indirect double antibody sandwich assays which require the use of viral antibody produced in two animal species.[70] Since chicken antibodies do not cross react with mammalian immunoglobulins,[71] there is no binding between the anti-virus chicken antibody and the conjugated anti-rabbit globulin antibody used in the assay (see Section III.C.2). Quails have also been found to be suitable animals for obtaining viral antibody from the egg yolk.[72]

Rabbits should be bled before immunization to obtain control serum. Many types of immunization procedures have been used successfully but there is little published information on the relative merits of different protocols.[33,73] Since the immune response of individual outbred animals submitted to the same immunizing protocol is highly variable, it is always advisable to inject several animals in order to be able to select the best antiserum produced. It is customary to inject antigen-adjuvant mixtures intramuscularly or subcutaneously. The antigen is usually emulsified in Freund's adjuvant, which is a mixture of mineral oil and an emulsifier. Freund's complete adjuvant contains, in addition, killed mycobacteria; without the bacteria it is referred to as Freund's incomplete adjuvant. A series of injections is given at intervals of 2 to 3 weeks and the animals are bled after the second and subsequent injections. Multisite injections given subcutaneously or intramuscularly tend to give a better immune response than single-site injections. Blood obtained from immunized animals is allowed to clot and the serum is separated from the clot. Antisera can be stored frozen at $-20°C$ or mixed with an equal volume of glycerol and kept in liquid form at $-20°C$. Detailed procedures for the immunization and handling of experimental animals are described in several reviews.[33,58,74-76,327]

B. PRODUCTION OF MONOCLONAL ANTIBODIES

Technical details of the production of Mabs are available in many reviews,[76-81] including the procedures that have been used for obtaining Mabs to plant viruses.[82,327] Only some of the main points will be briefly described here. The main steps in producing hybridoma can be summarized as follows:

1. Immunization of mice or rats[83]
2. Culturing of a mouse myeloma cell line

3. Extraction of spleen of immunized mouse
4. Fusion of spleen cells with myeloma cells
5. Fused cells (hybridoma) are grown in a selective medium that does not support the growth of nonfused cells
6. Supernatants of hybridoma cultures are screened for the presence of Mabs
7. Subcloning of the hybridoma to ensure that the cell lines are monoclonal
8. Production of ascitic fluid

1. Immunization of Mice

Several 6-week-old BALB/c mice are immunized with two to four intra-peritoneal injections (50 to 100 μg of purified virus) emulsified with an equal volume of complete Freund's adjuvant for the first injection and incomplete Freund's adjuvant for the others, at 7-day intervals. On day -4 and -3 before the fusion, the two mice giving the highest specific response in enzyme-linked immunosorbent assay (ELISA) are given intravenous booster injections with 10 μg of virus in saline each; 3 days after the second booster injection, the spleens are excised and used for fusion.

2. Preparation of Cell Suspensions

Myeloma cells — Myeloma cells, for instance, nonsecreting PAI cells,[84] are maintained in culture in supplemented Dulbecco's modified Eagle's medium (DMEM), containing glutamine-pyruvate antibiotic β-mercaptoethanol (GPAM) plus 10% fetal calf serum (FCS) at least 2 weeks before the fusion.

Feeder cells — Mouse peritoneal macrophages are obtained from BALB/c or OF1 mice by flushing the peritoneal cavities with about 5 ml of 0.34 M sucrose. The pooled fluids are centrifuged for 10 min at 400 g. The pellet is resuspended in DMEM-GPAM-HAT (hypoxanthine aminopterine thymidine) 10% FCS culture medium at a cell density of $2 \cdot 10^4$. Thymocytes are obtained from the thymus of 2- to 6-week-old mice. The thymus is removed and teased in DMEM-GPAM. After centrifugation (200 g, 10 min), the pellet is resuspended in DMEM-GPAM-HAT 10% FCS medium at a cell density of $3 \cdot 10^3$ cells/ml.

3. Cell Fusion and Culture

Spleen cells are dislodged by repeatedly injecting the intact spleen with serum-free medium through a 26-gauge needle. The cells are then mixed with nonsecreting myeloma cells at a ratio myeloma cell:spleen cell of 1:1 to 1:2 (e.g., $5 \cdot 10^7:10^8$) and co-pelleted (400 g) at room temperature for 10 min. A volume of 1 ml of polyethylene glycol 4000 at 1 mg/ml is added with gentle agitation for 1 min. After an additional 1 min of gentle agitation, 10 ml of DMEM-GPAM are added over a period of 2 min. After centrifugation (10 min, 50 g), the pellet is resuspended in DMEM-GPAM FCS-HAT. The cells are then distributed into 96-well Falcon plates (50 μl) containing 100 µl of feeder cells. The cells are placed into a 7% CO_2 incubator at 37°C; 10 to 15 days after the fusion, the supernatant from wells containing clones are tested for the presence of specific antibody by ELISA. Selected positive cultures are cloned by limiting dilution where the cells are distributed into the wells (containing as feeder cells thymocytes or macrophages) at a cell density of 0.5 cell/well. The wells where there is only one clone are selected and the supernatant is tested for the presence of specific antibody by ELISA. Cloning can also be achieved by growing cells in soft agar in which individual clones will appear as white spots in the agarose after 1 to 2 weeks.[85] Moderate amounts of specific Mabs can be obtained by maintaining positive subclones in culture while large quantities are obtained from ascitic fluid. Ascites are induced in pristane-primed BALB/c mice by injecting 10^6 to 10^7 hybridoma cells in 0.5 ml of DMEM into the peritoneal cavity. The ascitic fluid is harvested after 10 days by tapping the peritoneum with a 19-gauge syringe needle. The cells and debris are removed by centrifugation at 300 g for 15 min.

4. Screening of Hybridoma

One of the most critical steps in obtaining suitable Mabs lies in the screening of culture supernatants.[31,86] The most commonly used assay is some form of ELISA which should be adjusted to ensure maximum sensitivity of antibody detection. Although multilayered sandwich procedures are often the most sensitive,[87] it is advaisable to use the same type of ELISA for screening as will be used in the subsequent work with the Mabs. For instance, if microtiter plates coated with purified virus are used for screening, it is often found that the selected Mabs will not react in a test where virus is presented by a first layer of antibodies adsorbed to the microtiterplate. When plates are coated with virus at pH 9.6, it seems that dissociated coat protein subunits become preferentially adsorbed.[88] Under these conditions, Mabs specific for neotopes will not react[35,89] and hybridoma secreting such antibodies could be discarded inadvertently.

In view of the common phenomenon of heterospecificity,[33,34,90] it is advisable to test culture supernatants not only with the antigen used for immunization but also with related viruses that show close or distant serological relationships with the immunogen. If only the immunizing virus is used in the initial screening, i.e., before the subcloning step, a number of hybridoma useful for detecting serologically related viruses could be discarded.[31]

5. Monoclonal Antibodies to Plant Viruses

Since 1981, Mabs have been produced against 60 different plant viruses (Table 2). Mab reagents are now available for many viruses of economically important crops and these reagents are especially valuable for detecting viruses that can only be purified with difficulty and for which adequate supplies of polyclonal antisera have been difficult to obtain.[31]

Table 2. Monoclonal Antibodies Prepared Against Plant Viruses

Virus	Group	Ref.
Alfalfa mosaic virus		89, 53
African cassava mosaic virus	*Geminivirus*	91
Apple chlorotic leaf spot virus	*Closterovirus*	92
Apple mosaic virus	*Ilarvirus*	89
Arabis mosaic virus	*Nepovirus*	93, 94
Banana bunchy top virus	*Luteovirus*	95
Barley yellow dwarf virus	*Luteovirus*	96–100
Bean common mosaic virus	*Potyvirus*	101
Bean pod mottle virus	*Comovirus*	66
Bean yellow mosaic virus	*Potyvirus*	102
Beet necrotic yellow vein virus	*Tobamovirus*	103, 104
Beet western yellow virus	*Luteovirus*	86, 105
Carnation etched ring virus	*Caulimovirus*	106, 107
Carnation latent virus	*Carlavirus*	108
Carnation mottle virus	*Carmovirus*	108
Carnation necrotic fleck virus	*Closterovirus*	108
Citrus tristeza virus	*Closterovirus*	109, 110
Citrus variegation virus	*Ilarvirus*	111
Clover yellow vein virus	*Potyvirus*	102
Cowpea mosaic virus	*Comovirus*	112
Cowpea severe mosaic virus	*Comovirus*	112
Cucumber green mottle mosaic virus	*Tobamovirus*	113
Cucumber mosaic virus	*Cucumovirus*	114–117
Grapevine fanleaf virus	*Nepovirus*	118
Grapevine leafroll virus type III	*Potyvirus*	119, 120
Lettuce mosaic virus	*Potyvirus*	121
Maize dwarf mosaic virus	*Potyvirus*	121
Maize streak virus	*Geminivirus*	122
Odontoglossum ringspot virus	*Tobamovirus*	123
Papaya ringspot virus	*Potyvirus*	124
Pea mosaic virus	*Potyvirus*	102
Peanut clump virus	*Furovirus*	125
Peanut mottle virus	*Potyvirus*	126–128
Peanut stripe virus	*Potyvirus*	129
Plum pox virus	*Potyvirus*	130
Potato leafroll virus	*Luteovirus*	86, 131, 132
Potato virus A	*Potyvirus*	133, 134
Potato virus M	*Carlavirus*	135
Potato virus X	*Potexvirus*	136–139
Potato virus Y	*Potyvirus*	44, 140
Prune dwarf virus	*Ilarvirus*	141
Prunus necrotic ringspot virus	*Ilarvirus*	89
Raspberry bushy dwarf virus	—	142
Rice dwarf virus	*Luteovirus*	143
Rice ragged stunt virus	*Reovirus*	144
Rice stripe virus	*Tenuivirus*	145
Satsuma dwarf virus	*Neopovirus*	146, 147
Soilborne wheat mosaic virus	*Furovirus*	148
Southern bean mosaic virus	*Sobamovirus*	149
Soybean mosaic virus	*Potyvirus*	121
Subterranean clover red leaf virus	*Luteovirus*	150
Sweet clover necrotic mosaic virus	*Dianthovirus*	151
Tobacco etch virus	*Potyvirus*	40
Tobacco mosaic virus	*Tobamovirus*	152, 153
Tobacco necrotic dwarf virus	*Luteovirus*	154
Tobacco streak virus	*Ilarvirus*	89
Tomato mosaic virus	*Tobamovirus*	155, 156
Tomato ringspot virus	*Nepovirus*	157, 158

Table 2. Monoclonal Antibodies Prepared Against Plant Viruses (continued)

Virus	Group	Ref.
Tomato spotted wilt virus	*Tospovirus*	159–162
Tulip breaking virus	*Potyvirus*	97, 163
Watermelon mosaic virus II	*Potyvirus*	164
Zucchini yellow mosaic virus	*Potyvirus*	165, 166

When Mabs are produced for diagnostic work, it is important that they should detect the widest possible range of viral serotypes. This can be achieved by collecting as many serotypes of the virus as possible and selecting those Mabs that recognize the largest number of serotypes.[44] Although Mabs specific for a single virus strain can be readily obtained,[105,115,121,153,167,168] it is also possible to select for Mabs that recognize a wide range of different virus species within a plant virus group or genus.

The so-called "universal" Mab PTY-1 developed by Jordan and Hammond[45] has been found to react in ACP-ELISA (see Section III.C.2) with 48 distinct aphid-transmitted potyviruses. However, since seven potyvirus isolates were not recognized by this Mab[46] it is not a truly "universal" reagent. Mab PTY-1 also does not react with fungus-, mite-, or whitefly-transmitted potyviruses.

Mabs have also been produced to nonstructural potyviral proteins such as nuclear inclusions,[169] cylindrical inclusion proteins,[170] amorphous inclusions,[171] and helper components.

The most commonly used assay for detecting plant viruses by means of Mabs is ELISA. It should be noted, however, that not all Mabs are equally effective in all steps of an ELISA.[123,125,155] Some Mabs cannot be used as a capturing antibody because they lose their binding activity when adsorbed to the plastic surface of a microtiter plate. Other Mabs lose their activity when coupled to an enzyme. Because of their molecular homogeneity, Mabs retain their activity only within a limited range of experimental conditions (pH, temperature, etc.). As a result, Mabs may appear to be less stable than an antiserum containing a mixture of antibodies that are stable under different conditions. Because of its heterogeneity, a polyclonal antiserum will always retain its activity over a wider range of environmental conditions than a single Mab.

Mabs are in fact also assay specific, i.e., they may perform well in one assay and have no activity in another serological test. For instance, only some of the Mabs raised against a plant virus usually will be able to react in immunodiffusion tests[53,89] and it may be necessary to add 1 to 3% polyethylene glycol to the gel in order to obtain a visible precipitation with some Mabs.[172] Certain Mabs have been used successfully in passive hemagglutination,[151] latex agglutination,[83] infectivity neutralization,[97] and immunoelectron microscopy.[88,106]

Because of their specificity for a single epitope, Mabs have been found to be excellent reagents for analyzing the antigenic structure of plant viruses at the molecular level.[34,35,66,152,173-177] The availability of Mabs to a viral antigen also makes it possible to produce anti-idiotypic antibodies[178] which can be used, for instance, as a probe to identify putative cell surface receptors responsible for insect vector specificity.[179] An anti-idotypic antibody is an antibody directed to an epitope associated with the antigen-combining site of an immunoglobulin molecule.

C. PURIFICATION OF IMMUNOGLOBULINS AND ANTIBODY FRAGMENTS

In some assays such as ELISA and immunoelectron microscopy it is necessary to use purified immunoglobulins instead of whole antiserum. Some commonly used methods for purifying immunoglobulins will be described briefly.

1. Ammonium Sulfate Precipitation[33]

Ammonium sulfate precipitation is the most commonly used procedure for preparing a crude immunoglobulin fraction from whole serum. By adjusting the salt concentration to 1/3 to 1/2 saturation, the globulin are precipitated, whereas the albumin and many other serum proteins will remain in solution.

To 1 ml of antiserum add dropwise with constant stirring one volume of 4 M ammonium sulfate solution and adjust to pH 7.8 with 1 N NaOH. After standing 1 h at room temperature, the precipitate is collected by centrifugation (10 min, 8000 g) and dissolved in half the volume of original serum. Further purification may be achieved by a second and third precipitation. After dissolving the final precipitate, the suspension is dialyzed against phosphate-buffered saline, pH 7.8. After dialysis, the suspension is centrifuged to remove small amounts of insoluble material. The extinction coefficient $E^{0.1\%}$ of IgG is 1.4 at 280 nm. The immunoglobulins are stored frozen.

2. Rivanol Precipitation[69]

Rivanol (2-ethoxy-6,9-diaminoacridine lactate) is the soluble salt of an acridine base that is used for precipitating albumin and other serum proteins, while leaving the IgG in solution. The rivanol itself is removed from the IgG suspension by conversion to the insoluble bromide form, and the IgG is then precipitated with ammonium sulfate. A product of very high purity is obtained.[69]

The pH of the serum is adjusted to 8 to 8.5 using 0.1 N NaOH and the total volume measured. For each 1 ml of alkaline serum, 3.5 ml of 0.4% aqueous rivanol are added slowly with stirring. After centrifugation, the precipitate is reextracted with water. The supernatant plus the water used for extracting the precipitate are treated with excess saturated aqueous KBr to precipitate the rivanol (rivanol bromide forms a yellow precipitate). After centrifugation (15,000 rpm, 30 min), IgG contained in the supernatant is precipitated with 1 volume of 4 M ammonium sulfate and centrifuged at 10,000 rpm for 20 min. The precipitate is resuspended in phosphate-buffered saline (PBS) and dialyzed against PBS. The precipitate formed during dialysis is removed by centrifugation.

3. Caprylic Acid Precipitation

Under acidic conditions, the addition of caprylic acid to serum or ascitic fluid will precipitate most proteins, with the exception of the IgG molecules. Serum or ascitic fluid is diluted with 4 volumes of acetate buffer (60 mM, pH 4) and the pH adjusted to 4.5 (with 0.1 N NaOH). Caprylic acid at 25 μg/ml is added slowly with thorough mixing and the mixture is stirred for 30 min. The insoluble material (albumin and other non-IgG proteins) is removed by centrifugation (10,000 g for 30 min). The supernatant is filtered and the pH adjusted to 7.4 with NaOH, 1 M.[180] If necessary, the IgG can be concentrated and purified further by ammonium sulfate precipitation.[181]

4. DEAE Cellulose Chromatography

After ammonium sulfate precipitation, immunoglobulins can be further purified using the anion exchanger diethylaminoethyl (DEAE) cellulose. If the pH is kept at 6.5 (i.e., below the isoelectric point of most antibodies), only contaminants will bind to the DEAE cellulose. The separation of immunoglobulins can be effected by a batch procedure or on a column.[76,182] The cellulose is washed with 10 volumes of 5 mM sodium phosphate, pH 6.5, and the immunoglobulin in the same buffer is then added; 2 ml of wet DEAE matrix will bind the proteins found in 1 ml of serum. The unbound globulins can be separated by centrifuging the slurry or washing the column with the same phosphate buffer.

If the pH of the DEAE cellulose is raised to 8.5, the antibodies will bind to the cellulose and can then be eluted by increasing the ionic strength of the column buffer. The DEAE is first equilibrated with 10 mM Tris buffer, pH 8.5, and immunoglobulins in the same buffer are then added to the column and eluted with the Tris buffer containing increasing NaCl concentrations (50 to 200 mM).[76]

5. Protein A Chromatography

Since protein A from the cell wall of *Staphylococcus aureus* has a strong affinity for antibodies, this property has been used for purifying immunoglobulins.[76] There are two variations of this method of purification, depending on the affinity of the immunoglobulin for protein A. If the globulin has a high capacity for binding to protein A, it is added to the column in low salt. If the globulin has a low affinity for protein A (mouse IgG1), its affinity for protein A is raised by increasing the salt concentration and the strength of the hydrophobic bonds.[76,183]

In the first procedure, at low salt concentration, the pH of the crude antibody preparation is adjusted to 8 by adding 1/10 volume of 1.0 M Tris, pH 8. The antibody solution is passed through a column of protein A Sepharose beads (capacity 10 to 20 mg antibody per milliliter of wet beads). The beads are then washed with 10 column volumes of 100 mM Tris, pH 8, then with 10 column volumes of 10 mM, Tris pH 8. The column is eluted with 100 mM glycine, pH 3. The fractions containing immunoglobulin are identified by absorbance at 280 nm and are neutralized with 1 M Tris, pH 8.

In the second procedure at high salt concentration, the protein A Sepharose column is equilibrated with a borate buffer, pH 8.9, containing 1.5 M glycine and 3 M NaCl. After adding the immunoglobulin, the column is washed with the borate glycine NaCl buffer, pH 8.9.

The IgG1 is eluted with a citrate buffer 0.1 M, pH 6. The eluate containing the IgG1 is neutralized with 1 M Tris, pH 8. To elute the other proteins the column is washed with a citrate buffer 0.1 M, pH 3.5.

6. Preparation of Antibody Fragments

Fragments of immunoglobulins are useful reagents in some techniques, such as immunoelectron microscopy and certain ELISA formats, as well as for the analysis of virus-antibody interactions.[181,184-186] Fab fragments can be obtained by digesting IgG on a column of immobilized papain. The column is prepared by mixing excess papain with Affi-gel 10 (BioRad). The papain is activated with 150 mM phosphate buffer, pH 6.5, containing 10 mM β-mercaptoethanol and 2 mM ethylene diaminetetraacetate (EDTA). After removing the mercaptoethanol from the column with the phosphate buffer containing 2 mM EDTA, the antibody is cycled through the column. Digestion is stopped when whole antibody can no longer be detected by sodium dodecyl sulfate-polyacrylamide gel electrophoresis (SDS-PAGE). Fab fragments can be purified by passing the digest through a Sepharose 4B protein A column in phosphate buffer, pH 8.0. Fab concentration is determined using E_{280nm} = 15.3.[85]

To obtain F(ab')$_2$ fragments, purified IgG (1 to 5 mg/ml in 0.07 M sodium acetate, 0.05 M sodium chloride, pH 4) is digested by the addition of pepsin in distilled water (45 μg enzyme/mg IgG) and incubation at 37°C overnight. F(ab')$_2$ is separated from the enzyme and from the digestion products by dialysis against PBS. Monovalent Fab' fragments are obtained by reduction of F(ab')$_2$ fragments in the presence of 10 mM 2-mercaptoethylamine.[181]

7. Purification of Specific Antibody

Although specific viral antibody rarely accounts for more than 10% of the immunoglobulins present in antiserum, purified total immunoglobulins usually function adequately in most immunoassays designed to detect viral antigens. In some circumstances, however, it may be advantageous to use specific viral antibody, for instance, in binding assays[19,186] or for electron microscopy.

A high yield of viral antibody can be obtained by mixing virus with purified immunoglobulin under conditions of very low ionic strength at pH 7.8. After ultracentrifugation of the virus and virus-antibody complexes, the nonreactive immunoglobulin in the supernatant is discarded and the complexes, resuspended in water, are brought to pH 2.9 with 0.1 N HCl. The virus is then pelleted by ultracentrifugation and the supernatant containing the antibody is brought to pH 7.5 with 0.1 N NaCl.[69] Instead of using virus particles, it is also possible to mix the immunoglobulin with a virus-serum albumin insoluble polymer (prepared by aldehyde treatment) which can be sedimented at low speed. This reduces the time during which the antibody is kept under acid conditions.[69]

D. LABELING OF IMMUNOGLOBULINS
1. Enzyme Conjugates

The most commonly used enzyme for preparing immunoglobulin conjugates is alkaline phosphatase,[182] followed by peroxidase.[187] The presence of peroxidase and oxidizing substances in many plant extracts makes this enzyme less suitable in plant virology.[188] Other enzymes, such as penicillinase,[189] have been used occasionally.

Preparation of alkaline phosphatase-labeled globulins[190] — A mixture of 1 mg of globulin and 2 mg of alkaline phosphatase is dialyzed overnight at 4°C against PBS. Coupling is achieved by the addition of 0.05% glutaraldehyde (final concentration) is distilled water (w/v) for 4 h at room temperature.[191] Glutaraldehyde is removed by dialysis against PBS and the conjugate is stored at 4°C in the presence of 1% bovine serum albumin (BSA).

Preparation of peroxidase-labeled globulins[185] — Conjugation of IgG to horseradish peroxidase (HRP) by the periodate oxidation method links protein to HRP via carbohydrate moieties on the enzyme. HRP (4 mg/ml in distilled water) is activated by the addition of freshly prepared 0.1 M sodium periodate (0.2 ml/ml HRP solution). After shaking for 20 min at room temperature, ethylene glycol is added (1 drop/ml) and the HRP is dialyzed against 1 mM sodium acetate, pH 4.4. Activated HRP is then mixed with IgG in 0.01 M sodium carbonate, pH 9.5 (0.5 mg HRP to 1 mg Ig G in 1 ml), and 0.2 M sodium carbonate, pH 9.5, is added dropwise to raise the solution to about pH 9.5. After shaking for 2 h at room temperature, freshly prepared sodium borohydride solution (4 mg/ml in water) is added (0.1 ml/ml solution). After 2 h at 4°C the conjugate is dialyzed against PBS.

Preparation of Fab' peroxidase conjugates[181] — When Fab' fragments are conjugated to enzymes by means of maleimide compounds, the yield and binding activity of the conjugated antibody is higher than when other coupling agents are used. A single thiol group remote from the antibody combining site is present on the Fab' fragment in the hinge region of the immunoglobulin and is able to interact with the maleimide compound bound to the enzyme molecule. This coupling procedure leads to the formation of a monomeric

Fab'-enzyme conjugate that is more active than IgG conjugates, which have a number of randomly located enzyme molecules attached to each IgG molecule.

Fab' fragments are coupled to HRP activated by means of N-succinimidyl 4-(N-maleimidomethyl) cyclohexane-1-carboxylate and the enzyme-Fab' conjugates are purified by gel filtration.[181] Fab'-enzyme conjugates prepared by the maleimide method have been found to detect smaller quantities of viral antigen by direct or indirect ELISA than IgG enzyme conjugates prepared by other methods. Since very short substrate incubation times (15 min) can be used with these conjugates, the time needed for an assay is shortened.[181]

2. Biotin Conjugates[192]

Since biotin can be covalently linked to antibody without affecting its antigen-binding capacity,[192] biotin-labeled antibodies are a superior reagent compared to enzyme-labeled antibodies. The use of biotinylated viral antibody increases the sensitivity of ELISA[193-195] and overcomes the narrow serotype specificity of double antibody sandwich (DAS)-ELISA. It seems that the attachment of biotin to viral antibody is less detrimental to its capacity to recognize a broad range of serologically related viruses than is conjugation to an enzyme.[195] When biotinylated antibodies instead of enzyme-labeled viral antibodies are used in DAS-ELISA, an additional incubation step with enzyme-labeled avidin is needed. However, this can be circumvented by incubating the biotinylated antibody together with the enzyme-labeled avidin.

Biotinylated globulins are prepared by simply adding N-hydroxysuccinimidobiotin (0.2 mg) in dimethylformamide to 1 mg globulin in 1 ml PBS and incubating for 4 h at 25°C.

The reaction can be stopped by adding 20 μl of 1 M NH_4Cl followed by dialysis against PBS at 4°C.

3. Europium Conjugates

Antibody-europium conjugates are used in time-resolved fluoroimmunoassay to overcome the nonspecific fluorescence originating from the sample.[196] A Eu^{3+} chelate of N^1-(p-isothiocyanatobenzyl)-diethylene triamine-N^1, N^2,N^3,N^3-tetra-acetic acid (Wallac, Turku) is used in 30-fold molar excess to antibody for conjugation in the presence of phosphate buffer, pH 9.5. Following overnight incubation at 4°C, the conjugate is purified by two gel filtration steps, the first on a 1.5 × 15 cm column of Sephadex G50 and the second on a 1.5 × 30 cm column of Sephacryl S400 (Pharmacia). The elution buffer is 0.05 M Tris HCl, pH 7.7, containing 0.1% NaCl and 0.05% NaN_3. the Eu-IgG complex is stored at 4°C.[197]

4. Gold-Labeled Immunoglobulins

Gold-labeled anti-virus monoclonal antibodies are used to localize by electron microscopy the exact positions of antigenic sites on virus particles.[88,198]

Purified monoclonal antibody (100 μl at 1 mg/ml) in 2 mM borate buffer, pH 9.6, is mixed with 900 μl of a suspension of colloidal gold (size of particles:5 nm). After 2 min shaking, the mixture is centrifuged at 45,000 rpm for 20 min in a 50 Ti Beckman rotor. The pellet is resuspended in 1 ml borate buffer containing 0.05% Tween 20 and sodium azide.

E. COMPARATIVE VALUES OF MONOCLONAL AND POLYCLONAL ANTIBODIES

The advantages of Mab reagents over polyclonal antisera have been well documented,[31,199,200] and can be summarized as follows:

1. **Standardization** — Since Mabs are biochemically defined and homogeneous reagents, their use ensures that uniform results can be obtained in different laboratories. In contrast, the inherent variability of individual bleedings from an immunized animal often leads to discrepant results. This has been responsible for the poor reproducibility of serological data obtained with conventional antisera in the past.
2. **Ready availability** — Once a hybridoma has been developed, Mabs of the same specificity can be obtained in virtually unlimited quantities. Cell lines can be stored in liquid nitrogen indefinitely and new batches of the Mab can be produced when needed. In principle, Mab reagents could be distributed worldwide to all interested laboratories. This would ensure uniform results, for instance in quarantine screening and virus strain identification.
3. **Increased specificity** — Since each Mab is specific for a single epitope, only a small portion of the total antigenicity of a virus is analyzed at a time. In contrast, the reactivity of a polyclonal antiserum results from the simultaneous recognition of many separate epitopes by a large number of different antibodies. Those antibodies in an antiserum that recognize a change in one particular epitope of an antigenic variant are usually swamped by the more numerous antibodies that react normally with the unchanged epitopes. Since Mabs are epitope-specific rather than virus- or strain-specific, it is possible either to select Mabs that do not recognize an antigenically related virus, because the epitope is not shared between the two viruses, or, on the contrary, to select Mabs that do not differentiate between the two viruses because they recognize an identical epitope present in both. This means that Mabs can be selected to emphasize either what is common between two viruses or what is different between them. While a single amino acid substitution in an epitope may suffice to abolish the reactivity of the complementary Mab,[32,34] two viruses that differ in the sequence of their coat proteins by as much as 18% may not be distinguished by some Mabs.[153] This phenomenon is illustrated in Figure 2, in the case of Mabs raised to TMV. The very high specificity of Mabs makes it possible to demon-

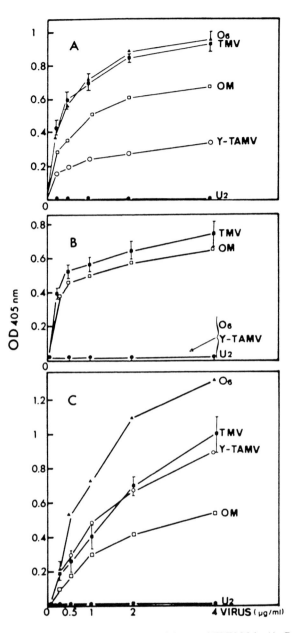

FIGURE 2. Cross-reactivity in indirect ELISA of three anti-TMV Mabs (A, B, and C) with related TMV strains and other tobamoviruses. U2 is a strain of tobacco mild green mosaic virus that is not recognized by the three Mabs. Y-TAMV, which is a strain of tomato mosaic virus with 18% coat protein sequence difference compared to TMV, could not be differentiated from TMV with Mab C and was not recognized at all by Mab B. Strain 06, which differs from TMV by three exchanges in the coat protein, could not be differentiated by Mab A. Strain 06 was not recognized at all by Mab B and was better recognized (heterospecific reaction) by Mab C than TMV. (Adapted from Briand, J. P. et al., *J. Virol. Meth.*, 5, 293, 1982.)

strate the selection of viral variants during replication in the host or transmission by vectors. This phenomenon of selection is particularly noticeable with viruses such as the luteoviruses that cannot be obtained as ''pure'' cultures since they can only be transferred by aphid transmission. For instance, an isolate of barley yellow dwarf virus (BYDV) will always comprise a mixture of variants that happen to be similarly transmissible by the vector used for passage. It was shown by Lister and Sward[201] that continued maintenance by vector transmission may lead to the selection of antigenic variants of BYDV identifiable with suitable Mabs but which remain undetected with polyclonal antiserum.

4. **Ease of immunization** — Mice and rats[83] can be immunized with as little as 50 µg of virus. If the virus preparation is contaminated with other viruses or plant antigens, it is nevertheless possible to select Mabs that will react only with the virus of interest.

5. **Selection of high-affinity reagents** — Mabs with a very high affinity constant can be selected which allow highly sensitive immunoassays to be developed [202-204] For instance, it is not unusual for ascitic fluids to be used in ELISA at a dilution of 10^{-6} or 10^{-7}.[125]

In spite of these advantages, it is clear that Mabs also possess certain limitations. The level of specificity and discrimination that can be achieved with an individual Mab may be excessive for the task at hand. Small alterations in the conformation of a viral polypeptide brought about by test conditions (pH, ionic strength, presence of detergents or various additives, adsorption to solid phase, etc.) may alter the epitope to the extent that it is no longer recognized by a Mab. If the antigen preparation used in the assay is not carefully standardized or experimental conditions are slightly varied, the reactivity of Mabs may appear to be erratic. Moreover, small antigenic differences between mutants could be emphasized although they may have no relevance when all other biological and physicochemical properties of the virus remain unchanged.

Another limitation is that each Mab tends to be assay specific, i.e., it usually functions well in only a limited number of serological techniques. Some Mabs lose their activity when conjugated to a label or when they are adsorbed to plastic, while others are not able to precipitate viral antigens. In contrast, the activity of a polyclonal antiserum will be preserved over a wide range of experimental conditions. Finally, Mabs are much more expensive to produce that conventional antisera and the large investment in time and resources that is needed may be prohibitive for many laboratories.

In view of these disadvantages of Mabs, it is clear that polyclonal antisera remain useful reagents for studying plant viruses even today. When information is needed on the overall degree of antigenic relatedness between two viruses, it is in fact preferable to use polyclonal antibodies. An antiserum will reflect the average similarity of all the epitopes of the virus, whereas single Mabs will generally give a distorted view of the overall similarity.[153]

III. SEROLOGICAL TECHNIQUES

A. PRECIPITATION

Precipitation or precipitin reactions derive their name from the visible precipitates that are formed when adequate quantities of antigen and antibodies are allowed to combine. It is the oldest technique used for quantitating immunochemical reactions[205] and it also has been extensively used for analyzing the interaction between antibodies and plant viruses.[33,206,207,327] Precipitation and agglutination are distinguished on the basis of the size of the reacting antigen. Precipitation refers to the insolubilization of macromolecules and virus particles, whereas agglutination describes the clumping of cells or particles of similar size.

1. Precipitin Tests in Liquid Medium

Quantitative precipitin tests in liquid medium are usually performed in small tubes with a volume of about 0.5 ml of each of the two reagents. Twofold dilutions of antiserum and purified antigen are mixed and incubated in a water bath at 25 to 40°C. The highest dilution of antiserum that gives a visible precipitate is known as the antiserum precipitin titer and is recorded after 1 to 2 h of incubation. Reliable antiserum titers are obtained when a virus concentration of about 10 μg/ml is used. The visibility of precipitates is enhanced by holding the tubes over a light box in front of a black background.

Various factors that influence the results of precipitin tests have been discussed by Matthews.[208] An important factor is the size of the antigen. Filamentous virus particles of a length of 500 to 1000 nm require fewer antibody molecules to produce a visible precipitate than smaller isometric virus particles or dissociated protein subunits.[209] As a result, antiserum precipitin titers of 10^{-3} to 10^{-4} are commonly observed with filamentous viruses, whereas precipitin titers with dissociated viral subunits are seldom higher than 10^{-2}.

Liquid precipitin tests can also be performed in single drops (50 μl) of the mixed antigen and antiserum reactants deposited on a slide or on the bottom of a petri dish.[210,211] If glass surfaces are used, they should be rendered hydrophobic by a coat of silicone or 0.1% formvar in chloroform. When dishes are used, drops can be covered with a layer of mineral oil to prevent evaporation. Slides should be kept in a moist chamber. The precipitates can be observed after 30 to 60 min at room temperature under a dark-field microscope at a magnification of 10 to 100 ×. If the drops are kept at 4°C, a longer incubation time (6 h) is needed. A complete grid titration can be performed in a single petri dish. The main advantage of microprecipitin tests in drops is that it is economical in its use of reagents.

Another variation of precipitin tests is the interface ring test. This test is based on the appearance of a ring of precipitation at the interface between superimposed layers of antibody and antigen preparations. Antiserum diluted in 10 to 30% glycerin in saline is placed in the bottom of a small tube and the antigen preparation is carefully layered onto the surface to form a sharp interface. A positive reaction takes the form of a precipitin ring at the interface. Compared to other precipitin tests, the main advantage of the interface ring test is that the relative concentration of the reactants is less critical. The test has been used successfully with many different plant viruses.[33,212,327]

Precipitin tests can be carried out with purified virus preparations or with clarified extracts of infected plant tissue. When viruses occur at very low concentration in their hosts, it may be necessary to concentrate them by ultracentrifugation of clarified plant sap.

The main application of liquid precipitin tests is the characterization of viruses by means of antiserum titers and the detection of serological relationships between large viruses that cannot be studied by immunodiffusion tests. It is customary to express the degree of antigenic relatedness between two viruses by a serological differentiation index (SDI), which corresponds to the average number of twofold dilution steps separating homologous and heterologous titers.[213] The homologous titer refers to the titer of an antiserum with respect to the antigen used for immunizing the animal, while the heterologous titer refers to the titer with respect to another related antigen.

The SDI is a reliable measure of serological relationship only if it represents the average value calculated from a large number of bleedings from different immunized animals. This is because individual antisera obtained from different animals or from the same animal at different times often showed considerable variation in the amount of cross-reactive antibodies they contain.[214] Average SDIs should be determined with antisera to the two viruses that are being compared. SDI values obtained in such reciprocal tests have been shown to agree closely and to correlate with the degree of sequence similarity of the coat proteins of the two viruses.[214] In the case of tymoviruses, the grouping of individual virus species obtained on the basis of reciprocal SDIs[215] was found to agree better with sequence relationships of the individual coat proteins[216,217] than with coat protein composition[218] or cDNA-RNA hybridization tests.[219]

2. Agglutination Tests

In agglutination reactions, either the antigen or the antibody is attached to the surface of red blood cells or of carrier particles such as latex. Because of the size of the cells and latex particles, a visible serological clumping can be induced with lower concentrations of reactants than is necessary for visible precipitation.

In passive hemagglutination, virus particles or antibodies are coupled to erythrocytes by various chemical treatments,[33,220] whereas in the latex test the virus or antibody is simply attached to the latex particles by adsorption.[221] Antibody molecules can also be attached to the latex via an intermediate layer of protein A.[222] The sensitivity of agglutination tests is two to three orders of magnitude higher than precipitin or immunodiffusion tests.[33]

Gelatin particles (3 μm in diameter) made of gelatin and gum arabic of various colors have also been used successfully for detecting plant viruses.[223]

B. IMMUNODIFFUSION TESTS

Immunodiffusion tests are precipitin tests that are carried out in gel instead of free liquid. The most commonly used format is the double diffusion technique in which antigen and antibody diffuse towards each other into a gel which initially contains neither of them. As diffusion progresses the two reactants meet and a band of precipitation occurs at a position where so-called optimal proportions are reached.[33,224] When the reagents are present in optimal proportions in the wells prior to the start of diffusion, the position and width of the formed precipitin line will not change with increasing time of diffusion. The position of the precipitin line with respect to the starting fronts of diffusion will provide information on the size of the antigen.[33] If one of the reactants is initially present in excess of the other, the precipitin band will broaden and move with time toward the reservoir containing the less concentrated reagent. Since the visibility of the precipitin line will then be less than at optimal proportions, antiserum titers will be underestimated under these conditions.

Double diffusion tests can be performed in petri dishes or on microscope slides. The gel consists of 0.5 to 1.5% agar or agarose in a buffer suitable for the antigen being tested. Usually a buffer concentration not higher than 0.1 M is used. In the case of elongated viruses, the electrolyte concentration in the gel can influence the diffusion process and the formation of precipitin lines.[225]

Wells are formed in the gel by positioning templates on the plate before pouring the agar or by using gel cutters after the agar has set. Agar plugs are removed by suction. Different well patterns suitable for various types of analysis have been described.[226] Commonly used patterns consist of a central well of 4 to 7 mm diameter surrounded by six or eight peripheral wells of the same size at a distance of 3 to 6 mm from the central well. After the reactants have been introduced in the wells, diffusion is allowed to proceed for 24 to 72 h, depending on the diffusion rate of the antigen. Evaporation is prevented by keeping the petri dish in a high-humidity incubation box or by pouring a layer of light mineral oil over the gel surface. Precipitin lines are clearly seen by examining the dish against a dark background over a box with a circular light source. Records of precipitin lines can be obtained by simple contact printing onto photographic paper or by using a polaroid camera.

1. Precipitation Patterns

When cylindrical wells and a balanced ratio of reactants are used, the position and curvature of precipitin lines give an indication of the size of the antigen. When the diffusion coefficient of the antigen is of the same order of magnitude as that of the antibody (5×10^{-7} cm^2 s^{-1}), the precipitin bands will be straight and located midway between the two wells. Virus particles have lower diffusion coefficients (0.3 to 1.6×10^{-7} cm^2 s^{-1}) and will form bands near the antigen well and curve around it.[33]

When two antigens diffuse from neighboring wells toward the same antibody source, three different precipitation patterns can develop at the position where the lines meet. These patterns, known as coalescence, spur formation, and crossing of precipitin lines, are usually interpreted in terms of serological identity, relatedness, or nonrelatedness of the two antigens. The main value of immunodiffusion tests lies in the ability of these patterns to provide a visible demonstration of the relationships that exist between antigens. The formation of a spur immediately indicates that the two antigens that are compared possess common as well as distinct epitopes. In view of the lack of sufficient precision of quantitative liquid precipitin tests and of ELISA, it is much more difficult to demonstrate by these tests that two serologically closely related antigens possess a small antigenic difference. Small differences between two closely related viruses are sometimes more easily demonstrated with an heterologous antiserum to a third, more distantly related, virus.[29,227]

It should be stressed that useful information concerning the degree of relationship between two antigens can be obtained only by testing them in double diffusion tests against a single antiserum. The reciprocal test where two antisera prepared against related antigens are tested against only one of the two antigens always gives rise to a pattern of coalescence of lines from which it would be erroneous to conclude that the two antigens are identical.[33,228]

Owing to their length and tendency to aggregate, most filamentous viruses do not diffuse readily into 0.5 to 1.0% agar gel. In order to make elongated particles diffuse in the gel, the particles may be broken by ultrasonication[212,229] or they may be chemically degraded by alkaline buffers, detergents, or reagents such as pyrrolidine or sodium dodecylsulfate.[54,212] Dissociated coat protein subunits diffuse readily and this approach makes it possible to apply immunodiffusion analysis to viruses of any size. Since dissociated subunits of different viruses tend to be more closely related serologically than the corresponding intact virions,[230] additional cross-reactions between viruses will be revealed by this approach (see Section III.C.1).

2. Intragel Cross-Absorption

This method of serological cross-absorption is very convenient for demonstrating the presence of distinct epitopes in related viruses.[29,33] It represents the simplest way of identifying a particular viral serotype.[231,232] In this technique, the cross-reacting antigen used for absorption is allowed to diffuse into the gel from a central well. In so doing, it establishes a concentration gradient in the gel around the well. After 24 to 48 h, the antiserum is allowed to diffuse from the same well, which leads to a ring of precipitation by the cross-reacting antibodies all around the central well. Unadsorbed antibodies will continue to diffuse freely and will form a precipitin band with the homologous antigen diffusion from a neighboring well. The completeness of absorption must be controlled by the absence of a similar band when the cross-absorbing antigen is allowed to diffuse from a neighboring well.[33]

3. Immunoelectrophoresis

This technique combines electrophoresis in a gel medium with immunodiffusion and is a powerful analytical tool for resolving complex mixtures of antigens. It is, however, only rarely used in diagnostic work. The technique and some of its applications in plant virology have been described elsewhere.[33,327] Rocket immunoelectrophoresis is a quantitative assay in which the antigen is allowed to migrate electrophoretically in a gel containing antiserum. It is a useful technique for measuring virus concentration and was used, for instance, to detect 0.025 to 10 μ of virus contained in milligram quantities of leaf tissue.[233,327]

C. ELISA AND OTHER SOLID-PHASE ASSAYS
1. General Principles

Enzyme-linked immunosorbent assays (ELISA) are solid-phase assays in which each successive reactant is immobilized on a plastic surface and the reaction is detected by means of enzyme-labeled antibodies. Because of its greater sensitivity and economic use of reagents, ELISA has replaced precipitin and immunodiffusion assays as the most popular serological test used in plant virology.

ELISA methods can be divided into direct and indirect procedures. In direct procedures, the specific antivirus antibody is itself labeled with an enzyme, while in indirect procedures the enzyme conjugate is an anti-immunoglobulin reagent. The conjugation of an enzyme to anti-virus antibody molecules tends to reduce their affinity and, as a result, conjugated antibodies show a lower degree of serological cross-reactivity with related viral serotypes than unconjugated antibodies. In effect, the conjugated antibody becomes more specific for the homologous serotype used to raise the antibody.

In contrast, in indirect procedures the activity of the cross-reacting antibody which is not labeled is kept intact.[70,234] When direct and indirect ELISA were compared for their ability to detect cross-reactions between related plant viruses, it was found that viruses that differed by an SDI greater than 3 (measured in precipitin tests) could no longer be detected in the direct assay format.[70,232,235] However, viruses that differed by an SDI of 7 could still be detected in the indirect procedure. These comparisons were done by the double antibody sandwich (DAS)-ELISA format in which the wells of the microtiter plate are first coated with virus antibody. The antigen is then captured by the immobilized antibody and is subsequently detected by a labeled virus antibody (direct ELISA) or by an unlabeled antibody followed by an anti-immunoglobulin conjugate. In order to prevent the anti-immunoglobulin conjugate from reacting with the first capturing antibody on the plastic, the first and second anti-virus antibody required in the indirect method must be prepared in different animal species, such as rabbit and chicken. Since chicken immunoglobulins used as first antibody do not cross react with mammalian immunoglobulins,[71] they will not be detected by the goat anti-rabbit conjugate used to reveal the second anti-virus antibody.

Some workers consider the need for two antisera prepared in different species to be a major inconvenience of the indirect DAS-ELISA. The need for antibodies from two animal species can be circumvented by coating the microtiter plates either with $F(ab')_2$ fragments of rabbit immunoglobulins[184,185] or with the C1q component of bovine complement.[236,237] However, neither of these approaches has become widely used, probably because of the additional work that is involved. Since enough purified virus material is usually available, it may in fact be simpler to immunize a rabbit and a chicken rather than to use the more complicated strategies that allow indirect DAS-ELISA to be done with only one virus antiserum.

Another advantage of indirect ELISA is that a simple, commercially available enzyme conjugate (for instance enzyme-labeled goat anti-rabbit immunoglobulin) can be used for detecting any number of different viruses, thus eliminating the need to prepare different conjugates for each virus system. Instead of using an anti-immunoglobulin reagent it is also possible to use a general purpose conjugate of protein A and enzyme.[238]

In many virus-host combinations, crude plant extracts contain enough virus to allow ELISA plates to be coated directly with sap from infected plants. An indirect assay will then provide a simple and rapid means of detecting and identifying viruses in crude plant extracts.[239,240] However, the presence of a first layer of capturing antibody in DAS-ELISA allows the viral antigen present in crude plant sap to become preferentially trapped on the solid phase. It also ensures that the viral antigen becomes trapped in a standardized manner and prevents it from being disrupted or denatured by direct adsorption to the plastic surface. It is well known that proteins become at least partly denatured when they are adsorbed to a layer of plastic during a solid-phase assay.[241-244]

When virus particles are directly adsorbed to the solid phase, usually in the presence of carbonate buffer, pH 9.6, it seems that it is mainly dissociated coat protein subunits that become preferentially adsorbed. This was clearly demonstrated in the case of TMV by a procedure in which ELISA is carried out on electron microscope grids and gold-labeled antibodies are used to visualize the structure of the antigen to which the antibodies bind.[88] It was found that at the virus concentration (1 μg/ml) normally used for coating plates in ELISA, no intact TMV particles became adsorbed to the plastic at pH 9.6. Only protein subunits that became dissociated at this pH were adsorbed and detected by Mabs specific for cryptotopes. A similar preferential adsorption of dissociated coat protein instead of intact virions on microtiter plates is probably responsible for the changes in antigenicity that have been observed with many other viruses such as alfalfa mosaic virus,[89] tobacco etch virus,[40] barley yellow dwarf virus,[245,246] potato virus X,[174] and bean pod mottle virus.[66] Since many of the Mabs raised against plant viruses are specific for neotopes found only on the surface of intact virions, these antibodies will react only in a DAS-ELISA format in which intact virus particles are trapped by a first layer of antibodies.

It has been claimed that when a virus is tested by direct DAS-ELISA using only Mabs as reagents, it is imperative to use as coating and labeled antibody two Mabs that are specific for different epitopes of the virus.[121,245] However, ELISA results obtained with Mabs to tomato mosaic virus,[155] *Odontoglossum* ringspot virus,[123] cucumber mosaic virus,[115] and peanut clump virus[125] clearly demonstrated that the same Mab can be used both as capturing and as biotinylated antibody. Since virions possess a large number of identical epitopes, there are always enough epitopes available on the particles to react with the labeled Mab after some of the epitopes have reacted with the same Mab used as capturing antibody. However, since certain Mabs may lose some or all of their activity after absorption to plastic or after conjugation to a label, tests involving the use of the same antibody in both positions in DAS-ELISA may appear to be inferior simply because the antibody activity is reduced in one or both positions.[31,83] Another design for dual Mab DAS-ELISA which avoids the need to label the second antibody consists in using an IgM Mab as capture antibody and an IgG Mab as second antibody. An anti-γ-chain enzyme conjugate that does not bind to IgM is used to detect antigen-bound IgG.[247]

The precision of ELISA readings makes this type of assay particularly suitable for quantitative measurements,[248] e.g., for estimating virus concentration[249,250] or measuring the degree of serological relationships between viruses.[251-253] SDI values obtained by indirect ELISA have been found to agree with the values measured in precipitin tests.[251] Since ELISA titers are usually much higher than precipitin titers of the same antiserum (usually 10^{-6} insted of 10^{-3}), ELISA is able to detect much more distant serological relationships than precipitin tests. When virus particles are directly adsorbed to the solid phase at pH 9.6, the SDI values measured in ELISA correspond in fact to the degree of antigenic relatedness between the coat protein subunits. Subunits from different viruses tend to be more closely related serologically than the corresponding intact virions. The reason for this is that a larger portion of the polypeptide chain is antigenically expressed in the subunit and the newly exposed surfaces that are instrumental in subunit polymerization tend to be structurally and functionally more conserved. Therefore the extent of cross-reactivity between subunits measured with antigen-coated plates is greater than when virions are compared in antibody-coated plates.[251] For the same reason, Mabs directed to cryptotopes are more likely to reveal antigenic relationships between viruses than Mabs to neotopes.[164]

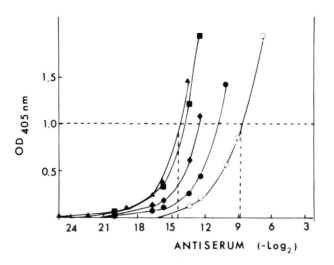

FIGURE 3. Quantitative measurement of antigenic cross-reactivity between several tobamo-viruses assessed by indirect ELISA on antigen-coated plates. ▲, Homologous reaction; ■, ♦, ●, and ○, heterologous reactions with increasingly distantly related viruses. The SDI value separating the homologous (▲) from the most distant heterologous reaction (○) is $14.5 - 8.8 = 5.7$. (Adapted from Jaegle, M. and Van Regenmortel, M. H. V., *J. Virol. Meth.*, 11, 189, 1985.)

When ELISA is used to measure distant serological relationships, the concentration of the reactants should be adjusted to allow the most sensitive detection of antibody. A grid titration of antiserum vs. homologous and heterologous antigens is used to select suitable conditions for the assay. The concentration of enzyme conjugate and the substrate hydrolysis time must be chosen so as to give a rapidly increasing absorbance curve with no trace of a plateau.[251] SDI values are calculated by comparing the antiserum dilutions that lead to the same absorbance (e.g., 0.5 or 1.0) for the homologous and heterologous antigens (Figure 3).

The maximum sensitivity of antigen detection by ELISA is of the order of 1 ng/ml. As a result, ELISA procedures have been found particularly suitable for detecting plant viruses in their insect vectors.[254-259] It was possible, for instance, to demonstrate by ELISA that maize stripe virus multiplies in its insect vector.[260]

2. ELISA Procedures

ELISA is usually performed in 96-well polystyrene or polyvinyl microtiter plates. The first coating step consists in incubating antigen or antibody in the wells, usually for 2 to 3 h at 37°C or for 16 h at 4°C. If unsatisfactory results are obtained with the pH 9.6 carbonate buffer, a variety of other buffers can be used.[261-263] After each subsequent step of reagent addition, the wells of the microtiter plate are washed with a buffer such as 10 mM phosphate, 150

Table 3. ELISA Procedures Used in Plant Virology

	Ref.
1. Ab^R,Ag,AB^R-E	190
2. Ag,Ab^R-E	—
3. Ag,AB^R,anti-R^G-E	234
4. Ab^C,Ag,Ab^R,anti-R^G-E	70
5. $F(ab')_2^R,Ag,Ab^R$,anti-Fc-E	185
6. $F(ab')_2^R,Ag,Ab^R$,PA-E	184
7. $Ab^C,Ag,(Ab^R$,anti-R^G-E)	274
8. $F(ab')_2^R,Ag,(Ab^R$,anti-Fc-E)	275
9. Ab^C,Ag,Ab^M,anti-M^R,anti-R^G-E	87
10. PA,Ab^R,Ag,Ab^R,PA-E	238
11. $Ab^{C/M},Ag,Ab^{R/M}$-biot,Avidin-E	195

Key: Ab, antibody; Ag, antigen; R, rabbit; E, enzyme label; G, goat; C, chicken; M, mouse; PA, protein A; biot, biotin.

mM NaCl, 2.7 mM KCl, 0.05% Tween 20, pH 7.4 (PBST), to remove unbound molecules. After adsorption of the first layer of reagent on the plate, remaining sites on the plastic are saturated with 1 to 2% BSA, ovalbumin, or gelatin in PBST.

Sometimes, this blocking step is incomplete and nonspecific binding of subsequent reagents used in the assay can occur. A more efficient medium for saturating the plates is 1% defatted milk incubated for 1 h at 37°C. Defatted milk was found to be particularly useful to prevent the nonspecific reactions that are prevalent when high concentrations of antigen (1 to 10 μg/ml) or antibody (for instance Mab culture supernatant containing 0.1 to 1 mg/ml protein) are used in the assay.[264] Such nonspecific reactions were found to be responsible for the alleged serological cross-reactivity reported to occur between unrelated plant viruses.[265] When milk instead of BSA was used as blocking agent these spurious cross-reactions were abolished.[264] The alleged cross-reaction between TMV and ribulose-1.5-biphosphate carboxylase[266] was also found to be artifactual and was abolished by using skimmed milk as blocking agent.[267]

It is known that electrostatic interactions between antigen and antibody can give rise to nonspecific binding, especially when serological reactants are used at relatively high concentration or when the antigen is fairly basic or contains basic domains.[268] The polyanion heparin was found to be able to abolish the electrostatic binding of antibodies to the basic domain of viral subunits.[269] In sensitive immunoassays, such as ELISA or immunoblotting, high reagent concentration should be avoided. In most cases such high concentrations are superfluous anyway, as the very sensitivity of these techniques allows genuine, specific reactions to be observed at low reagent concentrations.

A wide variety of ELISA procedures have been used in plant virology.[182,270,327] Eleven of the most useful procedures are listed in Table 3 and will be briefly described. It should be noted that the incubation times, the temperature, and the reagent concentrations can be varied in different systems and that optimal conditions should be established in each particular case.[87,255,271,272] Recently, it was shown that the use of antibody-coated magnetic beads could reduce the time needed to detect plant viruses by ELISA to 30 to 45 min.[273]

Procedure 1: *Direct ELISA procedure using antibody-coated plates*[182,190]

The most commonly used form of ELISA in plant virology is the double antibody sandwich (DAS) format. In this method, the wells are first coated with antivirus immunoglobulins and the virus in the test sample is then captured by the adsorbed antibody. The presence of virus is revealed by an enzyme-labeled antivirus immunoglobulin conjugate.

1. Microtiter plates are coated by incubation of 200 μl of the immunoglobulins (1 to 10 μg/ml) purified from antivirus antiserum and diluted in carbonate buffer, pH 9.6 (15 mM Na$_2$CO$_3$; 35 mM NaHCO$_3$; 0.2 g/l NaN$_3$). Incubation time: 2 to 6 h at 37°C.
2. Three washing steps with phosphate buffered saline, pH 7.4, containing 0.05% Tween-20 (PBST).
3. Blocking of plastic surface with 1 to 2% BSA in PBST (this step may not be necessary in all systems).
4. Incubation with 200 μl of purified viral antigen (0.1 to 2 μg/ml) or crude plant extract diluted in PBST, for 2 h at 37°C or overnight at 6°C.
5. Three washing steps with PBST.
6. Incubation with 200 μl enzyme-labeled antivirus immunoglobulin (1/200 to 1/2000) in PBST, 3 to 4 h at 37°C.
7. Three washing steps with PBST.
8. Incubation at room temperature with 200 μl of the enzyme substrate (30 min to 2 h). When alkaline phosphatase-labeled globulins are used, the substrate is 1 mg/ml *p*-nitrophenyl phosphate in diethanolamine buffer, pH 9.8. Absorbance at 405 nm is read after 30 min to 3 h of incubation of the substrate. It is customary to consider the results positive if the absorbance is twice that found with healthy controls or if it is 2 to 3 standard deviation units higher than the mean of negative control samples.

Procedure 2: *Direct ELISA procedure using antigen-coated plates*

Microtiter plates are coated by incubation of 200 μl of virus antigen (1 to 50 μg/ml purified virus or plant sap diluted 1/2 to 1/100) in carbonate buffer, pH 9.6, 2 to 3 h at 37°C. After washing, the antivirus conjugate and the enzyme substrate are incubated as described above.

Procedure 3: *Indirect ELISA procedure using antigen-coated plates*[234]
1. Microtiter plates are coated by incubation of 200 µl of purified virus preparation (1 to 50 µg/ml) or diluted plant sap in carbonate buffer, 2 h at 37°C.
2. Three washing steps with PBST.
3. Incubation with virus antiserum (1/500 to 1/5000) in PBST containing 1 to 2% polyvinylpyrrolidone (PBST-PVP), 18 h at 4°C.
4. Three washing steps with PBST.
5. Incubation with enzyme-labeled goat anti-rabbit-IgG immunoglobulin (1/1000 to 1/5000) in PBST-PVP, 4 h at 37°C.
6. Three washing steps with PBST.
7. Subsequent steps as in Procedure 1.

Procedure 4: *Indirect ELISA using antibody-coated plates (Indirect DAS procedure)*[70]
In this test, virus antiserum prepared in two animal species is required. A good combination is rabbit and chicken antibodies, since avian immunoglobulins do not cross react with mammalian Ig. Antivirus chicken immunoglobulins are readily obtained from the eggs of immunized hens.[67]

1. Coating of microtiter plates with antivirus chicken immunoglobulin (1 to 20 µg/ml) diluted in carbonate buffer (see Procedure 1).
2. Three washing steps with PBST.
3. Blocking of plastic surface by incubation with 200 µl of 1 to 2% BSA in PBST, 30 min to 1 h at 37°C. This step may not be necessary in all systems.
4. Washing steps.
5. Incubation with purified virus antigen (0.1 to 2 µg/ml) or diluted plant sap in PBST, 2 h at 37°C.
6. Washing steps.
7. Incubation with antivirus rabbit immunoglobulins (50 ng to 1 µg/ml) or diluted rabbit antiserum (1/1000 to 1/10,000) in 1% BSA for 2 h at 37°C.
8. Washing steps.
9. Incubation with goat anti-rabbit IgG enzyme conjugate. Washing steps and incubation with substrate.

Procedure 5: *Indirect ELISA using plates coated with specific F(ab')₂*[185]
The assay requires only a single virus-specific antiserum. $F(ab')_2$ fragments, prepared as described in Section II.C.6, are used to coat the plates and specific intact globulin molecules are used as second antibody.

The $F(ab')_2$ assay is similar to the indirect DAS Procedure 4. Microtiter plates are coated with $F(ab')_2$ fragment from rabbit IgG in carbonate buffer, pH 9.6, instead of with chicken immunoglobulins. After incubation with antigen, the virus-specific intact IgG is added to react with captured antigen. An enzyme conjugate reactive only with the Fc portion of rabbit IgG is used to visualize the reaction.

Donkey antiserum to whole rabbit immunoglobulin G can be rendered specific to the Fc region of rabbit IgG by treatment with an immunosorbent gel consisting of F(ab')$_2$ fragments of rabbit IgG with BSA as a carrier protein. The anti-Fc immunoglobulin is conjugated to the enzyme as described in Section II.D.1.

Procedure 6[184]

This method is based on the ability of protein A to bind only to Fc fragments and not to the Fab portion of IgG. The assay is similar to Procedure 5, the only difference being that the enzyme conjugate is replaced by a protein A-enzyme conjugate. Protein A (0.5 mg) is conjugated by the periodate oxidation method[182] simultaneously to 0.75 mg of HRP and to 1 mg of BSA used as a carrier protein.

Procedures 7 and 8[274,275]

In these shortened versions of indirect ELISA, the specific antivirus antibody and the enzyme conjugate are incubated simultaneously at room temperature and the mixture is applied to the plates on which the antigen has been trapped by antivirus chicken antibodies (Procedure 7) or rabbit F(ab')$_2$ (Procedure 8). The assay is less sensitive when the serum containing the detecting antibodies is present in the preincubated detection mixture at a low dilution. However, when high serum dilutions are used, the shortened procedures are sometimes more sensitive than the corresponding conventional methods.

Variations of short procedures have been described,[275,276] where either virus together with specific antibody or virus antibody together with anti-immunoglobulin antibody are added at the same time on the microtiter plate. Such simultaneous incubation steps reduce the time needed to complete the assay but in general diminish the sensitivity of the assay. If the antigen concentration is not limiting, the gain in time may justify the adoption of such procedures.

Procedure 9[87]

As in Procedure 4, this procedure uses avian antibody as capturing antigen. The second antibody is a Mab which is detected by a rabbit anti-mouse antiserum (incubated 2 h at 37°C) followed by goat anti-rabbit enzyme conjugate (2 h at 37°C). The additional layer of goat anti-rabbit conjugate provides an amplification of the reaction.

Procedure 10[238]

This method uses protein A at two positions in a sandwich assay. A first layer of protein A is adsorbed to the plate to trap the coating antibody layer.

After incubation with virus and a second antibody, a final layer of enzyme-conjugated protien A is used to detect the second antibody.

The protein A coating is done in carbonate buffer, pH 9.6, and the other reactants are diluted in PBST. Optimal incubation conditions must be established empirically for each system. The main advantage of this assay is that it requires only a single virus antiserum. However, difficulties are sometimes experienced in reducing the nonspecific background while maintaining sufficient assay sensitivity.

Procedure 11: *Direct biotin-avidin ELISA*[195]

When compared to standard ELISA procedures that use antibodies labeled with enzyme, the biotin-avidin system increases the assay sensitivity and allows a wider range of related viral serotypes to be detected. The attachment of biotin molecules to viral antibody is less detrimental to antibody activity than is conjugation to an enzyme. This method is particularly suitable for dual Mab assays.[123,155,277]

1.   Coating of plates with 2 μg/ml rabbit immunoglobulins in carbonate buffer (3 h at 37°C).
2.   Washing and blocking with 1% BSA in PBST.
3.   Incubation with antigen and washings as in ELISA procedures above.
4.   Incubation with biotin-labeled antibodies (usually about 1 μg/ml) for 2 h.
5.   Washing and incubation for 2 h at 37°C with avidin labeled with alkaline phosphatase (diluted 1/4000 in 1% BSA in PBST).
6.   Incubation with enzyme substrate.

Additional techniques based on signal amplification for instance with fluorogenic substrates,[278] have been discussed by Cooper & Edwards.[279]

3. Radioimmunoassay

Radioimmunoassay (RIA) procedures in which the antibody is labeled with a radioactive isotope instead of with an enzyme have been used only rarely in plant virology.[33] Solid-phase RIA using microtiter plates[280,281] or plastic beads[121] has been used occasionally for detecting plant viruses, but since it requires expensive counting equipment and uses potentially hazardous radioactive materials, it has never been popular with plant virologists. In fact, RIA offers no major advantage over ELISA procedures.

4. Time-Resolved Fluoroimmunoassay

A time-resolved fluoroimmunoassay (TRFIA) technique based on the measurement of the fluorescence of antibody-europium conjugates has been developed which completely avoids the nonspecific background fluorescence originating from the sample.[196]

Some lanthanides, especially europium, form highly fluorescent chelates with many different organic ligands. The sensitized fluorescence results from the ligand absorbing light, the energy of which is transferred to the chelated metal ion. The metal ion emits the energy as narrow-banded line-type fluorescence with a long Stokes shift (over 250 nm) and a long fluorescence decay time (0.1 to 1 msec).[282] Measurement of the europium conjugate is started 0.4 ms after excitation. After this delay period the nonspecific background fluorescence has decayed and only the fluorescence coming from the europium chelate is measured. The TRFIA gives a higher specific activity than conventional radioactive labels but without the hazards and drawbacks of radioactivity.[283]

Assays are performed in the wells of polystyrene microtitration strips previously coated with antibodies for 18 h at 37°C in carbonate buffer, pH 9.6, and blocked as in ELISA. Prior to use the strips are washed twice with TBS-T (20 mM Tris-HCl, 150 mM NaCl, pH 7.5, containing 0.05% Tween 20).

Two assay procedures have been described:

1. One-step incubation procedure — Antigen (50 μl) and the europium conjugate are incubated together for 1 h at 37°C in polystyrene microtiter strips previously coated with Mabs or IgG. The Eu conjugate is suspended in a 0.05 M Tris-HCl buffer, pH 7.7, containing 0.9% NaCl, 0.5% gelatin, 20 μM diethylenetriamine N^1,N^1,N^2,N^3-pentaacetic acid, 0.01% Tween 40, and 0.05% NaN$_3$ in 0.05 M Tris HCl buffer, pH 7.7. The strips are incubated for 1 h at 37°C and washed with 20 mM Tris-HCl buffer, pH 7.5, containing 150 mM NaCl. An enhancement solution containing 15 μM 2-naphtoyltrifluoroacetone, 50 μM tri-n-octylphosphine oxide, and 0.1% Triton X-100 in 0.1 M acetate phthalate buffer, pH 3.2, is then added. After shaking for 10 min the fluorescence in the wells is measured in an Arcus 1230 Fluorometer. Results are presented as counts per second.
2. Two-step incubation procedure, in which the antigen (100 μl) is first incubated for 1 h at 37°C in antibody-coated strips. After washing, the Eu conjugate is incubated also for 1 h at 37°C.

TRFIA has been used to detect potato viruses in leaf and tuber extracts.[197,283,284] In the case of purified potato virus X, TRFIA was found to be 5 to 100 times more sensitive than conventional DAS-ELISA. The lowest concentration of potato virus M detectable in TRFIA was found to be 0.5 ng/ml compared to 10 ng/ml in ELISA. Two different potato viruses have been detected simultaneously by using appropriate Mabs labeled with two lanthanides, europium and samarium, that have different emission wavelengths.[135]

D. IMMUNOBLOTTING

Recent reviews of immunoblotting techniques are available[285,286] and their application in plant virology has been described by Koenig and Burgermeister.[287,327] Immunoblotting techniques comprise such methods as electroblot immunoassay, dot blot immunoassay, and tissue blotting.

1. Electroblot Immunoassay

In this method the components of a mixture are first fractionated by gel electrophoresis before they are electroblotted onto nitrocellulose paper and revealed by labeled antibodies.[288-290] Purified virus or virus in plant extract is denatured during 10 min at 95°C in one volume of disruption buffer (125 mM Tris-HCl, pH 6.8, 10% SDS, 10% β-mercaptoethanol, 15% glycerol) and fractionated by SDS-PAGE.[291]

Comparative levels of reactivity with antibody can be assessed by incubation with serially diluted antiserum. The method has been used, for instance, to calculate SDI values for potyvirus proteins.[292]

Electroblotting is performed by the method of Towbin et al.[293] Resolving gels are laid upon wetted nitrocellulose sheets (0.45-μm pore, Schleicher et Schuell), and sandwiched between wetted filter paper sheets (Whatman 3MM). The gel sandwiches are laid upon Scotch-Brite scouring pads and the transfer of protein to nitrocellulose is done in transfer buffer (25 mM Tris, 112 mM glycine, 20% [v/v] methanol, pH 8.3). The anode is placed nearest the nitrocellulose and a current of 0.6 to 1 A for 4 to 10 h is used.

After electroblotting, blots are soaked 3 to 4 h at 37°C or overnight at 22°C in 10 mM Tris-HCl/saline, pH 7.4, 1% BSA (Tris-saline-BSA buffer) to saturate free protein-binding sites. Rabbit antisera are diluted in the same buffer and incubated with the blots in individual closed containers on a shaker at 22°C for 1 to 2 h. Blots are washed with buffer for 10 min or a shaker four times and are incubated with anti-rabbit HRP conjugate on a shaker for 1 to 2 h at 22°C. After washing, the enzyme substrate (25 μg/ml *o*-dianisidine, 0.01% H_2O_2, 10 mM Tris-HCl, pH 7.4) is added and kept at 22°C for 30 min. Color development is stopped by washing in tap water.

2. Dot Blot Immunoassay

In dot blot immunoassay, purified antigen or crude sap from infected plants (2 μl) is applied directly to a nitrocellulose membrane and air dried without prior electrophoretic separation.[294,295] After saturation with BSA, the blots are incubated successively with specific anti-virus antibody, an anti-immunoglobulin enzyme conjugate, and substrate, as described for the electroblot immunoassay. The technique has been used successfully with both polyclonal and monoclonal antibodies.[296-298] Because of the small volumes that are used, as little as 0.5 pg of virus can be detected.[299]

3. Tissue Blotting

This technique is similar to dot blot but does not require mechanical disruption of tissues for the extraction of antigen. The application of samples on the nitrocellulose occurs simply by pressing the freshly cut tissue surface on nitrocellulose membrane.[300]

E. IMMUNOELECTRON MICROSCOPY

As a separate chapter of this text is devoted to electron microscopy techniques, only some brief remarks will be made here. The most useful method of immunoelectron microscopy for detecting and identifying plant viruses is the technique known as immunosorbent electron microscopy (ISEM). In this method, an electron microscope grid is coated for 15 min with antiserum diluted at least 1/1000, and after rinsing, the grid is incubated for 15 to 60 min with crude sap from infected plants. Antibodies adsorbed to the grid specifically trap virus particles while the absorbed serum constitutents inhibit the binding of plant material. In effect, the procedure results in selective immunopurification of the virus with concomitant improved visualization on the grid.[301]

This approach was introduced by Derrick,[302] who called the technique "serologically specific electron microscopy". However, since all forms of electron microscopy that involve the use of antibodies can be said to be "serologically specific" (not only those relying on antibody-coated grids), it seems preferable to use the term immunosorbent electron microscopy for this technique.[303] The sensitivity of virus detection by ISEM is very similar to that achieved by ELISA. It was found, for instance, that 1 to 5 particles of turnip yellow mosaic virus were visible per micrograph when a 50-μl drop containing 5 ng/ml virus (5×10^7 virions) was deposited on a grid coated with a 10^{-4} dilution of antiserum.[304] When the grid was first coated with a layer of protein A, the sensitivity of virus detection was slightly improved.[305-307]

Depending on the stability of the virus, it may be necessary to test buffers of different pH and ionic strength for optimal results.[308,309] Improved detection can be obtained by combining ISEM with a so-called decoration step, whereby the virus particles are coated with a layer of antibody molecules.[310] Further improvement can be achieved by a second decoration with a gold-labeled antiglobulin antibody, provided the first antibody used to coat the grid was obtained from a different animal species than the second antibody used for decoration.[88] It is also possible to use F(ab')$_2$ fragments for coating the grids and to reveal the second antibody with gold-labeled protein A.[311] ISEM using gold-labeled antibodies has been found useful for the screening of antivirus Mabs during hybridoma production.[130]

By combining ELISA with immunoelectron microscopy using gold-labeled antibodies, it is possible to visualize which type of viral antigen is reacting in ELISA. It so happens that the usual electron microscope grids fit neatly into the wells of microtiter plates. It is feasible, therefore, to carry out the first adsorption step of ELISA on the grid, and the subsequent steps with the grid placed in the microtiter well. Using duplicate grids, the one grid is incubated with enzyme-labeled antibodies and read in an ELISA reader (keeping the grid in the well does not interfere with the reading) while the other grid is incubated with gold-labeled antibodies and examined by electron microscopy.[88] Using this approach, it could be shown that at the virus concentration (1 µg/ml) normally used to coat ELISA plates, no intact TMV particles became adsorbed to the solid phase at pH 9.6. Unexpectedly, only dissociated viral subunits became preferentially adsorbed and could be detected by Mabs specific for cryptotopes. This type of particle degradation during adsorption to the solid phase is no doubt responsible for the inability of antineotope Mabs to react with the viral antigen present on so-called virus-coated plates, since the adsorbed antigen is actually dissociated viral protein.[87,88,174,245]

When TMV particles were trapped on grids coated with antibody, ISEM showed that antineotope Mabs decorated the entire surface of the virus, whereas all antimetatope Mabs reacted only with one of the two extremities of viral rods.[88] To establish if the antimetatope antibodies were binding to the extremity containing the 3' or 5' end of the RNA, virus particles were partially uncoated by the action of 6 *M* urea, a procedure that uncovers RNA tails at the 5' terminus. It was found that all antimetatope antibodies recognized the extremity of the particle containing the 5' end of the RNA and thus that they reacted with the bottom face of the coat protein subunit containing the two α-helices made up of residues 73 to 89 and 111 to 135.[15] The same antibodies were also shown to bind to both ends of the viral protein aggregates known as stacked disks,[312] indicating that the same face of the subunit was exposed at both ends of the stacked disks.[15] This implies that the disks cannot be converted directly into helical rods,[313] as previously believed.[314]

The reason why so many of the Mabs obtained from a single fusion experiment[25] recognized the same TMV subunit surface was also explained by these results. Since the mice used for hybridoma production had been immunized with a concentrated TMV protein preparation which must have contained stacked disks, the presence of the same subunit surface on both ends of the disks explains the predominance of antimetatope antibodies recognizing this particular surface.[15]

The phenomenon of antimetatope antibodies reacting with only one extremity of TMV rods appears not to be general since, with beet necrotic yellow vein virus, Mabs have been obtained that react with epitopes on both extremities of the particles.[198]

Additional information on immunoelectron microscopy techniques is available in several reviews.[301,310,315]

F. USE OF DIFFERENT TECHNIQUES IN DIAGNOSIS OF PLANT VIRUS DISEASES

For more than a decade, ELISA has been the most widely used method for the diagnosis of plant virus diseases.[316] ELISA procedures using Mabs have been used very successfully in virus-free certification programs for potatoes[133] and for eliminating viruses from nursery stocks in the case of vegetatively propagated fruit crops.[199] Whenever individual Mabs recognize only a restricted number of serotypes of a particular virus, this limitation can be overcome by using a cocktail of different Mabs.[317] Although ELISA is less sensitive than nucleic acid hybridization methods,[318,319] it is much more suitable for routine, large-scale testing of field samples. Biological assays on sensitive indicator plants are also more sensitive than ELISA, but they usually take several days or weeks to complete and are also less amenable to large-scale indexing.[320]

The level of viral antigen concentration found in infected plant material is usually sufficient to allow testing of grouped samples followed, if necessary, by retesting of the individuals in those groups found to be infected. When very large numbers of samples have to be indexed for the presence of virus, it may be necessary to store them prior to testing by ELISA. Although several storage procedures have been used,[33,321,322] one of the most effective ones is to homogenize the samples in 1% K_2HPO_4, 0.1% Na_2SO_3 (1 g leaf in 1 ml buffer) and storing them at $-20°C$.[323] Storing large numbers of samples as leaf pieces is inconvenient, since they must then all be homogenized on the same day at the time of the test.

With virus infections of fruit trees, the sampling conditions, the type of material tested (young or mature leaves, buds, or flowers), and the testing period greatly influence the outcome of serological indexing tests.[322,324] The ELISA methods most suitable for detecting viruses in ornamental plants have been reviewed by Luisoni[325] while the procedure applicable for testing seed samples have been discussed by Lange.[326]

Since control test samples from healthy plants will always give a certain level of background reading in ELISA, it is necessary to set a threshold value of absorbance for regarding samples as negative or positive. A common practice is to regard samples as infected if their absorbance value exceeds the mean value of a series of negative controls by 2 to 3 standard deviations.

After ELISA, the second most popular serological technique for diagnosis is immunodiffusion. The different patterns of precipitin lines obtainable in this technique provide a visible discrimination between homologous and heterologous antigens and greatly facilitates the comparison between different viral strains and species.[227] The intragel cross-absorption technique is especially useful for demonstrating the presence of small antigenic differences between individual viral serotypes.[232] A third technique that is being used increasingly is dot blot immunoassay which is well adapted to large-scale testing of viruses in crude plant sap.[294,316]

When the number of samples to be examined in small, ISEM is also very useful, since it combines serological specificity with the high sensitivity of electron microscopy. However, the need for an expensive instrument limits the application of ISEM in diagnostic work to a small number of well-equipped laboratories.

REFERENCES

1. **Sela, M.,** Antigenicity: some molecular aspects, *Science,* 166, 1365, 1969.
2. **Atassi, M. Z. and Smith, J. A.,** A proposal for the nomenclature of antigenic sites in peptides and proteins, *Immunochemistry,* 15, 609, 1978.
3. **Minor, P. D.,** Humoral immune response to poliovirus, in *Immune Responses, Virus Infections and Disease,* Dimmock, N. J. and Minor, P. D., Eds., IRL Press, Oxford, 1989, 35.
4. **Acharya, R., Fry, E., Stuart, D., Fox, G., Rowlands, D., and Brown, F.,** The three-dimensional structure of foot-and-mouth disease virus at 2.9 Å resolution, *Nature,* 337, 709, 1989.
5. **Jemmerson, R.,** Antigenicity and native structure of globular proteins: low frequency of peptide reactive antibodies, *Proc. Natl. Acad. Sci. U.S.A.,* 84, 9180, 1987.
6. **Laver, W. G., Air, G. M., Webster, R. G., and Smith-Gill, S. J.,** Epitopes on protein antigens: misconceptions and realities, *Cell,* 61, 553, 1990.
7. **Amit, A. G., Mariuzza, R. A., Phillips, S. E. V., and Poljak, R. J.,** Three-dimensional structure of an antigen-antibody complex at 2.8 Å resolution, *Science,* 233, 747, 1986.
8. **Sheriff, S., Silverton, E. W., Padlan, E. A., Cohen, G. H., Smith-Gill, S., Finzel, B. C., and Davies, D. R.,** Three-dimensional structure of an antibody-antigen complex, *Proc. Natl. Acad. Sci. U.S.A.,* 84, 8075, 1987.
9. **Colman, P. M., Air, G. M., Webster, R. G., Varghese, J. N., Baker, A. T., Lentz, M. R., Tulloch, P. A., and Laver, W. G.,** How antibodies recognize virus proteins, *Immunol. Today,* 8, 323, 1987.
10. **Padlan, E. A., Silverton, E. W., Sheriff, S., and Cohen, G. H.,** Structure of an antibody-antigen complex: crystal structure of the HyHEL-10 Fab-lysozyme complex, *Proc. Natl. Acad. Sci. U.S.A.,* 86, 5938, 1989.
11. **Tulip, W. R., Varghese, J. N., Webster, R. G., Air, G. M., Laver, W. G., and Colman, P. M.,** Crystal structures of neuraminidase-antibody complexes, *Cold Spring Harbor Symp. Quant. Biol.,* 54, 257, 1989.
12. **Geysen, H. M., Mason, T. J., and Rodda, S. J.,** Cognitive features of continuous antigenic determinants, *J. Mol. Recognition,* 1, 32, 1988.
13. **Jerne, N. K.,** Immunological speculations, *Annu. Rev. Microbiol.,* 14, 341, 1960.
14. **Van Regenmortel, M. H. V.,** Plant virus serology, *Adv. Virus Res.,* 12, 207, 1966.
15. **Dore, L., Ruhlmann, P., Oudet, P., Cahoon, M., Caspar, D. L. D., and Van Regenmortel, M. H. V.,** Polarity of binding of monoclonal antibodies to tobacco mosaic virus rods and stacked disks, *Virology,* 176, 25, 1990.
16. **Van Regenmortel, M. H. V.,** The structure of viral epitopes, in *Immunochemistry of Viruses. II. The Basis for Serodiagnosis and Vaccines,* Van Regenmortel, M. H. V. and Neurath, A. R., Eds., Elsevier, Amsterdam, 1990, 1.
17. **Benjamini, E.,** Immunochemistry of the tobacco mosaic virus protein, in *Immunochemistry of Proteins,* Atassi, M. Z., Ed., Plenum Press, New York, 1977, 265.

18. **Van Regenmortel, M. H. V.**, Tobacco mosaic virus. Antigenic structure, in *The Plant Viruses,* Vol. 2, *The Rod-Shaped Plant Viruses,* Van Regenmortel, M. H. V. and Fraenkel-Conrat, H., Eds., Plenum Press, New York, 1986, 79.
19. **Van Regenmortel, M. H. V. and Hardie, G.**, Immunochemical studies of tobacco mosaic virus. II. Univalent and monogamous bivalent binding of the IgG antibody, *Immunochemistry,* 13, 503, 1976.
20. **Van Regenmortel, M. H. V. and Lelarge, N.**, The antigenic specificity of different states of aggregation of tobacco mosaic virus protein, *Virology,* 52, 89, 1973.
21. **Anderer, F. A.**, Recent studies on the structure of TMV, *Adv. Protein Chem.,* 18, 1, 1963.
22. **Benjamini, E., Young, J. D., Shimizu, M., and Leung, C. Y.**, Immunochemical studies on the tobacco mosaic virus protein. I. The immunological relationship of the tryptic peptides of the tobacco mosaic virus protein to the whole protein, *Biochemistry,* 3, 1115, 1964.
23. **De Milton, L. R. C. and Van Regenmortel, M. H. V.**, Immunochemical studies of tobacco mosaic virus. III. Demonstration of five antigenic regions in the protein subunit, *Mol. Immunol.,* 16, 179, 1979.
24. **Altschuh, D., Hartman, D., Reinbolt, J., and Van Regenmortel, M. H. V.**, Immunochemical studies of tobacco mosaic virus. V. Localization of four epitopes in the protein subunit by inhibition tests with synthetic peptides and cleavage peptides from three strains, *Mol. Immunol.,* 20, 271, 1983.
25. **Al Moudallal, Z., Briand, J. P., and Van Regenmortel, M. H. V.**, A major part of the polypeptide chain of tobacco mosaic virus protein is antigenic, *EMBO J.,* 4, 1231, 1985.
26. **Trifilieff, E., Dubs, M. C., and Van Regenmortel, M. H. V.**, Antigenic cross-reactivity potential of synthetic peptides immobilized on polyethylene rods, *Mol. Immunol.,* 28, 889, 1991.
27. **Westhof, E., Altschuh, D., Moras, D., Bloomer, A. C., Mondragon, A., Klug, A., and Van Regenmortel, M. H. V.**, Correlation between segmental mobility and the location and antigenic determinants in proteins, *Nature,* 311, 123, 1984.
28. **Van Regenmortel, M. H. V., Altschuh, D., and Klug, A.**, Influence of local structure on the location of antigenic determinants in tobacco mosaic virus protein. Synthetic peptides as antigens, *Ciba Found. Symp.,* 119, 76, 1986.
29. **Van Regenmortel, M. H. V.**, Serological studies on naturally occurring strains and chemically induced mutants of tobacco mosaic virus, *Virology,* 31, 467, 1967.
30. **De Milton, L. R. C., Milton, S. C. F., Von Wechmar, M. V., and Van Regenmortel, M. H. V.**, Immunochemical studies of tobacco mosaic virus. IV. Influence of single amino acid exchanges on the antigenic activity of mutant coat proteins and peptides, *Mol. Immunol.,* 17, 1205, 1980.
31. **Van Regenmortel, M. H. V.**, The potential for using monoclonal antibodies in the detection of plant viruses, in *Developments and Applications in Virus Testing,* Jones, R. A. C. and Torrance, L., Eds., Association of Applied Biologists, Wellesbourne, England, 1986, 89.
32. **Van Regenmortel, M. H. V.**, Molecular dissection of antigens by monoclonal antibodies, in *Hybridoma Technology in Agricultural and Veterinary Research,* Stern, N. J. and Gamble, H. R., Eds., Rowan and Allanheld, Totowa, NJ, 1984, 43.
33. **Van Regenmortel, M. H. V.**, *Serology and Immunochemistry of Plant Viruses,* Academic Press, New York, 1982.
34. **Al Moudallal, Z., Briand, J. P., and Van Regenmortel, M. H. V.**, Monoclonal antibodies as probes of the antigenic structure of tobacco mosaic virus, *EMBO J.,* 1, 1005, 1982.
35. **Altschuh, D., Al Moudallal, Z., Briand, J. P., and Van Regenmortel, M. H. V.**, Immunochemical studies of tobacco mosaic virus. VI. Attempts to localize viral epitopes with monoclonal antibodies, *Mol. Immunol.,* 22, 329, 1985.

36. **Brandes, J. and Bercks, R.,** Gross morphology and serology as a basis for the classification of elongated plant viruses, *Adv. Virus Res.,* 11, 1, 1965.
37. **Moghal, S. M. and Francki, R. I. B.,** Towards a system for the identification and classification of potyviruses. I. Serology and amino acid composition of six distinct viruses, *Virology,* 73, 350, 1976.
38. **Ward, C. W. and Shukla, D. D.,** Taxonomy of potyviruses: current problems and some solutions, *Intervirology,* 32, 269, 1991.
39. **Allison, R. F., Dougherty, W. G., Parks, T. D., Willis, L., Johnston, R. E., Kelly, M. E., and Armstrong, F. B.,** Biochemical analysis of the capsid protein gene and capsid protein of tobacco etch virus: N-terminal amino acids are located on the virion's surface, *Virology,* 147, 309, 1985.
40. **Dougherty, W. G., Willis, L., and Johnston, R. E.,** Topographic analysis of tobacco etch virus capsid protein epitopes, *Virology,* 144, 66, 1985.
41. **Shukla, D. D., Strike, P. M., Tracy, S. L., Gough, K. H., and Ward, C. W.,** The N and C termini of the coat proteins of potyviruses are surface-located and the N termini contains the major virus-specific epitopes, *J. Gen. Virol.,* 69, 1497, 1988.
42. **Geysen, H. M., Rodda, S. J., Mason, T. J., Tribbick, G., and Schoofs, P. G.,** Strategies for epitope analysis using peptide synthesis, *J. Immunol. Meth.,* 102, 259, 1987.
43. **Shukla, D. D., Tribbick, G., Mason, T. J., Hewish, D. R., Geysen, H. M., and Ward, C. W.,** Localization of virus-specific and group-specific epitopes of plant potyviruses by systematic immunochemical analysis of overlapping peptide fragments, *Proc. Natl. Acad. Sci. U.S.A.,* 86, 8192, 1989.
44. **Gugerli, P. and Fries, P.,** Characterization of monoclonal antibodies to potato virus Y and their use for virus detection, *J. Gen. Virol.,* 64, 2471, 1983.
45. **Jordan, R. and Hammond, J.,** Comparison and differentiation of potyvirus isolates and identification of strain-, virus-, subgroup-specific and potyvirus group-common epitopes using monoclonal antibodies, *J. Gen. Virol.,* 72, 25, 1991.
46. **Jordan, R.,** Potyviruses, monoclonal antibodies and antigenic sites, in *Potyvirus Taxonomy, Arch. Virol. Suppl.,* Barnett, O. W., Ed., Springer-Verlag, Berlin, 1992, 81.
47. **Rochow, W. F., Aapola, A. I. E., Brakke, M. K., and Carmichael, L. E.,** Purification and antigenicity of three isolates of barley yellow dwarf virus, *Virology,* 46, 117, 1971.
48. **Rowland, G. F., Engelbrecht, D. J., Pool, E. J., Schmollgruber, E. C., Thompson, G. J., and Van der Merwe, K. J.,** The use of peroxidase anti-peroxidase (PAP) complexes in the detection of plant viruses by ELISA, *J. Virol. Methods,* 25, 259, 1989.
49. **Ikegami, M. and Francki, R. I. B.,** Presence of antibodies to double stranded RNA in sera of rabbits immunized with rice dwarf and maize rough dwarf viruses, *Virology,* 56, 404, 1973.
50. **Von Wechmar, M. B. and Van Regenmortel, M. H. V.,** Serological studies on bromegrass mosaic virus and its protein fragments, *Virology,* 34, 36, 1968.
51. **Habili, N. and Francki, R. I. B.,** Comparative studies on tomato aspermy and cucumber mosaic viruses. IV. Immunogenic and serological properties, *Virology,* 64, 421, 1975.
52. **Rybicki, E. P. and Von Wechmar, M. B.,** Serology and immunochemistry, in *The Plant Viruses.* Vol. 1, *Polyhedral Virions with Tripartite Genomes,* Francki, R. I. B., Ed., Plenum Press, New York, 1985, 207.
53. **Hajimorad, M. R., Dietzgen, R. G., and Francki, I. B.,** Differentiation and antigenic characterization of closely related alfalfa mosaic virus strains with monoclonal antibodies, *J. Gen. Virol.,* 71, 2809, 1990.
54. **Purcifull, D. E. and Batchelor, D. L.,** Immunodiffusion tests with sodium dodecyl sulfate (SDS)-treated plant viruses and plant viral inclusions, *Agric. Exp. Stn. Inst. Food Agric. Sci. Bull. Univ. Fla.,* 788, 1, 1977.
55. **Carroll, R. B., Goldfine, S. M., and Melero, J. A.,** Antiserum to polyacrylamide gel-purified simian virus 40 antigen, *Virology,* 87, 194, 1978.

56. **Hiebert, E., Purcifull, D. E., and Christie, R. G.,** Purification and immunological analyses of plant viral inclusion bodies, *Meth. Virol.,* 8, 225, 1984.
57. **Van Regenmortel, M. H. V., Briand, J. P., Muller, S., and Plaue, S.,** Synthetic polypeptides as antigens, *Lab. Tech. Biochem. Mol. Biol.,* 19, 1, 1988.
58. **Mayer, R. J. and Walker, J. H.,** *Immunochemical Methods in Cell and Molecular Biology,* Academic Press, London, 1990.
59. **Pellequer, J. L., Westhof, E., and Van Regenmortel, M. H. V.,** Predicting the location of continuous epitopes in proteins from their primary structures, *Meth. Enzymol.,* 203, 176, 1991.
60. **Plaue, S., Muller, S., Briand, J. P., and Van Regenmortel, M. H. V.,** Recent advances in solid-phase peptide synthesis and preparation of antibodies to synthetic peptides, *Biologicals,* 18, 147, 1990.
61. **Ziegler, V., Laquel, P., Guilley, H., Richards, K., and Jonard, G.,** Immunological detection of cauliflower mosaic virus gene V protein produced in engineered bacteria or infected plants, *Gene,* 36, 271, 1985.
62. **Berna, A., Briand, J. P., Stussi-Garaud, C., and Godefroy-Colburn, T.,** Kinetics of accumulation of the three non-structural proteins of alfalfa mosaic virus in tobacco plants, *J. Gen. Virol.,* 67, 1135, 1986.
63. **Van Pelt-Heerchap, H., Verbeek, H., Huisman, M. J., Loesch-Fries, L. S., and Van Vloten-Doting, L.,** Non-structural proteins and RNAs of alfalfa mosaic virus synthesized in tobacco and cowpea protoplasts, *Virology,* 161, 190, 1987.
64. **Jaegle, M., Briand, J. P., Burckard, J., and Van Regenmortel, M. H. V.,** Accessibility of three continuous epitopes in tomato bushy stunt virus, *Ann. Inst. Pasteur Virol.,* 139, 39, 1988.
65. **Dubs, M. C. and Van Regenmortel, M. H. V.,** Odontoglossum ringspot virus coat protein: sequence and antigenic comparisons with other tobamoviruses, *Arch. Virol.,* 115, 239, 1990.
66. **Joisson, C. and Van Regenmortel, M. H. V.,** Influence of the C terminus of the small protein subunit of bean pod mottle virus on the antigenicity of the virus determined using monoclonal antibodies and anti-peptide antiserum, *J. Gen. Virol.,* 72, 2225, 1991.
67. **Polson, A., Von Wechmar, M. B., and Van Regenmortel, M. H. V.,** Isolation of viral IgY antibodies from yolks of immunized hens, *Immunol. Commun.,* 9, 475, 1980.
68. **Polson, A., Coetzer, T., Kruger, J., Von Maltzahn, E., and Van der Merwe, K. J.,** Improvements in the isolation of IgY from the yolks of eggs laid by immunized hens, *Immunol. Invest.,* 14, 323, 1985.
69. **Hardie, G. and Van Regenmortel, M. H. V.,** Isolation of specific antibody under conditions of low ionic strength, *J. Immunol. Meth.,* 15, 305, 1977.
70. **Van Regenmortel, M. H. V. and Burckard, J.,** Detection of a wide spectrum of tobacco mosaic virus strains by indirect enzyme immunoassay (ELISA), *Virology,* 106, 327, 1980.
71. **Leslie, G. A. and Clem, L. W.,** Phylogeny of immunoglobulin structure and function. III. Immunoglobulins of the chicken, *J. Exp. Med.,* 130, 1337, 1969.
72. **Somowiyarjo, S., Sako, N., and Nonaka, F.,** Production of avian antibodies to three potyviruses in coturnix quail, *J. Virol. Meth.,* 28, 125, 1990.
73. **Lister, R. M., Hammond, J., and Clement, D. L.,** Comparison of intradermal and intramuscular injection for raising plant virus antisera for use in ELISA, *J. Virol. Meth.,* 6, 179, 1983.
74. **Clausen, J.,** *Immunochemical Techniques for the Identification and Estimation of Macromolecules,* 3rd ed., Elsevier, New York, 1989.
75. **Herbert, W. J. and Kristensen, F.,** Laboratory animal techniques for immunology, in *Handbook of Experimental Immunology.* Vol. 4, *Applications of Immunological Methods in Biomedical Sciences,* Weir, D. M., Ed., Blackwell Scientific, Oxford, 1986, chap. 133.

76. **Harlow, E. and Lane, D.,** *Antibodies. A Laboratory Manual,* Cold Spring Harbor Laboratory, Cold Spring Harbor, NY, 1988.
77. **Fazekas de St. Groth, S. and Scheidegger, D.,** Production of monoclonal antibodies: strategy and tactics, *J. Immunol. Meth.,* 35, 1, 1980.
78. **Galfrè, G. and Milstein, C.,** Preparation of monoclonal antibodies: strategies and procedures, *Meth. Enzymol.,* 73, 1, 1981.
79. **Goding, J. W.,** in *Monoclonal Antibodies: Principles and Practice,* 2nd ed., Academic Press, London, 1986, 126.
80. **Campbell, A. M.,** *Monoclonal Antibody Technology,* Elsevier, Amsterdam, 1984.
81. **Liddell, J. E. and Cryer, A.,** *A Practical Guide to Monoclonal Antibodies,* John Wiley & Sons, New York, 1991.
82. **Sander, E. and Dietzgen, R. G.,** Monoclonal antibodies against plant viruses, *Adv. Vir. Res.,* 29, 131, 1984.
83. **Torrance, L. and Pead, M. T.,** The application of monoclonal antibodies to routine tests for two plant viruses, in *Developments and Applications in Virus Testing,* Jones, R. A. C. and Torrance, L., Eds., Association of Applied Biologists, Wellesbourne, 1986, 103.
84. **Stocker, J. W., Foster, H. K., Miggiano, V., Stahli, C., Staiger, G., Takacs, B., and Staehelin, Th.,** Generation of two new myeloma cell lines "PAI" and "PAI-O" for hybridoma production, *Res. Disclosure,* 217, 155, 1982.
85. **Hudson, L. and Hay, F.,** *Practical Immunology,* 2nd ed., Blackwell, Oxford, 1980, 318.
86. **Ohshima, K. and Shikata, E.,** On the screening procedures of ELISA for monoclonal antibodies against three luteoviruses, *Ann. Phytopathol. Soc. Jpn.,* 56, 219, 1990.
87. **Al Moudallal, Z., Altschuh, D., Briand, J. P., and Van Regenmortel, M. H. V.,** Comparative sensitivity of different ELISA procedures for detecting monoclonal antibodies, *J. Immunol. Meth.,* 68, 35, 1984.
88. **Dore, I., Weiss, E., Altschuh, D., and Regenmortel, M. H. V.,** Visualization by electron microscopy of the location of tobacco mosaic virus epitopes reacting with monoclonal antibodies in enzyme immunoassay, *Virology,* 162, 279, 1988.
89. **Halk, E. L., Hsu, H. T., Aebig, J., and Franke, J.,** Production of monoclonal antibodies against three ilarviruses and alfalfa mosaic virus and their use in serotyping, *Phytopathology,* 74, 367, 1984.
90. **Underwood, P. A.,** Theoretical considerations of the ability of monoclonal antibodies to detect antigenic differences between closely related variants, with particular reference to heterospecific reactions, *J. Immunol. Meth.,* 85, 295, 1985.
91. **Thomas, J. E., Massalski, P. R., and Harrison, B. D.,** Production of monoclonal antibodies to African cassava mosaic virus and differences in their reactivities with other whitefly-transmitted geminiviruses, *J. Gen. Virol.,* 67, 2739, 1986.
92. **Poul, F. and Dunez, J.,** Production and use of monoclonal antibodies for the detection of apple chlorotic leaf spot virus, *J. Virol. Meth.,* 25, 153, 1989.
93. **Dietzgen, R. G.,** Monoklonale Antikörper gegen Pflanzenviren: Herstellung, Reinigung und Charakterisierung, Ph.D. Thesis, University of Tübingen, Germany, 1983.
94. **Tirry, L., Welvaert, W., and Samyin, G.,** Differentiation of the hop strain from other Arabis mosaic virus isolates by means of monoclonal antibodies, *Meded. Fac. Landbouwwet. Rijksuniv. Gent,* 53, 423, 1988.
95. **Wu, R. Y. and Su, H. J.,** Production of monoclonal antibodies against banana bunchy top virus and their use in enzyme-linked immunosorbent assay, *J. Phytopathol.,* 128, 203, 1990.
96. **Diaco, R., Lister, R. M., Durand, D. P., and Hill, J. H.,** Production of monoclonal antibodies against three isolates of barley yellow dwarf virus, *Phytopathology,* 73, 788, 1983.
97. **Hsu, H. T., Aebig, J., and Rochow, W. F.,** Differences among monoclonal antibodies to barley yellow dwarf viruses, *Phytopathology,* 74, 600, 1984.

98. **D'Arcy, C. J., Murphy, J. F., and Miklasz, S. D.,** Murine monoclonal antibodies produced against two Illinois strains of barley yellow dwarf virus: production and use for virus detection, *Phytopathology,* 80, 377, 1990.

99. **Torrance, L., Pead, M. T., Larkins, A. P., and Butcher, G. W.,** Characterization of monoclonal antibodies to a U.K. isolate of barley yellow dwarf virus, *J. Gen. Virol.,* 67, 549, 1986.

100. **Pead, M. T. and Torrance, L.,** Some characteristics of monoclonal antibodies to a British MAV-like isolate of barley yellow dwarf virus, *Ann. Appl. Biol.,* 113, 639, 1988.

101. **Wang, W. Y., Mink, G. I., and Silbernagel, M. J.,** A broad-spectrum monoclonal antibody prepared against bean common mosaic virus, *Phytopathology,* 75, 1352, 1985.

102. **Scott, S. W., McLaughlin, M. R., and Ainsworth, A. J.,** Monoclonal antibodies produced to bean yellow mosaic virus, clover yellow vein virus, and pea mosaic virus which cross-react among the three viruses, *Arch. Virol.,* 108, 161, 1989.

103. **Merten, O. W., Heberle-Bors, E., Himmler, G., Reiter, S., Messner, P., and Katinger, H.,** Monoclonal ELISA for the determination of BNYV-virus, *Dev. Biol. Stand.,* 60, 451, 1985.

104. **Torrance, L., Pead, M. T., and Buxton, G.,** Production and some characteristics of monoclonal antibodies against beet necrotic yellow vein virus, *Ann. Appl. Biol.,* 113, 519, 1988.

105. **D'Arcy, C. J., Torrance, L., and Martin, R. R.,** Discrimination among luteoviruses and their strains by monoclonal antibodies and identification of common epitopes, *Phytopathology,* 79, 869, 1989.

106. **Hsu, H. T. and Lawson, R. H.,** Comparison of mouse monoclonal antibodies and polyclonal antibodies of chicken egg yolk and rabbit for assay of carnation etched ring virus, *Phytopathology,* 75, 778, 1985.

107. **Hsu, H. T. and Lawson, R. H.,** Detection of carnation etched ring virus using mouse monoclonal antibodies and polyclonal chicken and rabbit antisera, *Acta Hortic.,* 164, 199, 1985.

108. **Jordan, R.,** Successful production of monoclonal antibodies to three carnation viruses using an admixture of only partially purified virus preparations as immunogen, *Phytopathology,* 79, 1213, 1989.

109. **Vela, C., Cambra, M., Cortes, E., Moreno, P., Miguet, J. G., Perez de San Roman, C., and Sanz, A.,** Production and chracterization of monoclonal antibodies specific for citrus tristeza virus and their use for diagnosis, *J. Gen. Virol.,* 67, 91, 1986.

110. **Permar, T. A., Garnsey, S. M., Gumpf, D. J., and Lee, R. F.,** A monoclonal antibody that discriminates strains of citrus tristeza virus, *Phytopathology,* 80, 224, 1990.

111. **Hsu, H. T., Halk, E. L., and Lawson, R. H.,** Monoclonal antibodies in plant virology, Proc. 4th Int. Cong. Plant Pathology, Melbourne, Australia, 1983, 25.

112. **Kalmar, G. B. and Eastwell, K. C.,** Reaction of coat proteins of two comoviruses in different aggregation states with monoclonal antibodies, *J. Gen. Virol.,* 70, 3451, 1989.

113. **Takahashi, Y., Kameya-Iwaki, M., Shohara, K., and Toriyama, S.,** Detection of watermelon strain of cucumber green mottle mosaic virus by using monoclonal antibodies, *Ann. Phytopathol. Soc. Jpn.,* 55, 369, 1989.

114. **Maeda, T., Sako, N., and Inouye, N.,** Production of monoclonal antibodies to cucumber mosaic virus and their use in ELISA, *Ann. Phytopathol. Soc. Jpn.,* 54, 600, 1988.

115. **Porta, C., Devergne, J. C., Cardin, L., Briand, J. P., and Van Regenmortel, M. H. V.,** Serotype specificity of monoclonal antibodies to cucumber mosaic virus, *Arch. Virol.,* 104, 271, 1989.

116. **Haase, A., Richter, J., and Rabenstein, F.,** Monoclonal antibodies for detection and serotyping of cucumber mosaic virus, *J. Phytopathol.,* 127, 129, 1989.

117. **Cai, W. Q., Wang, R., Qin, B. Y., Ma, S. F., Tian, B., Zhang, C. L., Chen, J., and Li, L.,** Production of CMV specific monoclonal antibodies by electrofusion, *Acta Microbiol. Sinica,* 29, 444, 1989 (in Chinese).

118. **Huss, B., Muller, S., Sommermeyer, G., Walter, B., and Van Regenmortel, M. H. V.,** Grapevine fanleaf virus monoclonal antibodies: their use to distinguish different isolates, *J. Phytopathol.,* 119, 358, 1987.

119. **Hu, J. S., Boscia, D., and Gonsalves, D.,** Use of monoclonal antibodies in the study of closteroviruses associated with grape leafroll disease, *Phytopathology,* 79, 1189, 1989.

120. **Zimmermann, D., Sommermeyer, G., Walter, B., and Van Regenmortel, M. H. V.,** Production and characterisation of monoclonal antibodies specific to closterovirus-like particles associated with grapevine leafroll disease, *J. Phytopathol.,* 130, 277, 1990.

121. **Hill, E. K., Hill, J. H., and Durand, D. P.,** Production of monoclonal antibodies to viruses in the potyvirus group: use in radioimmunoassay, *J. Gen. Virol.,* 65, 525, 1984.

122. **Dekker, E. L., Pinner, M. S., Markham, P. G., and Van Regenmortel, M. H. V.,** Characterization of maize streak virus isolates from different plant species by polyclonal and monoclonal antibodies, *J. Gen. Virol.,* 69, 983, 1988.

123. **Dore, I., Dekker, E. L., Porta, C., and Van Regenmortel, M. H. V.,** Detection by ELISA of two tobamoviruses in orchids using monoclonal antibodies, *J. Phytopathol.,* 20, 317, 1987.

124. **Baker, C. A. and Purcifull, D. E.,** Antigenic diversity of papaya ringspot virus (PRSV) isolates detected by monoclonal antibodies to PRSV-W, *Phytopathology,* 79, 214, 1989.

125. **Huguenot, C., Givord, L., Sommermeyer, G., and Van Regenmortel, M. H. V.,** Differentiation of peanut clump virus serotypes by monoclonal antibodies, *Res. Virol.,* 140, 87, 1989.

126. **Sherwood, J. L., Sanborn, M. R., and Keyser, G. C.,** Production of monoclonal antibodies to peanut mottle virus and wheat streak mosaic virus, *Phytopathology,* 75, 1358, 1985.

127. **Sherwood, J. L., Sanborn, M. R., and Melouk, H. A.,** Use of monoclonal antibodies (MCA) for detection of peanut mottle virus (PMV), *Proc. Am. Peanut Res. Educ. Soc.,* 18, 60, 1986.

128. **Sherwood, J. L., Sanborn, M. R., and Keyser, G. C.,** Production of monoclonal antibodies to peanut mottle virus and their use in enzyme-linked immunosorbent assay and dot-immunobinding assay, *Phytopathology,* 77, 1158, 1987.

129. **Culver, J. N. and Sherwood, J. L.,** Detection of peanut stripe virus in peanut seed by an indirect ELISA using a monoclonal antibody, *Plant Dis.,* 72, 676, 1988.

130. **Himmler, G., Brix, U., Steinkellner, H., Laimer, M., Mattanovich, D., and Katinger, H. W. D.,** Early screening for anti-plum pox virus monoclonal antibodies with different epitope specificities by means of gold-labelled immunosorbent electron microscopy, *J. Virol. Meth.,* 22, 351, 1988.

131. **Martin, R. R. and Stace-Smith, R.,** Production and characterization of monoclonal antibodies specific to potato leaf roll virus, *Can. J. Plant Pathol.,* 6, 206, 1984.

132. **Ohshima, K., Uyeda, I., and Shikata, E.,** Production and characteristics of monoclonal antibodies to potato leafroll virus, *J. Fac. Agric. Hokkaido Univ.,* 63, 373, 1988.

133. **Gugerli, P.,** Use of enzyme immunoassay in phytopathology, in *Immunoenzymatic Techniques,* Avrameas, S., Druet, P., Masseyeff, R., and Feldmann, G., Eds., Elsevier, Amsterdam, 1983, 369.

134. **Bonnekamp, P. M., Pomp, H., and Gussenhoven, G. C.,** Production and characterization of monoclonal antibodies to potato virus A, *J. Phytopathol.,* 128, 112, 1990.

135. **Saarma, M., Järvekülg, L., Hemmilä, I., Siitari, H., and Sinijärv, R.,** Simultaneous quantification of two plant viruses by double-label time-resolved immunofluorometric assay, *J. Virol. Meth.,* 23, 47, 1989.

136. **Torrance, L., Hutchins, A., Larkins, A. P., and Butcher, G. W.,** Characterization of monoclonal antibodies to potato virus X and assessment of their potential for use in large-scale surveys for virus infection, Proc. Sixth Int. Congress of Virology, Sendai, Japan, 1984, 27.

137. **Sanz, A., Cortes, E., Miguet, J., Cambra, M., Perez de San Roman, C., Moreno, P., Lavina, A., and Vela, C.,** Preparation of monoclonal antibodies of the potato virus PVX and PVY, *Centro Invest. Tecn. Agrarias,* 1988, 316, 1988 (in Spanish).

138. **Lizarraga, C. and Fernandez-Northcote, E. M.,** Detection of potato viruses X and Y in sap extracts by a modified indirect enzyme-linked immunosorbent assay on nitrocellulose membranes (NCM-ELISA), *Plant Dis.,* 73, 11, 1989.

139. **Xiao, X. W., Gou, J., Chai, S., H., Lui, J., Feng, L. X., and Hsu, H. T.,** Some properties of potato virus X specific monoclonal antibodies and the hybridoma cell lines, *Chin. J. Biotechnol.,* 5, 108, 1989 (in Chinese).

140. **Rose, D. G. and Hubbard, A. L.,** Production of monoclonal antibodies for the detection of potato virus Y, *Ann. Appl. Biol.,* 109, 317, 1986.

141. **Jordan, R.,** Evaluating the relative specific activities of monoclonal antibodies to plant viruses, in *Hybridoma Technology in Agricultural and Veterinary Research,* Stern, N. J. and Gamble, H. R., Eds., Rowman and Allanheld, Totowa, NJ, 1984, 259.

142. **Martin, R. R.,** The use of monoclonal antibodies for virus detection, *Can. J. Plant Pathol.,* 9, 177, 1987.

143. **Harjosudarmo, J., Ohshima, K., Uyeda, I., and Shikata, E.,** Production and characterization of monoclonal antibodies against rice dwarf virus, *J. Fac. Agric. Hokkaido Univ.,* 64, 107, 1990.

144. **Ohshima, K., Harjosudarmo, J., Ishikawa, Y., and Shikata, E.,** Relationship between hybridoma screening procedures and the characteristics of monoclonal antibodies for use in direct double antibody sandwich ELISA for the detection of plant viruses, *Ann. Phytopathol. Soc. Jpn.,* 56, 569, 1990.

145. **Omura, T., Takahashi, Y., Shohara, K., Minobe, Y., Tsuchizaki, T., and Nozu, Y.,** Production of monoclonal antibodies against rice stripe virus for the detection of virus antigen in infected plants and viruliferous insects, *Ann. Phytopathol. Soc. Jpn.,* 52, 270, 1986.

146. **Nozu, Y., Usugi, T., and Nishimori, K.,** Production of monoclonal antibodies to a plant virus, *Seikagaku,* 55, 837, 1983.

147. **Nozu, Y., Usugi, T., and Nishimori, K.,** Production of monoclonal antibodies to satsuma dwarf virus, *Ann. Phytopathol. Soc. Jpn.,* 52, 86, 1986.

148. **Bahrani, Z., Sherwood, J. L., Sanborn, M. R., and Keyser, G. C.,** The use of monoclonal antibodies to detect wheat soil-borne mosaic virus, *J. Gen. Virol.,* 69, 1317, 1988.

149. **Tremaine, J. H. and Ronald, W. P.,** Serological studies of sobemovirus antigens, *Phytopathology,* 73, 794, 1983.

150. **Hewish, D. R., Shukla, D. D., Johnstone, G. R. and Sward, R. J.,** Monoclonal antibodies to luteoviruses, Proc. 4th Int. Congress of Plant Pathology, Melbourne, Australia, 1983, 116.

151. **Hiruki, C., Figueiredo, G., Inoue, M., and Furuya, Y.,** Production and characterization of monoclonal antibodies specific to sweet clover necrotic mosaic virus, *J. Virol. Meth.,* 8, 301, 1984.

152. **Dietzgen, R. G. and Sander, E.,** Monoclonal antibodies against a plant virus, *Arch. Virol.,* 74, 197, 1982.

153. **Briand, J. P., Al Moudallal, Z., and Van Regenmortel, M. H. V.,** Serological differentiation of tobamoviruses by means of monoclonal antibodies, *J. Virol. Meth.,* 5, 293, 1982.

154. **Ohshima, K., Uyeda, I., and Shikata, E.,** Characterization of monoclonal antibodies against tobacco necrotic dwarf virus, *Ann. Phytopathol. Soc. Jpn.,* 55, 420, 1989.

155. **Dekker, E. L., Dore, I., Porta, C., and Van Regenmortel, M. H. V.,** Conformational specificity of monoclonal antibodies used in the diagnosis of tomato mosaic virus, *Arch. Virol.,* 94, 191, 1987.

156. **Takahashi, Y., Kameya-Iwaki, M., and Shihara, K.,** Characteristics of epitopes on the tomato strain of tobacco mosaic virus detected by monoclonal antibodies, *Ann. Phytopathol. Soc. Jpn.,* 55, 179, 1989.

157. **Powell, C. A. and Marquez, E. D.,** The preparation of monoclonal antibody to tomato ringspot virus, Proc. 4th Int. Congress of Plant Pathology, Melbourne, Australia, 1983, 121.

158. **Powell, C. A.,** Detection of tomato ringspot virus with monoclonal antibodies, *Plant Dis.,* 7, 904, 1990.

159. **Sherwood, J. L., Sanborn, M. R., Keyser, G. C., and Myers, L. D.,** Use of monoclonal antibodies in detection of tomato spotted wilt virus, *Phytopathology,* 79, 61, 1989.

160. **Huguenot, C., Van den Dobbelsteen, G., De Haan, P., Wagemakers, C. A. M., Osterhaus, A. D. M. E., and Peters, D.,** Detection of tomato spotted wilt virus using monoclonal antibodies and riboprobes, *Arch. Virol.,* 110, 47, 1990.

161. **Hsu, H. T., Wang, Y. C., Lawson, R. H., Wang, M., and Gonsalves, D.,** Splenocytes of mice with induced immunological tolerance to plant antigens for construction of hybridomas secreting tomato spotted wilt virus-specific antibodies, *Phytopathology,* 80, 158, 1990.

162. **Adam, G., Lesemann, D. E., and Vetten, H. J.,** Monoclonal antibodies against tomato spotted wilt virus: characterisation and application, *Ann. Appl. Biol.,* 118, 87, 1991.

163. **Hsu, H. T., Lawson, R. H., Beijersbergen, J. C. M., and Derks, A. F. L. M.,** Murine hybridomas secreting antibodies specific to carnation etched ring and tulip breaking viruses, Proc. 4th Int. Congress of Plant Pathology, Melbourne, Australia, 1983, 25.

164. **Somowiyarjo, S., Sako, N., and Nonaka, F.,** Production and characterization of monoclonal antibodies to watermelon mosaic virus 2, *Ann. Phytopathol. Soc. Jpn.,* 56, 541, 1990.

165. **Somowiyarjo, S., Sako, N., and Nonaka, F.,** The use of monoclonal antibody for detecting zucchini yellow mosaic virus, *Ann. Phytopathol. Soc. Jpn.,* 54, 436, 1988.

166. **Wisler, G. C., Baker, C. A., Purcifull, D. E., and Hiebert, E.,** Partial characterization of monoclonal antibodies to zucchini yellow mosaic virus ZYMV and watermelon mosaic virus-2 WMV-2, *Phytopathology,* 79, 1213, 1989.

167. **Torrance, L., Larkins, A. P., and Butcher, G. W.,** Characterization of monoclonal antibodies against potato virus X and comparison of serotypes with resistance groups, *J. Gen. Virol.,* 67, 57, 1986.

168. **Halk, E. L.,** Serotyping plant viruses with monoclonal antibodies, *Meth. Enzymol.,* 118, 766, 1986.

169. **Slade, D. E., Johnston, R. E., and Dougherty, W. G.,** Generation and characterization of monoclonal antibodies reactive with the 49-kDa proteinase of tobacco etch virus, *Virology,* 173, 499, 1989.

170. **Baker, C. A. and Purcifull, D. E.,** Reactivity of two monoclonal antibodies to the cylindral inclusion protein of papaya ringspot virus type-W, *Phytopathology,* 80, 1033, 1990.

171. **Baker, C. A. and Purcifull, D. E.,** Reactivity of a monoclonal antibody to the amorphous inclusion protein of papaya ringspot virus type-W (PRSV-W), *Phytopathology,* 78, 1537, 1988.

172. **Massalski, P. R. and Harrison, B. D.,** Properties of monoclonal antibodies to potato leafroll luteovirus and their use to distinguish virus isolates differing in aphid transmissibility, *J. Gen. Virol.,* 68, 1813, 1987.

173. **Tremaine, J. H., MacKenzie, D. J., and Ronald, W. P.,** Monoclonal antibodies as structural probes of southern bean mosaic virus, *Virology,* 144, 80, 1985.

174. **Koenig, R. and Torrance, L.,** Antigenic analysis of potato virus X by means of monoclonal antibodies, *J. Gen. Virol.,* 67, 2145, 1986.

175. **MacKenzie, D. J. and Tremaine, J. H.,** The use of a monoclonal antibody specific for the N-terminal region of southern bean mosaic virus as a probe of virus structure, *J. Gen. Virol.,* 67, 727, 1986.

176. **Dore, I., Altschuh, D., Al Moudallal, Z., and Van Regenmortel, M. H. V.,** Immunochemical studies of tobacco mosaic virus. VII. Use of comparative surface accessibility of residues in antigenically related viruses for delineating epitopes recognized by monoclonal antibodies, *Mol. Immunol.,* 24, 1351, 1987.

177. **Sober, J., Järvekülg, L., Toots, I., Radavsky, J., Villems, R., and Saarma, M.,** Antigenic characterization of potato virus X with monoclonal antibodies, *J. Gen. Virol.,* 69, 1799, 1988.

178. **Slaoui, M., Franssen, J. D., Hérion, P., and Urbain, J.,** Idiotypic studies of monoclonal anti-tobacco mosaic virus antibodies from one mouse, *Mol. Immunol.,* 20, 1073, 1983.

179. **Hu, J. S. and Rochow, W. F.,** Anti-idiotypic antibodies against an anti-barley yellow dwarf virus monoclonal antibody, *Phytopathology,* 78, 1302, 1988.

180. **McKinney, M. M. and Parkinson, A.,** A simple, non-chromatographic procedure to purify immunoglobulins from serum and ascites fluid, *J. Immunol. Meth.,* 96, 271, 1987.

181. **Weiss, E. and Van Regenmortel, M. H. V.,** Use of rabbit Fab'-peroxidase conjugates prepared by the maleimide method for detecting plant viruses by ELISA, *J. Virol. Meth.,* 24, 11, 1989.

182. **Clark, M. F., Lister, R. M., and Bar-Joseph, M.,** ELISA techniques, *Meth. Enzymol.,* 118, 742, 1986.

183. **Ey, P. L., Prowse, S. J., and Jenkin, C. R.,** Isolation of pure IgG1, IgG2a and IgG2b immunoglobulins from mouse serum using protein A-sepharose, *Immunochemistry,* 15, 429, 1978.

184. **Adams, A. N. and Barbara, D. J.,** The use of F(ab')2-based ELISA to detect serological relationships amont carlaviruses, *Ann. Appl. Biol.,* 101, 495, 1982.

185. **Barbara, D. J. and Clark, M. F.,** A simple indirect ELISA using F(ab')2 fragments of immunoglobulin, *J. Gen. Virol.,* 58, 315, 1982.

186. **Azimzadeh, A., Weiss, E., and Van Regenmortel, M. H. V.,** Measurement of affinity of viral monoclonal antibodies using Fab'-peroxidase conjugate. Influence of antibody concentration on apparent affinity, *Mol. Immunol.,* in press.

187. **Polak, J. and Kristek, J.,** Use of horseradish-peroxidase labelled antibodies in ELISA for plant virus detection, *J. Phytopathol.,* 122, 200, 1988.

188. **Jones, A. T. and Mitchell, M. J.,** Oxidising activity in root extracts from plants inoculated with virus or buffer that interferes with ELISA when using the substrate 3,3', 5,5'-tetramethylbenzidine, *Ann. Appl. Biol.,* 111, 359, 1987.

189. **Sudarshana, M. R. and Reddy, D. V. R.,** Penicillinase-based enzyme-linked immunosorbent assay for the detection of plant viruses, *J. Virol. Meth.,* 26, 45, 1989.

190. **Clark, M. F. and Adams, A. N.,** Characteristics of the microplate method of enzyme linked immunosorbent assay for the detection of plant viruses, *J. Gen. Virol.,* 34, 475, 1977.

191. **Avrameas, S.,** Coupling of enzymes to proteins with glutaraldehyde. Use of the conjugates for the detection of antigens and antibodies, *Immunochemistry,* 6, 43, 1969.

192. **Guesdon, J. L., Ternynck, T., and Avrameas, S.,** Use of avidin-biotin interaction in immunoenzymatic techniques, *J. Histochem. Cytochem.,* 27, 1131, 1979.

193. **Diaco, R., Hill, J. H., Hill, E. K., Tachibana, H., and Durand, D. P.,** Monoclonal antibody-based biotin-avidin ELISA for the detection of soybean mosaic virus in soybean seeds, *J. Gen. Virol.,* 66, 2089, 1985.

194. **Hewish, D. R., Shukla, D. D., and Gough, K. H.,** The use of biotin-conjugated antisera in immunoassays for plant viruses, *J. Virol. Meth.,* 13, 79, 1986.

195. **Zrein, M., Burckard, J., and Van Regenmortel, M. H. V.,** Use of the biotin-avidin system for detecting a broad range of serologically related plant viruses of ELISA, *J. Virol. Meth.,* 13, 121, 1986.

196. **Siitari, H., Lövgren, T., and Halonen, P.,** Detection of viral antigens by direct one-incubation time-resolved fluoroimmunoassay, in *Developments and Applications in Virus Testing,* Vol. 1, Jones, R. A. C. and Torrance, L., Eds., Association of Applied Biologists, Wellesbourne, 1986, 155.

197. **Sinijärv, R., Järvekülg, L., Andreeva, E., and Saarma, M.,** Detection of potato virus X by one incubation europium time-resolved fluoroimmunoassay and ELISA, *J. Gen. Virol.,* 69, 991, 1988.

198. **Lesemann, D. E., Koenig, R., Torrance, L., Buxton, G., Bonnekamp, P. M., Peters, D., and Schots, A.,** Electron microscopical demonstration of different binding sites for monoclonal antibodies on particles of beet necrotic yellow vein virus, *J. Gen. Virol.,* 71, 731, 1990.

199. **Halk, E. L. and De Boer, S. H.,** Monoclonal antibodies in plant-disease research, *Annu. Rev. Phytopathol.,* 23, 321, 1985.

200. **Van Regenmortel, M. H. V.,** Monoclonal antibodies in plant virology, *Microbiol. Sci.,* 1, 73, 1984.

201. **Lister, R. M. and Sward, R. J.,** Anomalies in serological and vector relationships of MAV-like isolates of barley yellow dwarf virus from Australia and the USA, *Phytopathology,* 78, 766, 1988.

202. **Yolken, R. H.,** Solid phase immunoassays for the detection of viral diseases, in *Immunochemistry of viruses. The basis for serodiagnosis and vaccines,* Van. Regenmortel, M. H. V. and Neurath, A. R., Eds., Elsevier, Amsterdam, 1985, 121.

203. **Azimzadeh, A. and Van Regenmortel, M. H. V.,** Antibody affinity measurements, *J. Mol. Recogn.,* 3, 108, 1990.

204. **Azimzadeh, A. and Van Regenmortel, M. H. V.,** Measurement of affinity of viral monoclonal antibodies by ELISA titration of free antibody in equilibrium mixtures, *J. Immunol. Meth.,* 141, 199, 1991.

205. **Heidelberger, M. and Kendall, F. E.,** The precipitin reaction between type III penumococcus polysaccharide and homologous antibody III. A quantitative study and a theory of the reaction mechanisms, *J. Exp. Med.,* 61, 563, 1935.

206. **Matthews, R. E. F.,** *Plant Virus Serology,* University Press, Cambridge, 1957.

207. **Kleczkowski, A.,** Serological properties of viruses and their fragments, in *Viruses of Plants,* Beemster, A. B. R. and Dykstra, J., Eds., North-Holland, Amsterdam, 1966, 196.

208. **Matthews, R. E. F.,** Serological techniques for plant viruses, in *Methods in Virology,* Vol. 3, Maramorosch, K. and Koprowski, H., Eds., Academic Press, New York, 1967, 199.

209. **Shepard, J. F. and Shalla, T. A.,** An antigenic analysis of potato virus X and its degraded protein. I. Evidence for and degree of antigenic disparity, *Virology,* 42, 825, 1970.

210. **Wetter, C.,** Serology in virus-disease diagnosis, *Annu. Rev. Phytopathol.,* 3, 19, 1965.

211. **Noordam, D.,** Identification of Plant Viruses: Methods and Experiments, Center Agric. Publ. Document, Wageningen, The Netherlands, 1973, 207.

212. **Koenig, R.,** Serology and immunochemistry, in *The Plant Viruses,* Vol. 4, Milne, R. G., Ed., Plenum Press, New York, 1988, 111.

213. **Van Regenmortel, M. H. V. and Von Wechmar, M. B.,** A reexamination of the serological relationship between tobacco mosaic virus and cucumber virus 4, *Virology,* 41, 330, 1970.

214. **Van Regenmortel, M. H. V.,** Antigenic relationships between strains of tobacco mosaic virus, *Virology,* 64, 415, 1975.

215. **Koenig, R.,** A loop-structure in the serological classification system of tymoviruses, *Virology,* 72, 1, 1976.

216. **Ding, S. W., Keese, P., and Gibbs, A.,** The nucleotide sequence of the genomic RNA of kennedya yellow mosaic tymovirus-Jervis Bay isolate: relationships with potex- and carlaviruses, *J. Gen. Virol.,* 71, 925, 1990.

217. **Rybicki, E. P.**, The use of serological differentiation indices for the phylogenetic analysis of plant virus relationships, *Arch. Virol.*, 119, 83, 1991.

218. **Paul, H. L., Gibbs, A., and Wittmann-Liebold, B.**, The relationships of certain tymoviruses assessed from the amino acid composition of their coat proteins, *Intervirology*, 13, 99, 1980.

219. **Blok, J., Gibbs, A., and McKenzie, A.**, The classification of tymoviruses by cDNA-RNA hybridisation and other measures of relatedness, *Arch. Virol.*, 96, 225, 1987.

220. **Rajeshwari, R., Reddy, D. V. R., and Iizuka, N.**, Improvements in the passive hemagglutination technique for serological detection of plant viruses, *Phytopathology*, 71, 306, 1981.

221. **Koenig, R. and Bode, O.**, Sensitive detection of Andean potato latent and Andean potato mottle viruses in potato tubers with serological latex tests, *Phytopathol. Z.*, 92, 275, 1978.

222. **Torrance, L.**, Use of protein A to improve sensitisation of latex particles with antibodies to plant viruses, *Ann. Appl. Biol.*, 96, 45, 1980.

223. **Natsuaki, K. T., Nishimura, Y., Ikeda, M., and Tomura, K.**, Use of gelatine particle agglutination test for detection of plant viruses, *Ann. Phytopathol. Soc. Jpn.*, 54, 548, 1988.

224. **Ouchterlony, Ö.**, *Handbook of Immunodiffusion and Immunoelectrophoresis*, Ann. Arbor Scientific, Ann Arbor, Michigan, 1968.

225. **Wetter, C.**, Immunodiffusion of tobacco mosaic virus and its interaction with agar, *Virology*, 31, 498, 1967.

226. **Crowle, A. J.**, *Immunodiffusion*, 2nd ed., Academic Press, New York, 1973.

227. **Jones, A. T., Koenig, R., Lesemann, D. E., Hamacher, J., Nienhaus, F., and Winter, S.**, Serological comparison of isolates of cherry leaf roll virus from diseased beech and birch trees in a forest decline area in Germany with other isolates of the virus, *J. Phytopathol.*, 129, 339, 1990.

228. **Grogan, R. G., Taylor, R. H., and Kimble, K. A.**, The effect of placement of reactants on immunodiffusion precipitin patterns, *Phytopathology*, 54, 163, 1964.

229. **Tomlinson, J. A. and Walkey, D. G. A.**, Effects of ultrasonic treatment on turnip mosaic virus and potato virus X, *Virology*, 32, 267, 1967.

230. **Shepard, J. F., Segor, G. A., and Purcifull, D. E.**, Immunochemical cross-reactivity between the dissociated capsid proteins of PVY group plant viruses, *Virology*, 58, 464, 1974.

231. **Wetter, C. and Bernard, M.**, Identifizierung, Reinigung, und serologischer Nachweis von Tabakmosaikvirus und Para-Tabakmosaikvirus aus Zigaretten, *Phytopathol. Z.*, 90, 257, 1977.

232. **Wetter, C., Conti, M., Altschuh, D., Tabillion, R., and Van Regenmortel, M. H. V.**, Pepper mild mottle virus, a tobamovirus infecting pepper cultivars in Sicily, *Phytopathology*, 74, 405, 1984.

233. **Hagen, T. J., Taylor, D. B., and Meagher, R. B.**, Rocket immunoelectrophoresis assay for cauliflower mosaic virus, *Phytopathology*, 72, 239, 1982.

234. **Koenig, R.**, Indirect ELISA methods for the broad specificity detection of plant viruses, *J. Gen. Virol.*, 55, 53, 1981.

235. **Devergne, J. C., Cardin, L., Burckard, J., and Van Regenmortel, M. H. V.**, Comparison of direct and indirect ELISA for detecting antigenically related cucumoviruses, *J. Virol. Meth.*, 3, 193, 1981.

236. **Torrance, L.**, Use of bovine C1q to detect plant viruses in an ELISA-type assay, *J. Gen. Virol.*, 51, 229, 1981.

237. **Torrance, L.**, Use of C1q enzyme-linked immunosorbent assay to detect plant viruses and their serologically different strains, *Ann. Appl. Biol.*, 99, 291, 1981.

238. **Edwards, M. L. and Cooper, J. I.**, Plant virus detection using a new form of indirect ELISA, *J. Virol. Meth.*, 11, 309, 1985.

239. **Lommel, S. A., McCain, A. H., and Morris, T. J.,** Evaluation of indirect enzyme-linked immunosorbent assay for the detection of plant viruses, *Phytopathology,* 72, 1018, 1982.

240. **Mowat, W. P. and Dawson, S.,** Detection and identification of plant viruses by ELISA using crude sap extracts and unfractionated antisera, *J. Virol. Meth.,* 15, 233, 1987.

241. **Soderquist, M. E. and Walton, A. G.,** Structural changes in proteins adsorbed on polymer surfaces, *J. Colloid Interface Sci.,* 75, 386, 1980.

242. **Kennel, S. J.,** Binding of monoclonal antibody to protein antigen in fluid phase or bound to solid supports, *J. Immunol. Meth.,* 55, 1, 1982.

243. **McCullough, K. C., Crowther, J. R., and Butcher, R. N.,** Alteration in antibody reactivity with foot-and-mouth disease virus (FMDV) 146S antigen before and after binding to a solid phase or complexing with specific antibody, *J. Immunol. Meth.,* 82, 91, 1985.

244. **Darst, S. A., Robertson, C. R., and Berzofsky, J. A.,** Adsorption of the protein antigen myoglobin affects the binding of conformation-specific monoclonal antibodies, *Biophys. J.,* 53, 533, 1988.

245. **Diaco, R., Lister, R. M., Hill, J. H. and Durand, D. P.,** Demonstration of serological relationships among isolates of barley yellow dwarf virus using polyclonal and monoclonal antibodies, *J. Gen. Virol.,* 67, 353, 1986.

246. **Murphy, J. F. and D'Arcy, C. J.,** Influence of ELISA conditions on detection of serological relationships among luteoviruses, *J. Virol. Meth.,* 31, 263, 1991.

247. **Jones, F. E., Hill, J. H., and Durand, D. P.,** Detection and differentiation of maize dwarf mosaic virus, strains A and B, by use of different class immunoglobulins in a double-antibody sandwich enzyme-linked immunosorbent assay, *Phytopathology,* 78, 1118, 1988.

248. **Burrows, P. M., Scott, S. W., Barnett, O. W., and McLaughlin, M. R.,** Use of experimental designs with quantitative ELISA, *J. Virol. Meth.,* 8, 207, 1984.

249. **Cardin, L., Devergne, J. C., and Pitart, M.,** Dosage immunoenzymatique (ELISA) du virus de la mosaique du concombre. I. Aspect methodologique, *Agronomie,* 4, 125, 1984.

250. **Vilker, V. L. and Verduin, B. J. M.,** Analysis and application of substrate hydrolysis rates in indirect ELISA of a purified plant virus, *J. Virol. Meth.,* 19, 141, 1988.

251. **Jaegle, M. and Van Regenmortel, M. H. V.,** Use of ELISA for measuring the extent of serological cross-reactivity between plant viruses, *J. Virol. Meth.,* 11, 189, 1985.

252. **Clark, M. F. and Barbara, D. J.,** A method for the quantitative analysis of ELISA data, *J. Virol. Meth.,* 15, 213, 1987.

253. **Hughes, J. D'A. and Thomas, B. J.,** The use of protein A-sandwich ELISA as a means for quantifying serological relationships between members of the tobamovirus group, *Ann. Appl. Biol.,* 112, 117, 1988.

254. **Tamada, T. and Harrison, B. D.,** Quantitative studies on the uptake and retention of potato leafroll virus by aphids in laboratory and fields conditions, *Ann. Appl. Biol.,* 98, 261, 1981.

255. **Hewings, A. D. and D'Arcy, C.,** Maximizing the detection capability of a beet western yellows virus ELISA system, *J. Virol. Meth.,* 9, 131, 1984.

256. **Govier, D. A.,** Purification and partial characterisation of beet mild yellowing virus and its serological detection in plants and aphids, *Ann. Appl. Biol.,* 107, 439, 1985.

257. **Caciagli, P. and Casetta, A.,** Maize rough dwarf virus (Reoviridae) in its planthopper vector Laodelphax striatellus in relation to vector infectivity, *Ann. Appl. Biol.,* 109, 337, 1986.

258. **Torrance, L., Plumb, R. T., Lennon, E. A., and Gutteridge, R. A.,** A comparison of ELISA with transmission tests to detect barley yellow dwarf virus-carrying aphids, in *Developments and Applications in Virus Testing,* Jones, R. A. C. and Torrance, L., Eds., Association of Applied Biologists, Wellesbourne, 1986, 165.

259. **Cho, J. J., Mau, R. F. L., Hamasaki, R. T., and Gonsalves, D.,** Detection of tomato spotted wilt virus in individual thrips by enzyme-linked immunosorbent assay, *Phytopathology,* 78, 1348, 1988.

260. **Falk, B. W. and Tsai, J. H.,** Serological detection and evidence for multiplication of maize mosaic virus in the planthopper, Peregrinus maidis, *Phytopathology,* 75, 852, 1985.

261. **Tijssen, P.,** *Practice and Theory of Enzyme Immunoassays,* Burdon, P. H. and van Knippenberg, P. H., Eds., Elsevier, Amsterdam, 1985.

262. **Butler, J. E.,** *Immunochemistry of Solid-Phase Immunoassay,* CRC Press, Boca Raton, FL, 1991.

263. **Kemeny, D. M.,** *A Practical Guide to ELISA,* Pergamon Press, New York, 1991.

264. **Zimmermann, D. and Van Regenmortel, M. H. V.,** Spurious cross-reactions between plant viruses and monoclonal antibodies can be overcome by saturating ELISA plates with milk proteins, *Arch. Virol.,* 106, 15, 1989.

265. **Dietzgen, R. G.,** Characterization of antigenic structures on arabis mosaic virus with monoclonal antibodies, *Arch. Virol.,* 91, 163, 1986.

266. **Dietzgen, R. G. and Zaitlin, M.,** Tobacco mosaic virus coat protein and the large subunit of the host protein ribulose 1.5-biphosphate carboxylase share a common antigenic determinant, *Virology,* 155, 262, 1986.

267. **Dietzgen, R. G. and Zaitlin, M.,** Alleged common antigenic determinant of tobacco mosaic virus coat protein and the host protein rubulose-1,5-biphosphate carboxylase is an artifact of indirect ELISA and western blotting, *Virology,* 184, 397, 1991.

268. **Sluyser, M., Rümke, P., and Hekman, A.,** Antigenicity of histones: comparative studies on histones with very high lysine content from various sources, *Immunochemistry,* 6, 494, 1969.

269. **Dietzgen, R. G. and Francki, R. I. B.,** Nonspecific binding of immunoglobulins to coat proteins of certain plant viruses in immunoblots and indirect ELISA, *J. Virol. Meth.,* 15, 159, 1987.

270. **Clark, M. F. and Bar-Joseph, M.,** Enzyme immunosorbent assays in plant virology, *Meth. Virol.,* 7, 51, 1984.

271. **Korpraditskul, P., Casper, R., and Lesemann, D. E.,** Evaluation of short reaction times and some characteristics of the enzyme-conjugation in enzyme-linked immunosorbent assay (ELISA), *Phytopathol. Z.,* 96, 281, 1979.

272. **McLaughlin, M. R., Barnett, O. W., Burrows, P. M., and Baum, R. H.,** Improved ELISA conditions for detection of plant viruses, *J. Virol. Meth.,* 3, 13, 1981.

273. **Banttari, E. E., Clapper, D. L., Hu, S. P., Daws, K. M., and Khurana, S. M. P.,** Rapid magnetic microsphere enzyme immunoassay for potato virus X and potato leafroll virus, *Phytopathology,* 81, 1039, 1991.

274. **Koenig, R. and Paul, H. L.,** Variants of ELISA in plant virus diagnosis, *J. Virol. Meth.,* 5, 113, 1982.

275. **Van Regenmortel, M. H. V.,** Strategy for control of plant virus diseases: advances in serodiagnosis, in *Control of Virus Diseases,* Kurstak, E., Ed., Marcel Dekker, New York, 1984, 405.

276. **Stobbs, L. W. and Barker, D.,** Rapid sample analysis with a simplified ELISA, *Phytopathology,* 75, 492, 1985.

277. **Dekker, E. L., Porta, C., and Van Regenmortel, M. H. V.,** Limitations of different ELISA procedures for localizing epitopes in viral coat protein subunits, *Arch. Virol.,* 105, 269, 1989.

278. **Torrance, L. and Jones, R. A. C.,** Increased sensitivity of detection of plant viruses obtained by using a fluorogenic substrate in enzyme-linked immunosorbent assay, *Ann. Appl. Biol.,* 101, 501, 1982.

279. **Cooper, J. I. and Edwards, M. L.,** Variations and limitations of enzyme-amplified immunoassays, in *Developments in Applied Biology: Developments and Applications in Virus Testing,* Jones, R. A. C. and Torrance, L., Eds., Association of Applied Biologists, Wellesbourne, 1986, 139.

280. **Bryant, G. R., Durand, D. P., and Hill, J. H.**, Development of a solid-phase radioimmunoassay for detection of soybean mosaic virus, *Phytopathology,* 73, 623, 1983.
281. **Ghabrial, S. A., Li, D., and Shepherd, R. J.**, Radioimmunosorbent assay for detection of lettuce mosaic in lettuce seed, *Plant Dis.,* 66, 1037, 1982.
282. **Hemmilä, I., Dakubu, S., Mukkala, V. M., Siitari, H., and Lövgren, T.**, Europium as a label in time-resolved immunofluorometric assays, *Anal. Biochem.,* 137, 335, 1984.
283. **Siitari, H. and Kurppa, A.**, Time-resolved fluoroimmunoassay in the detection of plant viruses, *J. Gen. Virol.,* 68, 1423, 1987.
284. **Järvekülg, L., Söber, J., Sinijärv, R., Toots, I., and Saarma, M.**, Time-resolved fluorimmunoassay of potato virus M with monoclonal antibodies, *Ann. Appl. Biol.,* 114, 279, 1989.
285. **Stott, D. I.**, Immunoblotting and dot blotting, *J. Immunol. Meth.,* 119, 153, 1989.
286. **Harper, D. R., Ming-Liu, K., and Kangro, H. O.**, Protein blotting: ten years on, *J. Virol. Meth.,* 30, 25, 1990.
287. **Koenig, R. and Burgermeister, W.**, Applications of immuno-blotting in plant virus diagnosis, in *Developments and Applications in Virus Testing. Developments in Applied Biology,* Jones, R. A. C. and Torrance, L., Eds., Association of Applied Biologists, Wellesbourne, 1986, 121.
288. **O'Donnell, I. J., Shukla, D. D., and Gough, K. H.**, Electro-blot radioimmunoassay of virus-infected plant sap. A powerful new technique for detecting plant viruses, *Phytopathology,* 4, 19, 1982.
289. **Rybicki, E. P. and Von Wechmar, M. B.**, Enzyme-assisted immune detection of plant virus proteins electroblotted onto nitrocellulose paper, *J. l. Meth.,* 5, 267, 1982.
290. **Ehlers, U. and Paul, H. L.**, Characterization of the coat proteins of different types of barley yellow mosaic virus by polyacrylamide gel electrophoresis and electro-blot immunoassay, *J. Phytopathol.,* 115, 294, 1986.
291. **Laemmli, U. K.**, Cleavage of structural proteins during the assembly of the head of bacteriophage T4, *Nature,* 227, 680, 1970.
292. **Suzuki, N., Shirako, Y., and Ehara, Y.**, Isolation and serological comparison of virus-coded proteins of three potyviruses infecting cucurbitaceous plants, *Intervirology,* 31, 43, 1990.
293. **Towbin, H., Staehelin, T., and Gordon, J.**, Electrophoretic transfer of proteins from polyacrylamide gels to nitrocellulose sheets: procedure and some applications, *Proc. Natl. Acad. Sci. U.S.A.,* 76, 4350, 1979.
294. **Banttari, E. E. and Goodwin, P. H.**, Detection of potato viruses S, X and Y by ELISA on nitrocellulose membranes (Dot-ELISA), *Plant Dis.,* 69, 202, 1985.
295. **Hibi, T. and Saito, Y.**, A dot immunobinding assay for the detection of tobacco mosaic virus in infected tissues, *J. Gen. Virol.,* 66, 1191, 1985.
296. **Williamson, C., Rybicki, E. P., Kasdorf, G. G. F., and Von Wechmar, M. B.**, Characterization of a new picorna-like virus isolate from aphids, *J. Gen. Virol.,* 69, 787, 1988.
297. **Rocha-Pena, M. A., Lee, R. F., and Niblett, C. L.**, Development of a dot-immunobinding assay for detection of citrus tristeza virus, *J. Virol. Meth.,* 34, 297, 1991.
298. **Somowiyarjo, S., Sako, N., and Nonaka, F.**, Dot-immunobinding assay for zucchini yellow mosaic virus using polyclonal and monoclonal antibodies, *Ann. Phytopathol. Soc. Jpn.,* 55, 56, 1989.
299. **Berger, P. H., Thornbury, D. W., and Pirone, T. P.**, Detection of picogram quantities of potyviruses using a dot blot immunobinding assay, *J. Virol. Meth.,* 12, 31, 1985.
300. **Lin, N. S., Hsu, Y. H., and Hsu, H. T.**, Immunological detection of plant viruses and a mycoplasmalike organism by direct tissue blotting on nitrocellulose membranes, *Phytopathology,* 80, 824, 1990.
301. **Katz, D. and Kohn, A.**, Immunosorbent electron microscopy for detection of viruses, *Adv. Virus Res.,* 29, 169, 1984.

302. **Derrick, K. S.**, Quantitative assay for plant viruses using serologically specific electron microscopy, *Virology,* 56, 652, 1973.

303. **Roberts, I. M., Milne, R. G., and Van Regenmortel, M. H. V.**, Suggested terminology for virus/antibody interactions observed by electron microscopy, *Intervirology,* 18, 147, 1982.

304. **Nicolaieff, A., Katz, D., and Van Regenmortel, M. H. V.**, Comparison of two methods of virus detection by immunosorbent electron microscopy (ISEM) using protein A, *J. Virol. Meth.,* 4, 155, 1982.

305. **Shukla, D. D. and Gough, K. H.**, The use of protein A from Staphylococcus aureus in immune electron microscopy for detecting plant virus particles, *J. Gen. Virol.,* 45, 533, 1979.

306. **Lesemann, D. E. and Paul, H. L.**, Conditions for the use of protein A in combination with the Derrick method of immuno electron microscopy, *Acta Hortic.,* 110, 119, 1980.

307. **Van Regenmortel, M. H. V., Nicolaïeff, A., and Burckard, J.**, Detection of a wide spectrum of virus strains by indirect ELISA and serological trapping electron microscopy (STREM), *Acta Hortic.,* 110, 107, 1980.

308. **Cohen, J., Loebenstein, G., and Milne, R. G.**, Effect of pH and other conditions on immunosorbent electron microscopy of several plant viruses, *J. Virol. Meth.,* 4, 323, 1982.

309. **Roberts, I. M.**, Practical aspects of handling, preparing and staining samples containing plant virus particles for electron microscopy, in *Developments and Applications in Virus Testing,* Jones, R. A. C. and Torrance, L., Eds., Association of Applied Biologists, Wellesbourne, 1986, 213.

310. **Milne, R. G.**, Quantitative use of the electron microscope decoration technique for plant virus diagnostics, *Acta Hortic.,* 234, 321, 1988.

311. **Louro, D. and Lesemann, D. E.**, Use of protein A-gold complex for specific labelling of antibodies bound to plant viruses. I. Viral antigens in suspensions, *J. Virol. Meth.,* 9, 107, 1984.

312. **Bloomer, A. C. and Butler, P. J. G.**, Tobacco mosaic virus. Structure and self assembly, in *The Plant Viruses,* Vol. 2, *The Rod-Shaped Viruses,* Van Regenmortel, M. H. V. and Fraenkel-Conrad, H., Eds., Plenum Press, New York, 1986, 19.

313. **Raghavendra, K., Kelley, J. A., Khairallah, L., and Schuster, T. M.**, Structure and function of disk aggregates of the coat protein of tobacco mosaic virus, *Biochemistry,* 27, 7583, 1988.

314. **Durham, A. C. H., Finch, J. T., and Klug, A.**, States of aggregation of tobacco mosaic virus protein, *Nature,* 229, 37, 1971.

315. **Milne, R. G. and Lesemann, D. E.**, Immunosorbent electron microscopy in plant virus studies, *Meth. Virol.,* 8, 85, 1984.

316. **Barnett, O. W.**, Application of new test procedures to surveys: merging the new with the old, in *Developments and Applications in Virus Testing,* Jones, R. A. C. and Torrance, L., Eds., Association of Applied Biologists, Wellesbourne, 1986, 247.

317. **Takahashi, Y. and Shohara, K.**, Practical detection of wasabi strain of tobacco mosaic virus by using "cocktail" monoclonal antibodies, *Ann. Phytopathol. Soc. Jpn.,* 56, 621, 1990.

318. **Herrbach, E., Lemaire, O., Ziegler-Graff, V., Lot, H., Rabenstein, F., and Bouchery, Y.**, Detection of BMYV and BWYV isolates using monoclonal antibodies and radioactive RNA probes, and relationships among luteoviruses, *Ann. Appl. Biol.,* 118, 127, 1991.

319. **Sakamoto, H., Lemaire, O., Merdinoglu, D., and Guesdon, J. L.**, Comparison of enzyme-linked immunosorbent assay (ELISA) with dot hybridization using 32P- and 2-acetylaminofluorene (AAF)-labelled cDNA probes for the detection and characterization of beet necrotic yellow vein virus, *Mol. Cell. Probes,* 3, 159, 1989.

320. **Rose, D. G., McCarra, S., and Mitchell, D. H.,** Diagnosis of potato virus YN: a comparison between polyclonal and monoclonal antibodies and a biological assay, *Plant Pathol.,* 36, 95, 1987.

321. **Singh, R. P. and Somerville, T. H.,** Effect of storage temperatures on potato virus infectivity levels and serological detection by enzyme-like immunosorbent assay, *Plant Dis. Rep.,* 67, 1133, 1983.

322. **Torrance, L. and Dolby, C. A.,** Sampling conditions for reliable routine detection by enzyme-linked immunosorbent assay of three ilarviruses in fruit trees, *Ann. Appl. Biol.,* 104, 267, 1984.

323. **Ward, C. M., Walkey, D. G. A., and Phelps, K.,** Storage of samples infected with lettuce or cucumber mosaic viruses prior to testing by ELISA, *Ann. Appl. Biol.,* 110, 89, 1987.

324. **Torrance, L.,** Use of forced buds to extend the period of serological testing in surveys for fruit tree viruses, *Plant Pathol.,* 30, 213, 1981.

325. **Luisoni, E.,** Diagnosis of viruses in ornamental plants with special reference to serological methods: new developments, *Bull. OEPP/EPPO Bull.,* 19, 27, 1989.

326. **Lange, L.,** The practical application of new developments in test procedures for the detection of viruses in seed, in *Developments and Applications in Virus Testing,* Jones, R. A. C. and Torrance, L., Eds., Association of Applied Biologists, Wellesbourne, 1986, 269.

327. **Hampton, R., Ball, E., and DeBoer, S., Eds.,** *Serological Methods for Detection and Identification of Viral and Bacterial Plant Pathogens,* American Phytopathological Society Press, St. Paul, Minnesota, 1990.

Chapter 8

ELECTRON MICROSCOPY OF *IN VITRO* PREPARATIONS

Robert G. Milne

TABLE OF CONTENTS

0-8493-4284-8/93/$0.00 + $.50

I. INTRODUCTION

In this chapter I will attempt to review electron microscope methods used in plant virus diagnostics, with reference to *in vitro* preparations, as opposed to thin sections. The general path we shall follow is to: obtain an aqueous extract containing the virus in question (purified or not), and adsorb the particles to filmed electron microscope grids, which may, if appropriate, first be antibody-coated; again if appropriate, decorate the particles using specific antibodies and optional gold labeling; apply negative stain, and examine the results by transmission electron microscopy.

All this sounds simple, and mostly it is; however, there are not many laboratories where it is done effectively, and there are cautionary examples aplenty in the literature of how it has been done without much finesse.

The main advantages of negative staining are that it is quick, simple, and direct. From sample to result, processing time may be 5 min, or with immunosorbent or gold label steps, about 2 h, and perhaps overnight. The reagents used are stable and can be taken off the refrigerator shelf. The links in the chain of evidence are few, and most are actually visible, so relatively little is open to false interpretation. On the other hand, processing requires skill and is labor intensive, so that large numbers of samples cannot be handled easily. Though about as sensitive as ELISA, immunoelectron microscopy cannot now compete with the sensitivity becoming available using molecular hybridization methods, especially if these incorporate a polymerase chain reaction step.[1] On the other hand, suitable primer sequences or probes are not available for a great many viruses, especially the more difficult or less well known ones, although use of degenerate primers[1] partially overcomes this difficulty.

The rational solution is of course that, while in a sense these different approaches are competitive, they should be orchestrated into a system in which each technique supports the others. In many situations, electron microscopy is sufficient to solve simple problems, and is also very useful in the phase of "breaking in" a less direct technique such as ELISA, or in establishing the precise specificity of a monoclonal antibody.

Electron microscope methods for examining particulate preparations of viruses have been reviewed extensively.[2-20]

II. THE ELECTRON MICROSCOPE

A. IMAGING AND RESOLUTION

Electron microscopes are expensive to buy and run, but whoever does not have one in the laboratory may be able to share, and whoever cannot afford a new one can get a very good and much less expensive second-hand model through the regular manufacturers. Whoever does not have access to an electron microscope at least sometimes is in difficult straits, but grids *can* be mailed to laboratories equipped and willing to receive and examine them.

Those with access to a microscope may not find it running in good condition, and this may be because no single competent person of good will is dedicated to maintaining the instrument, or because there are too many users, or indeed because spares and maintenance are too costly. A simple test of image quality is to examine a support film at 100,000 × screen magnification through the binoculars and see if the "grain" in the image is clearly visible near focus. This is important because it should form the basis of astigmatism correction, and without constant attention to this factor, no good images, let alone micrographs, can result. Having obtained this grain structure, photograph it at 40,000 × and make a print enlarged by a further 12.5 ×. Can you easily distinguish pairs of grains (not of course the photographic grain) separated by less than 0.5 mm? This should be possible, and indicates that your instrument is resolving a modest 1-nm separation.

A harder test, but also an important one for those interested in negative staining of virus particles as an aid to diagnostics, is the following. Take almost any old cigarette end (still commonly available material!) and homogenize a fragment in about ten volumes of water; collect a sample on a support film, negatively stain, and photograph the abundant tobacco mosaic virus (TMV) particles at 30,000 to 40,000 ×. Can you resolve the 2.3-nm primary helix? Some tips are (1) you may need liquid nitrogen cooling for the specimen; (2) do not look at the particle first! By that time it will be a cinder; photograph it by some variation of the minimal exposure technique of Williams and Fisher;[21] (3) do not be fooled by the very occasionally visible stacked-disk arrangement of coat protein subunits only, which is more resistant to beam damage and has a repeat of 2.5 nm, so is more easily seen.[22]

B. SIZE CALIBRATION

Often it is important to know how large a virus particle is, and although estimates within 10% are not difficult to obtain, those within 5 or 3% are less easy. Apart from the relatively simple problem of calibrating the instrument magnification, presumably identical virus particles in a field of negative stain have an unhappy tendency to appear of different sizes, especially on different parts of the grid or grid square, according to the thickness of the layer of stain. In addition, virus particles deposited on support films that have been treated by glow-discharge in air (see Reference 11) have a strong tendency to flatten, and therefore appear with an increased diameter. Even without glow-discharge, flattening often occurs, and if the carbon layer on a carbon-plastic support film is too thin, stretching may also occur in the neighborhood of the virus particle, generating artifacts. Finally, if a stain, such as uranyl acetate, is used incorrectly, virus particles may appear positively stained and shrunken or "mummified". Possibilities for retaining the original shape of virus particles more faithfully using freeze-drying and freeze-substitution have been discussed by Roberts.[18]

Calibration of magnification can be done in a number of ways, each of which is subject to certain problems (see discussions in Hayat and Miller[23] and Milne[8]). Therefore, it is best to employ at least two methods.

1. Diffraction Grating Replicas

Square grating replicas of 2160 lines/mm are commercially available. These make useful magnification standards up to about 20,000 ×, but above this, too few lines can be photographed for a reliable measurement. However, once a magnification near, shall we say, 10,000 × has been calibrated, a given portion of the replica (or another stable specimen) can be photographed at successively higher magnifications, and the distances between recognized points can be compared.

2. Catalase

Beef liver catalase crystals already mounted on grids and negatively stained are commercially available and should be kept at hand. There has been some dispute about the exact sizes of the spacings,[23,8] but the generally agreed figures are 8.75 and 6.85 nm.

3. Tobacco Mosaic Virus (Modal Length)

If the lengths of about 100 TMV particles are measured, a sharply defined modal length should emerge, and this will be close to 300 nm, though it should be checked against other calibration standards. Once a convenient source of TMV has been so calibrated, it can be used as an internal standard, by mixing it with the preparation to be measured and photographing both types of particle together.

4. Particle Diameters of Certain Viruses

The particle diameters of TMV ($\approx$18 nm), alfalfa mosaic virus ($\approx$18 nm), a *Geminivirus* such as maize streak (shorter dimension, $\approx$18 nm) or tomato bushy stunt virus ($\approx$34 nm) also make convenient, if somewhat less accurate, internal standards. The point here is not so much to obtain an absolute measurement on the unknown virus (this may be illusory), but to be able to say, as for example with white clover cryptic virus 2 "If tomato bushy stunt virus is assumed to be 34 nm, then our virus is 38 nm in diameter",[24,25] or "Ourmia melon virus has a diameter indistinguishable from that of TMV, alfalfa mosaic virus, and digitaria streak *Geminivirus.*"[26] For the purpose of morphological or size comparisons, the importance of co-imaging the different viruses on the same photograph cannot be overemphasized,[27] though it is not often done.

III. GRIDS AND SUPPORT FILMS

A. GRIDS

For normal negative staining, 400-mesh grids are recommended. Grids with wider meshes render the support film more unstable or liable to breakage. These defects can, of course, be corrected by making thicker films, but this sacrifices contrast and resolution.

For preparations where wet material is in contact with the grid for not more than about 40 min at room temperature, and where the pH is not far from 7, copper grids can be used, though it is more convenient to use copper/ rhodium or copper/palladium grids that have easily distinguished "coppery" and "silvery" faces. When incubations take much longer or involve say Tris-HCl buffer at pH 9.6, the preparation may turn blue and the face of the experimenter red. Nickel grids can be substituted, and the fact that these are magnetic brings problems and benefits. Nonmagnetic forceps must be used, otherwise you cannot let go of the grids; on the other hand, such grids can be incubated in a humidified petri dish placed on top of a magnetic stirrer. The grids then flip or twitch, and diffusion path lengths (hence the necessary incubation times) can be appreciably reduced.[28]

With magnetic stirring, simply floating the grids on small drops (20 μl) placed on a Parafilm surface is insufficient, as the drops tend to migrate and coalesce. We confine the drops in small dimples made in sheets of dental wax, as follows: slide the dental wax into a transparent polyethylene envelope and place it over the wells of an enzyme-linked immunosorbent assay (ELISA) microtiter plate, on a light-box; use the rounded end of a small glass test tube to press out the dimples.

Grids can, if necessary, be recycled after use by brief shaking (preferably sonication) in a solvent for the plastic of the support film, followed by, or combined with, an agent such as a dilute acid, that will clean the metal. We use a solution of 70 parts acetone and 30 parts aqueous 0.1 N HCl (v/v). There follows a rinse in acetone, before the grids are spread out to dry. New grids sometimes benefit from similar cleaning before use.

B. SUPPORT FILMS
1. Preparation

Making support films is now a routine procedure that has been reviewed many times.[10,18,29-34] Opinions still vary over the best type of routine support film, but generally a plastic film (tough but liable to move or melt under the beam) is first prepared, and a carbon layer (brittle but rigid) is evaporated over this. The combination should be as thin as possible without leading to excessive breakage. Because carbon evaporation involves the use of expensive equipment, some workers attempt to do negative staining using plastic films alone; this nearly always gives bad results because of the instability of the film and because virus particles and negative stains tend to adsorb to such films very poorly. Our protocol for preparing films is as follows:

1. Polyvinyl formaldehyde (Formvar, Pioloform) is dried at 40°C overnight and a 0.7% solution is made in dichloroethane. This solution keeps well if the solvent is not allowed to evaporate.
2. A vessel about 15 cm in diameter and 10 cm deep (preferably painted matte black on the inside) is overfilled with deionized or distilled water, and the surface is swept clean with a glass rod just before use.
3. One drop of the Formvar solution from a Pasteur pipette is allowed to fall on the water surface from a few millimeters above. This spreads rapidly and the solvent evaporates, leaving a film. (Note that dichloroethane vapor is poisonous, but that too much draft from a fume hood can cause irregularities in the film.)

4.	At least some of the film should have a uniform pale-grey interference color, as seen by reflected fluorescent light. If the film is too thin (colorless), try placing 2 drops rapidly on the clean water surface; alternatively, increase the concentration of Formvar in the solution. If the film is too thick (bright silver color or gold), use a more dilute Formvar solution. Further adjustment can be made by using a pipette with a finer tip that delivers a smaller drop. If the film is unsatisfactory, sweep it away with the glass rod and try again; after two or three attempts, renew the water in the vessel.

5.	Cleaned and flat grids are lowered onto suitable areas of film and should be seen to adhere strongly.

6.	A free edge of the film is now pushed down with the end of a clean glass slide, so that the grids are trapped between the film and the glass. A U-shaped motion brings the slide out, and it is left to dry. Alternatively, Parafilm is lowered onto the film, then lifted off, bringing film and grids with it.

7.	The dried film and grids and their glass or Parafilm backing are placed in a carbon evaporator, and when the vacuum is better than 10^{-4} torr (0.013 Pascal), a thin layer of carbon is deposited. This should be a barely visible pale brown.

8.	The grids are now ready to use, but may benefit from conditioning by glow-discharge or other treatments.[11]

2. Glow-Discharge

A persistent problem encountered in negative staining, especially of purified virus preparations and especially in some laboratories, is that support films are too hydrophobic or become so with age, and both the preparation and especially the negative stain tend to roll off the grid. A convenient remedy for this condition is to subject the grids to glow-discharge.[11,23,31,35-40] This exposes the support film on the grid to a plasma of ionized air molecules under reduced pressure (around 0.1 torr) created using a high-voltage discharge (about 500 V ac) for 10 to 15 s. The surface of the support film becomes negatively charged (and is also probably cleaned or etched), with the result that it is rendered strongly hydrophilic for up to an hour and possibly longer, after treatment. Glow-discharge using gases other than air can produce positive charges (see Table 1 in Reference 11), though this is less interesting for our immediate purpose.

The required vacuum can be obtained using a simple rotary pump, and a transformer to produce the necessary voltage should not be difficult to obtain. Construction of a relatively inexpensive "home-made" apparatus based on a plastic desiccator, a rotary pump, and a Tesla coil has also been described.[35,38]

As noted above, glow-discharged grids are most useful when it is necessary to negatively stain purified preparations. However, with crude or partially purified preparations, another favorable effect is sometimes noted. We have found for example, with viruses as diverse as maize rough dwarf (Reoviridae), tomato yellow leaf curl (a *Geminivirus*) and with an *Ilarvirus* from tomato, that grids subjected to glow-discharge usually adsorb several times more virus particles than untreated grids. This is especially so where preparations are rich in the small disc-like particles of "f1 protein" (Rubisco); glow-discharge may largely eliminate these, and correspondingly increase the amount of virus on the grid.[41]

If glow-discharge facilities are not available, grids can be precoated with 0.5 μg/ml polylysine (mol wt 4000) for 5 min; this is sometimes a good low-technology solution to the problem of hydrophobic support films.[42]

IV. SAMPLE PREPARATION

Sample preparation is very important for successful detection and recognition of plant viruses in fresh material (leaves, roots, phloem, insect or nematode vectors, etc.); unfortunately, good results come as much from cunning, cookery, and folklore as from theory and high principles, and the cocktail used for extraction should vary according to the characteristics of both host and virus.

A. EXTRACTION METHODS

There are three ways to approach the problem of extraction, the commonest of which is the first described below.

1. Homogenization

A small piece of tissue (for example a square of leaf 2 × 2 mm) can be placed on a glass slide in a small drop (about 15 μl) of extraction cocktail (see the following section) and homogenized using a flat-ended rod 3 mm in diameter, made of plastic, glass, or nonporous wood. This is an effective low-technology approach.

A better system is based on the use of small Eppendorf-type tubes, and rods with an end-shape conforming to that of the bottom of the tube. In the simplest procedure which we use, both the end of the rod and the bottom of the tube are flat, but there is a simple method for making grinding rods of a desired shape.[17,18,43] A small amount (25 μl) of epoxy resin mixed with sand or carborundum is polymerized in the bottom of the Eppendorf tube with a wooden toothpick or cocktail stick held vertically in place. After polymerization, the tool is easily released from the tube used as a mold.

A convenient load for the Eppendorf tube is 50 μl of cocktail and a sample of tissue 1/5 or 1/10 of this by weight or volume; Paliwal[43] used 20 μl when homogenizing single aphids. A rotary rather than pounding motion is most efficient for homogenization, and a small electric drill has been recommended for the purpose,[18] though there appears to be some risk of overheating the sample with this method. If the sample is tough, or perhaps in any event, a very small amount of washed carborundum or fine sand may aid homogenization.

A small sample of the homogenate is conveniently taken up in a small glass capillary, and used directly. Even if this liquid is very turbid, rinsing the grid (see below) will usually ensure a satisfactory preparation. The sample in the Eppendorf tube is, however, easy to centrifuge at low speed for a few minutes to sediment the major debris, if this is found beneficial. The supernatant is then used for electron microscopy.

2. Crushing

Gentle crushing rather than grinding will help to preserve the integrity of filamentous virus particles, and also the characteristic "bunch of grapes" appearance of aggregates of tomato spotted wilt virus particles, much easier to recognize than single dispersed particles. Large numbers of samples can be processed with a Pollähne press (Meku, Wennigsen, Germany) or similar device that squeezes out the juice without excessive shearing of the sample. One possibility is to seal the sample plus buffer in a small polyethylene bag containing gauze to prevent slippage of leaf material, and run over this with ball bearings (Bioreba Basel, Switzerland).

3. Leaf Dip

This method, developed originally by Brandes,[44] is still sometimes useful. It does not extract particles very efficiently, but it does so rapidly, without breaking them and without generating a lot of cell detritus (and presumably releasing many enzymes) as is usual with the process of homogenization.

Essentially, a drop of negative stain such as phosphotungstate (PTA) (see Section V.B.1) is placed on the grid, and the freshly cut edge of a leaf is drawn through it so that the contents of cut cells are released. (Alternatively, a fragment of epidermis is stripped from a leaf and passed through the drop.) The drop is then drained, preferably from below, and allowed to dry without further treatment. (Originally, the cut leaf was drawn through a drop of water, which was dried down, and the grid was then given contrast by metal shadowing.)

4. Spreading[45]

A 9-cm Petri dish is filled with water or another solution such as 1% buffered glutaraldehyde. Immediately before use, the surface of the liquid is swept with a glass rod, and a few particles of talc are sprinkled on the surface to act as markers. A piece of tissue is crushed with the end of a glass rod which is immediately touched to the liquid surface. The lipids and proteins in the sample cause it to spread out as a surface film in which virus particles are trapped and, if necessary, chemically fixed. Meanwhile salts, sugars, and other interfering substances dissolve in the bulk of the liquid and are lost. A grid is touched to the surface within the area defined by the talc, and after rinsing if necessary, is negatively stained.

Very clean preparations can be obtained in this way, and labile particles can be fixed (e.g., with glutaraldehyde) within seconds of extraction from the host.

B. EXTRACTION COCKTAILS

An advantage of negative staining is that one can get results quickly, so feedback is rapid. In an untried situation (unknown virus, novel host plant), simple things should be tested first, followed by progressively more sophisticated alternatives. The following progression is recommended.

1. Extract the sample in distilled water. This often works very well.
2. Extract in 0.1 M phosphate, pH 7, or in 0.05 M Tris-HCl, pH 7.2; these are good standard extraction buffers. Where plants with acid sap (e.g., grapevines) are handled, special care should be taken to keep the pH of the extract not lower than 6.5.
3. For particular situations, special buffers may be better (i.e., result in more virus particles on the grid). For some detailed examples, see References 17 and 18. Cucumber mosaic virus particles should be extracted in 0.5 M phosphate, pH 7, or 0.5 M citrate, pH 6.5; geminiviruses should generally be extracted in 0.01 M Tris buffer, pH 8, containing 0.1 M ethylene diaminetetracetate (EDTA), although maize streak *Geminivirus* was most efficiently extracted in phosphate buffer, pH 5, or citrate, pH 4.8.[46] Potyviruses may be extracted more efficiently in 0.5 M phosphate buffer containing 0.01 to 0.1 M EDTA.
4. Glutaraldehyde at levels of 0.01 to 0.1% can often increase the numbers of particles obtained on the grid.[47] It is not clear whether this effect is related to fixation phenomena.
5. Sap from pelargoniums, fruit trees, grapevines, or other woody species, and seeds frequently contain substances that upon extraction form complexes with virus particles and prevent their detection on grids. Very often, addition of 1 to 3% polyethylene glycol (mol wt 6000), 2% polyvinylpyrrolidone (mol wt 24,000 to 45,000) or 2 to 4% nicotine makes dramatic improvements. With grapevine fanleaf virus in crude extracts of young grapevine bark, these three substances were very effective, whereas 1% sodium sulfite was ineffective, though it was the best additive for preventing oxidative browning of the sap.[48]
6. The reducing agents commonly used in virus purification (e.g., 0.02 M concentrations of ascorbic acid, dithiothreitol, 2-mercaptoethanol, or sodium sulfite) may be very useful, if excessive oxidation of the sap is suspected. However, reducing agents have been reported to abolish or alter antigenicity of some virus particles because of activation (or failed suppression) of proteolytic enzymes.[11,49-52]
7. If incubation at room temperature or above is prolonged (say, more than 2 h), sodium azide should be incorporated in the extract to prevent

bacterial growth. It is prepared as a 1/50 (w/v) stock solution (very poisonous!) and used at a further dilution of 1/100 (final dilution 1/5000).

8. If proteolytic degradation is suspected, protease inhibitors can be incorporated. All such substances are poisons! Many are available in catalogues of chemical and biochemical companies, and a commonly used one is phenylmethylsufonyl fluoride (PMSF), which is made up to 0.05 M in isopropanol and added to the cocktail at 1/250, giving a final concentration of 0.2 mM.[18,53]

9. Traces (0.01 to 0.1%) of Tween 20 may give cleaner preparations, and especially may abolish "pebbly" backgrounds often encountered with negatively stained crude preparations. Tween, however, can sometimes prevent adhesion of virus particles to the grid.

C. FIXATION

Some kinds of virus particle are unstable *in vitro*; some negative stains disrupt certain viruses; antibody decoration procedures may induce virus breakage. For all these reasons, it is sometimes useful to be able to fix the virus particles. Fixation has been discussed extensively by Hayat,[54] and is most commonly achieved using glutaraldehyde (GA) because this often preserves morphology without compromising antigenicity. However, for at least one virus (maize rough dwarf virus, Reoviridae), GA causes swelling and fails to fix the virus particles, whereas the bifunctional cross-linking agent dithiobis(succinimidylpropionate) (DSP) was found to be an effective fixative.[55] Many other such agents exist,[56,57] and could be tried in case of difficulty. Other alternatives to GA are formaldehyde and acrolein.[54]

All these fixatives may well leave antigenicity relatively unaffected, although poor results are sometimes obtained with certain monoclonal antibodies, which may fail to recognize antigens that are altered only slightly. Fixatives such as permanganate or osmium tetroxide should be used only where conservation of antigenicity is not important.

GA can be used in the extraction cocktail, or once the virus is adsorbed to the grid. If used in the cocktail, low levels of GA (0.01 to 0.1%) will have some fixative effect but may serve mainly to obtain more particles on the grid (Section IV.B). To obtain good fixation, levels of 0.2 to 1.0% should be used, for about 1 h with stirring at 4°C. A possible adverse result of incorporating GA into the extraction cocktail is that many small fragments of tissue debris may then adhere to the grid or even to the virus particles. giving an undesirable background.

Alternatively, the virus is extracted and adsorbed to the grid before fixation. After rinsing with buffer and draining, the grid is incubated with about 0.1% buffered GA for 15 min at room temperature. Other fixatives may be handled in a similar manner. The grid is again rinsed with buffer before proceeding. If residual free aldehyde groups are suspected of giving trouble

9(e.g., by causing nonspecific attachment of IgG molecules), the grid can at this point be rinsed in 0.02 *M* ammonium chloride, glycine, or Tris buffer.

Labile virus particles adsorbed to support films are often more stable than their counterparts in suspension.

V. NEGATIVE STAINING

A. PROBLEMS ENCOUNTERED

Hayat and Miller[23] have recently devoted a whole book to negative staining, and there have been many other reviews and commentaries.[8,17,18,31,33,58-68] This will therefore be a very simplified account concentrating on essentials. It is worth noting that an increasingly interesting alternative to negative staining, especially for high-resolution studies, is cryoelectron microscopy;[69] a good illustration of the method is the paper by Hewat et al.[70]

Negative staining can go wrong in a number of ways, but most of these problems have solutions. The following discussion may be helpful.

1. **Resolution and contrast are poor** — The support film is too thick; the preparation is too concentrated; rinsing of the grid has been insufficient; you may have been using ammonium molybdate stain (Section V.B).
2. **The film dilates and stretches under the virus particles** — This causes them to distort; typically, filamentous virus particles may become partly "inflated". The carbon part of the support film is not thick enough, and the plastic portion is stretching under thermal load.
3. **There is too much material on the grid** — Dilute the preparation, and especially rinse the grid more thoroughly before staining.
4. **There is too little material on the grid** — Apply immunosorbent electron microscopy (ISEM) (Section VI); try using glow-discharge; leave the preparation to adsorb to the grid for longer times; try varying the extraction buffer (molarity, pH, addition of EDTA or GA, etc., see Section IV.B); apply one or more purifiction and concentration steps prior to electron microscopy.
5. **The stain does not adhere to the support film** — Try glow-discharge or try precoating with polylysine (Section III.B.2); try incubating the grid for longer times, both with the preparation and with the stain. Very hydrophobic support films may indicate that oil vapor is contaminating the carbon-coating unit.
6. **Virus particles are damaged or destroyed** — Change the type of negtive stain (you have been using PTA!) (Section V.B); fix the particles (Section IV.C); incorporate into the preparation selected antioxidants or enzyme inhibitors (Section IV.B).

7. **Some particles appear positively stained, shrunk, and "mummi- fied", against a background free of stain** — You have been using uranyl acetate on a hydrophobic-grid-plus-purified-virus combination! Try glow-discharge or polylysine; seek other areas on the grid; rinse the grid with water more carefully.

8. **Black precipitates or crystalline deposits have ruined the grid** — You have used uranyl acetate or another sensitive stain without sufficient care to eliminate contamination from traces of sap or buffer, especially between the tips of the forceps (Section V.B); you have failed to rinse away traces of buffer, sucrose, cesium chloride, etc. before staining.

9. **The background is "pebbly" or covered in white spots** — There is no sure remedy known for this disease; try again, examine a different part of the grid; try glow-discharge; try incorporating 0.01% Tween 20 into the preparation or the stain.

B. NEGATIVE STAINS

A negative stain should give high contrast (except when you are trying to see gold particles!) and high resolution, while supporting the particle against flattening. It should be easy to prepare and use, be reliable, and should not degrade the specimen, but rather should protect its morphology against beam damage. Many kinds of heavy metal salt have been tested, but none has proved to possess all the desired qualities, though naturally each negative stain has its faithful band of devotees. The characteristics of a few of the best negative stains are outlined below.

1. Phosphotungstate (Phosphotungstic Acid, PTA)

Interestingly, when used as the acid, at pH values of 3 to 5, 2% PTA is a rather muddy negative stain of low contrast and seemingly low resolution, but it causes little or no damage to labile specimens.[71] When, as is more usual, the acid is brought to near pH 7 with KOH, it gains high contrast but destroys ribosomes, damages host membranes, and attacks a significant num- ber of kinds of plant virus, unless these are first fixed. Among those viruses damaged are alfalfa mosaic, ilarviruses, some closteroviruses, fabaviruses, nepoviruses (more top-component, i.e., empty particles are generated), reo- viruses, rhabdoviruses, tospoviruses, and some geminiviruses. However, aqueous neutral PTA is stable and easy to use, since it is tolerant of various buffers or the presence of sap in the preparation.

2. Sodium Silicotungstate (SST)

This stain, made up to 2% in water, may be prepared from the acid and neutralized with NaOH, or from the sodium salt (pH near 9) which is neu- tralized with hydrochloric, sulfuric, or acetic acids; the staining results are the same. SST has properties very similar to those of PTA, including its propensity to damage unfixed viruses,[11,46] and has been little used by plant virologists, although it has been favored for some high-resolution studies of antibodies and of antibody-antigen binding,[59,72,73] because it apparently gives finer resolution than PTA or uranyl acetate.[23,70] SST tends to produce images that give contrast to the upper side of virus particles, at least the larger ones such as adenovirus, whereas uranyl acetate (see below) tends to give a more confused image that combines contrast from both the upper and lower surfaces.[68]

3. Uranyl Acetate (UA)

This poisonous and slightly radioactive stain should be handled with care. It is not the easiest stain to use, especially for beginners, because it precipitates in the presence of even small amounts of sap constituents, phosphate, or buffer above about pH 5.5, so that careful washing of the grid is necessary, and measures should be taken to avoid contamination of the forceps holding the grid. In addition, it may, under certain conditions, produce positive staining (Section V.A), especially of nucleic acid cores of virus particles.

On the other hand, when handled correctly, UA gives a stain of high contrast and fine grain[58] (unless the specimen is overtoasted in the beam, when it can become grainy). The great majority of viruses are not damaged, but an exception is rhabdoviruses, whose morphology may be transformed into that of a fried egg[58] and whose lipid envelopes are often disrupted, in the absence of fixation. A 1% solution makes a good routine stain; this may be reduced to 0.1% if gold labeling is used (but see under ammonium molybdate).

4. Uranyl Formate

Uranyl formate is handled and behaves like UA, except that in solution it is unstable, especially in the light, and should be freshly made up. It gives slightly better resolution of viral substructure than UA.

5. Ammonium Molybdate

This is an easy and flexible stain to use and it causes little or no damage to virus particles; unfortunately, the level of contrast obtained is disappointing. A 2% aqueous solution, pH 5.5, can be adjusted with HCl or ammonia between pH 4 and 9, though our experience[74] suggests that most often pH 4 is best. The stain is stable and can be mixed with or added to preparations without previous washing (though some prior washing is generally beneficial). For giving support ("sustain") and a deliberately low level of contrast to gold-labeled preparations, 0.3 to 0.5% ammonium molybdate, pH 6.5, is very useful.[75]

C. NEGATIVE STAINING TECHNIQUE

If possible, the tips of forceps used to hold the grid while it is negatively stained should be slightly bent inwards and should be hydrophobic, so that liquid does not get trapped between them. The tips of our forceps are washed briefly with soap and water under a dribbling tap while being rubbed with the fingertips, and are then dried. This procedure maintains the necessary hydrophobic surface. With Swiss forceps it may be impossible to bend in the tips without fracture, but with the Italian brand we use, this is possible; moreover, the Italian steel is more permanently stainless and not prone to eventual rusting.

All operations on the grid should, if possible, be confined to the filmed side, while the other side remains dry. On the other hand, the operational face of the grid should never be allowed to dry out until negative staining is completed.

For negative staining, the grid can be held in forceps and the face rinsed with a few drops of water, followed by 5 drops of negative stain. The grid is then drained by holding absorbent paper to the edge, and allowed to dry. When using UA or UF, up to 30 drops of water should be used, and if liquid does get between the points of the forceps, these also need careful rinsing.

As a slightly more laborious but more foolproof alternative, the grid is briefly floated on a series of small water drops, with a rinse of the forceps between each change, before the negative stain is applied.

VI. IMMUNOSORBENT METHODS

A. INTRODUCTION

Immunosorbent electron microscopy (ISEM), also sometimes called solid-phase immunoelectron microscopy (SPIEM), is the procedure in which an EM support film is first coated with a layer of antibody, which then serves to trap the virus particles in a subsequently applied virus preparation. It is helpful to recognize that ISEM is very similar in principle, and even in method, to the first half of the traditional double antibody sandwich (DAS) ELISA, the second half of such an ELISA being more similar to electron microscope decoration (Section VIII).

ISEM is used for two main purposes. One is to trap increased numbers of virus particles on the grid, while reducing the amount of host material collected; in effect we are making use of an *in situ* immunopurification step. Obtaining more and cleaner virus on the grid can in turn be advantageous in three ways: it may simply improve detectability of the virus; it may produce grids that are better suited to the application of decoration methods; and it may allow trapping of more particles of an elongated virus with a view to improving estimates of modal length. However, for this last purpose, it has been noted[76] that there may be quite strong selection by ISEM for shorter particles and particle fragments, and this may compromise establishment of modal lengths.

The second main purpose of ISEM is to make use of the trapping response to estimate degrees of serological relationship between viruses. The number of virus particles trapped depends, among other things, on the degree of affinity of the viral surface epitopes for the antibody, and a grid coated with a given antibody will trap more homologous virus, less of a related but heterologous virus, and less still of an unrelated virus. Thus, a ratio can be derived, which is the number of virus particles adsorbed on the antibody-coated grid, divided by the number of particles from the same preparation adsorbed on a control grid (usually uncoated, but perhaps coated with preimmune serum). For a given antibody, this ratio can be obtained for preparations of different viruses, or different antibodies can be tested with the same virus. Examples of relationships being established between viruses using ISEM are found in References 46 and 77 to 87. Counting fields of particles trapped by ISEM seems, to this author, a laborious (though undoubtedly valid) way to measure serological differences between viruses, especially as variability in counts may be high. An often easier (though not quite equivalent) alternative is to trap the viruses, by ISEM if necessary, and then use quantitative decoration methods (Section VII.C) to measure such differences.

The first to introduce ISEM, under the name "serologically specific electron microscopy" (SSEM) was Derrick,[88] and there soon followed a number of key papers reporting either improvements in the steps involved or a deeper understanding of them.[15,40,43,46,77-86,89-98] The current situation has been reviewed a number of times[2-4,6-18,99,100] and it seems appropriate here only to summarize the present scene.

B. SUMMARY OF ISEM METHODS AND RESULTS
1. The Trapping Effect

In crude virus preparations incubated directly on support films, virus particles are in competition with host materials for a firm anchorage on the grid, and this reduces the number of virus particles retained, while giving a "dirty" background.

If the virus preparation is incubated on grids coated with an antibody layer, particles recognized by the antibody will, after a rinsing step, tend to be retained and other material will tend to wash off. The recognized virus particles are not in competition with "dirt" in the preparation for binding sites on the antibody layer, and depending on many cirumstances, antibody-coated grids can trap severalfold to over 10,000-fold more particles than untreated grids, while, as we have noted, coming out much cleaner.

As we can understand from the above, numbers of particles trapped from purified preparations may not be increased by antibody coating since there is no competing impurity to prevent efficient adsorption of virus in the first place. When there is an increase, this may be because the IgG adsorbs to the grid more efficiently than does the virus alone, under the conditions prevailing.

It should be added that unpurified virus preprations also may fail to respond to trapping by antibody. A major reason for this may be the presence of free coat protein subunits that bind to the antibody and block its capacity to trap the intact virus.

2. Antibody Dilutions

For crude antisera, optimal dilutions for coating are in the region of 1/1000. Using undiluted antisera or those diluted less than about 1/500 causes inhibition in the sense that numbers of virus particles trapped become less than maximal, almost certainly because serum proteins orther than IgGs successfully compete for sites on the grid surface, and either squeeze out or sterically impede the IgGs.

With purified IgG, an optimal concentration for coating may be 1 to 10 μg/ml or, alternatively, the same concentration as found best in ELISA. For purified IgG, the inhibitory effect encountered with crude serum largely disappears.

3. Antibody Dilution Buffers

It seems possible to vary quite widely the type, molarity, and pH of the buffer used to dilute the coating antibody, and still obtain good results. Phosphate and Tris have been particularly favored, over a range of 0.02 to 0.3 M and pH vaues from 6.5 to 8.0. Carbonate buffer, pH 9.5, has also been used successfully. Often, saline (0.85% NaCl) is added to dilution buffers, but it is not always clear that this addition improves the result.

4. Coating Times and Temperatures

Coating time can conveniently be 5 min at 25°C, and after this period nearly all of the effective coating that is possible appears to have taken place. Sometimes up to 60 min has been preferred, in the belief that variability in numbers of particles trapped may thus be reduced.

5. Virus Dilution Buffers

In Section IV.B we discussed suitable cocktails for extracting plant viruses, and much of this applies to ISEM. Perhaps the only ingredient that should *not* be included is glutaraldehyde or another fixative, as this will induce nonspecific cross-linkages. Often buffers such as 0.05 M Tris-HCl, pH 7.2, or 0.05 to 0.1 M phosphate, pH 6.5 to 7.0, are used, but for best results, type of buffer, pH, and molarity should be tested in each case. These parameters can have considerable effects on numbers of particles trapped, though the effects are not, so far, predictable.[86]

6. Virus Incubation Times and Temperatures

"Short" incubations — After coating, the grid is rinsed with buffer and drained but not dried. It can then be incubated with the virus preparation, at room temperature, for a short period, such as 15 min, usually long enough to trap considerable numbers of virus particles while eliminating much impurity. This may well be sufficient for the question in hand, in which case further incubation will serve no purpose. However, this system will not give the highest sensitivity in terms of numbers of particles trapped or ultimate dilutions from which virus particles can still be detected.

"Long" incubations — To maximize sensitivity, longer virus adsorption times are indicated, and these may be several hours at room temperature, or overnight, or even several days at 4°C.[4,7,13,17,99] With grapevine fanleaf nepovirus, maximal numbers of particles were trapped after 8 days of incubation (270 times more than after 15 min).[99] However, overlong trapping times may result in the degradation of particles, loss of those already adsorbed, or other deleterious reactions, such as oxidation. For example, Lesemann[7] observed that more particles of a *Geminivirus* were trapped after 1 h than after 4 h. When a virus is to be trapped using an antibody that is related but not homologous, and which therefore has lowered affinity for the antigen, long incubation times may be especially useful.

Incubation times mentioned above and in the literature refer to static incubation, but as noted (Section III.A),[28] use of nickel grids and magnetic stirring can appreciably shorten the incubation times required.

7. Protein A, Protein G

Protein A (PA), a wall protein from *Staphylococcus aureus,* binds up to two IgG molecules by the Fc portion, leaving the active Fab arms free. PA binds well to human and rabbit IgGs but less well to goat, rat, and mouse IgGs, and not to chicken IgGs.[101,102] Protein G, derived from human group G *Streptococcus* binds with greater affinity than PA to goat, rat, and mouse IgGs but otherwise has similar binding properties;[103] native protein G has a separate binding site for albumins but in engineered recombinant protein G this potentially confusing site has been eliminated.[104] The following refers only to PA.

The use of PA has been amply discussed.[4,7,11,12,14,17,18,40.85,87,94,105-107] Normally, a concentration of 10 µg/ml is coated for 5 min at room temperature, and after rinsing, the antiserum coating follows. Shukla and Gough[108] were the first to use PA to precoat grids before adding a layer of IgG, the aim being to concentrate the IgG fraction of crude serum on the grid, and also obtain a favorable orientation with the Fab arms extended. However, initial hopes that PA precoating would dramatically improve the sensitivity of ISEM in trapping virus particles were not realized and it is fair to say that many workers do not think PA precoating gives enough improvement to be worthwhile.

Nevertheless, PA can be useful on occasion. As noted above (Section VI.B.2), grids coated directly with crude antiserum will trap fewer than optimal numbers of virus particles if the serum is diluted less than about 1/500, and this means that an antiserum of low titer may not be able to trap particles at all. After PA precoating, this inhibitory effect is much reduced, so that antiserum dilutions of, say, 1/50 are practical, and the antiserum of low titer may now work. (Another option, as also noted in Section VI.B.1, is to purify the IgG fraction and use this for coating. This procedure largely gets around the problem discussed above.)

One further potentially useful point is that PA-precoated grids subsequently coated with antiserum diluted to about 1/50, and incubated for long times with the virus preparation, can trap a relatively wide spectrum of virus strains, compared to the narrower, more nearly homologous spectrum that would be trapped on grids simply coated with antiserum.[6,7,46,79,83,85,105,107]

VII. DECORATION METHODS

A. INTRODUCTION

Decoration of plant viruses has been widely reviewed,[2-4,6,8-12,15,95] so only some important or interesting points will be summarized here.

"Decoration" indicates the process whereby virus particles are first individually adsorbed to grids, and then, in the situation where the antigen is uniformly distributed on the viral surface, a layer of specific antibody is placed around each particle; after (usually) negative staining, this antibody halo is seen to have absorbed the stain and become electron dense. The dense layer is approximately 10 nm thick (the length of an IgG molecule), thus adding 20 nm to the diameter of the particle. With antibody that is progressively more dilute, or more distantly related to the antigen, the halo becomes thinner and finally can no longer be detected (though detection limits may be extended using gold); thus, pairs of viruses and antisera can be titrated. Alternatively, where different epitopes are exposed on the surface of the virus particle, these may be located, a procedure that is especially effective using a combination of monoclonal antibodies and gold (see Section VIII).

Decoration methods were introduced to plant virology by Milne and Luisoni,[95,109,110] though a similar approach had earlier been used with bacteriophages,[111-113] and of course the decoration phenomenon had been detected in 1941[114-116] and clearly illustrated, for example, by Ball and Brakke in 1968.[117]

The difference between decoration techniques and classical methods where antigen and antibody are mixed in suspension is that the antigen is first trapped on the grid, and antibody attachment to this antigen becomes a separate process. This means that virus particles can be decorated individually (not in clumps), and that each step can be conducted using conditions tailored to its requirements, instead of those that are the result of compromise. Thus, virus particles can be adsorbed to grids using the most appropriate buffer and protectants, and using ISEM as a possible option.[118] After adsorption, the virus particles can be fixed if required. Especially, they can be carefully rinsed, and this gives a cleaner background against which the decoration is to be judged.

The decorating antibody (not necessarily the same as that used in the optional preceding trapping phase) is then applied in the buffer of choice and at an appropriate dilution. The process is terminated by rinsing and negative staining, or alternatively the decoration is now accentuated with gold particles conjugated either to a secondary antibody or to PA (see Sections VI.B.7 and 8).

When a field of decorated virus particles is surveyed, one might expect to see each particle decorated to the same extent as the others, but this does not always happen; the reasons may be several.

First, uneven decoration, especially along the length of rod-shaped particles such as those of potyviruses, may indicate that partial degradation of the coat protein[49,51,52,119] has occurred, either in the plant or during purification and storage, or indeed during incubation on the EM grid.[120]

Second, as elegantly illustrated by Bourdin and Lecoq,[121] a mixed infection with two related viruses (in this case also, potyviruses) may lead to partial transcapsidation and, upon decoration with antibodies against one of the coat proteins, this shows up as uneven antibody coating along the length of the particles. (The interesting point of this paper is its demonstration that transcapsidation can enable a virus that has lost its ability to be aphid-transmitted to regain this ability in the presence of the second, aphid-transmissible, virus. Decoration studies of natural mixed infections had earlier shown that particles of poty- and closteroviruses can carry mixed coat proteins, with likely consequences for potentiating vector transmission in novel ways.[122-124])

Fukuda et al.[125] and Otsuki and colleagues[126,127] likewise elegantly and instructively demonstrated experimental transcapsidation using different strains of TMV, and identified the site of initiation of rod assembly.

Third, the presence of two or more distinct populations of particles clearly decorated to differing degrees probably indicates a mixture of viruses or virus strains; these may be a naturally occurring mixture or may have resulted from laboratory contamination. Such mixtures often go unsuspected until a decoration test is made. A nice example of the detection of mixtures is given by van Dijk et al.[128]

Fourth, especially where a few virus particles are not decorated, among a large number that are, undecorated particles may simply be those of the same virus, that were hidden behind a grid bar or in a fold of the support film during exposure to the decorating antibody; during subsequent washing and negative staining, they were dislodged into the open.

The following list summarizes the applications mentioned above.

1. Serological identification of virus particles
2. Detection of virus mixtures
3. Detection of partial degradation of the coat protein
4. Detection and localization of partial heteroencapsidation
5. Quantitative estimation of degrees of relationship between viruses

6. Measurement of antiserum (antibody) titers
7. Localization of particular antigens (viral gene products) on the viral surface; detection of the location of particular epitopes using monoclonal antibodies
8. Rendering the virus particle more conspicuous by increasing its size and electron density, as an aid to diagnostics

Some protocols for different decoration methods are suggested below.

B. SIMPLE DECORATION

1. Virus particles are adsorbed to the grid so that they are well separated but reasonably frequent (about 5 particles on the main screen at 40,000 × magnification is nice; 5 per 400-mesh grid square is enough if necessary). Where appropriate, control particles (such as those of an unrelated virus) are also adsorbed so that both kinds can be photographed on the same plate at the end. The viruses are adsorbed using the most convenient method, which may include ISEM (Section VI). If required, the preparation on the grid can be fixed, commonly with 0.1% buffered GA for 15 min at room temperature (Section IV.C).
2. The grid is carefully rinsed with buffer and drained (but never allowed to dry until it has been negative-stained). At this point the buffer may be changed from that used to adsorb the virus to that most suitable for the attachment of antibody. As noted in Section VI.B.3, a variety of buffers may be suitable. In our laboratory we have generally used 0.1 M phosphate, pH 7, but recently my colleagues and I titered (see Section VII.C, below) several antisera against poty- and carlaviruses in garlic by decoration, diluting the sera in the above buffer, in 0.05 M borate buffer, pH 8.1, or in distilled water; we were a little surprised to arrive at the same titers with each medium.[129]
3. The grid is incubated at room temperature for 15 min with a suitable dilution of the antibody. to give an easily recognized level of decoration, a dilution that is two or three twofold steps less than the titer should be used. However, when it is desired to demonstrate that a given antibody is *unrelated* to the antigen, undiluted antiserum or concentrated IgG can be used.
4. The grid is carefully rinsed with water.
5. Five drops of 1% aqueous UA are passed across the face of the grid, which is then drained at the edge with absorbent paper, and allowed to dry.

C. QUANTITATIVE DECORATION

As noted in Section VIII.A above, decoration can be used to titrate antisera or other antibody preparations against viruses. All that is necessary is to set up standard conditions of time and temperature (say 15 min of decorating time at 25°C), and to prepare serial dilutions of the antibody. The virus is adsorbed to the grid, which is then rinsed and drained but not allowed to dry. Each of the antibody dilutions is incubated with one grid, and a range of dilutions is used that brackets the expected endpoint. If the virus particles are well distributed on the grid and not too concentrated, the titer obtained is in practice independent of virus concentration. As an aid in detecting the endpoint, it is useful to have an unrelated virus of similar particle morphology also present on the grid. The titer, in our experience, generally reflects that obtained by classical gel-diffusion or microprecipitin tests, but is one or two twofold steps higher.

If a series of six dilutions is prepared, it is usually only necessary to examine four of these, spending less than 5 min on each grid; thus reading the results takes less than 30 min. Preparing and doing the text takes up a similar amount of time, so results can be obtained within a total elapsed time of 60 min — much faster than with alternative systems. Examples of the use of quantitative decoration to obtain information on antiserum titers have been cited in the literature.[10,24,26,55,82,122,129-135]

D. ISEM PLUS DECORATION

1. An ISEM procedure is followed (Section VI).
2. Instead of rinsing the grid with water, followed by negative stain, the grid is rinsed with buffer.
3. A decoration procedure is followed.
4. In this process it is important that the ISEM coating antibody is not applied in too thick a layer, and that rinsing is thorough. Otherwise some of the coating antibody can attach itself in significant amounts to virus particles, producing haloes. This is usually avoided in practice by using a comparatively high dilution of antibody (for an unelaborated antiserum of respectable titer, this dilution might be in the region of 1/5000).
5. The grid is rinsed with water and negative-stained.

VIII. GOLD LABELING

A. INTRODUCTION

Under many circumstances, and especially if the work is done cleanly, decoration of virus particles without additional gold labeling results in a preparation in which both the virus and the antibody are clearly seen, so interpretation of the image is not too difficult. However, fragments of virus particles or small amounts of decorating antibody are harder to recognize, and the particles of, for example, tospoviruses or rhabdodivuses are so large that antibody haloes tend to get lost in the pool of negative stain surrounding the particle. (With rhabdoviruses it is sometimes possible to remove the envelope and work instead with the nucleocapsid.[134,136,137]) Another problem

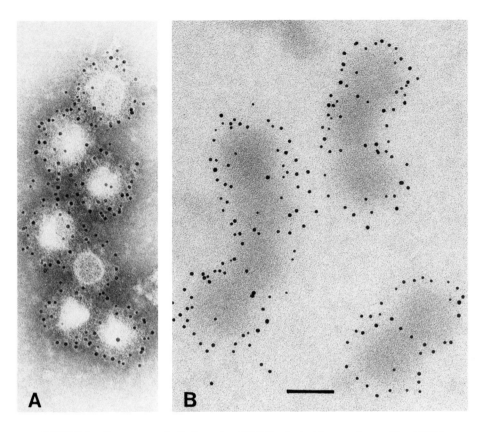

FIGURE 1. Tomato spotted wilt *Tospovirus* (TSWV) adsorbed from crude sap without ISEM, and decorated with rabbit antiserum to TSWV lettuce strain[162] adsorbed with healthy plant antigens (courtesy E. Luisoni) and diluted 1/500; probed with 5-nm gold-goat anti-rabbit IgG (Janssen Auroprobe) diluted 1/50. (A) Negatively stained in 0.5% ammonium molybdate, pH 6.8; (B) unstained. Bar = 100 nm. (TSWV antiserum courtesy of D. Gonsalves;[162] preparation and photo: Vera Masenga and Robert G. Milne.)

with enveloped particles is that the envelopes may possess spikes that can resemble attached antibody. With tospoviruses in particular (an increasingly important group whose particles every diagnostic lab must sooner or later learn to recognize) the particles may appear pleomorphic. For all these reasons, labeling the decorating antibody with gold can offer clear advantages in appropriate situations (Figure 1).

The term GLAD (gold label antibody decoration) was introduced by Pares and Whitecross in 1982[138] for the gold labeling of virus particles in suspension (as opposed to labeling of thin sections). Other early papers were those of Lin[139] and Giunchedi and Langenberg.[140] For plant virology at least, gold labeling techniques acquired some maturity with the papers of Louro and Lesemann[130] and van Lent and Verduin.[19] The subject has been reviewed[2,3,5,10-12,141-147] and there are many interesting papers.[148-154]

B. THE GOLD LABEL

Normally, the mean diameter of the gold particles employed is from 5 to 20 nm. In double labeling experiments (see below), labels of, for example, 7 and 16 nm[19,20] or 5 and 10 nm[152] have been used. The smallest practical gold label for normal transmission electron microscopy is 2 to 3 nm in diameter, as below this level there is insufficient contrast.[145] With a 5-nm gold probe, the label is easily visible in the EM and also on the photographic negative, but if it happens to be superimposed on a mass that is itself somewhat electron dense, making satisfactory prints can be a problem (this is illustrated using, not a virus, but a mycoplasma-like organism [MLO], in Figure 2). Generally, a smaller label will bind in greater amounts than a larger one, giving the former an advantage if other considerations are equal.

Gold can of course be attached directly to the primary (decorating) antibody, but more usually it is conjugated either to PA (Section VI.B.7) or to a secondary antibody — for example goat anti-rabbit IgG in the case where the primary antibody is rabbit anti-virus IgG, or goat anti-mouse IgG where the primary antibody is a mouse monoclonal. Reliable gold preparations can be purchased or can be made.[19,155,156]

For routine work, there seems to be little to choose between PA and the second-antibody systems for labeling, and both are used frequently. According to Baschong and Wrigley,[145] the PA combination places the center of the gold particle about 18 nm from the antigenic site, and the second-antibody system places it at a distance of about 28 nm, supposing the use of 5-nm gold particles. The smallest and most precise label for normal transmission electron microscopy currently obtainable is made by attaching a 2- to 3-nm gold particle to the papain-cleaved end of a single Fab fragment derived directly from the primary antibody: the distance from antigen to label is then only 8 nm.[145]

PA is most stably bound to gold at slightly acid or neutral pH,[157] and so a neutral phosphate buffer is a suitable dilution medium. With second antibody-gold, optimal binding of the gold to the antibody may occur at higher pH and, for example, Amersham-Janssen Biotech recommend 0.02 M Tris buffer, pH 8.2, for their products.

C. SUPPRESSION OF BACKGROUND LABEL

As with ELISAs, Western blots, or sections destined for immunostaining, it is generally necessary to block sites on the surface of the support film that remain unoccupied after the antigen (e.g., virus particles) has been adsorbed. This is usually done by incubating the grids with 1% buffered bovine serum albumin (BSA), which can also optionally contain 0.1% Tween 20. Instead of BSA, one can use normal goat serum (NGS) diluted 1/30. Thereafter, all solutions except the final rinse and negative stain can contain the blocking agent. However, if PA-gold is used, it is best not to include BSA in this step, since it may interact to lower the intensity of the specific label.[130] NGS should also be avoided as it may react with the PA.

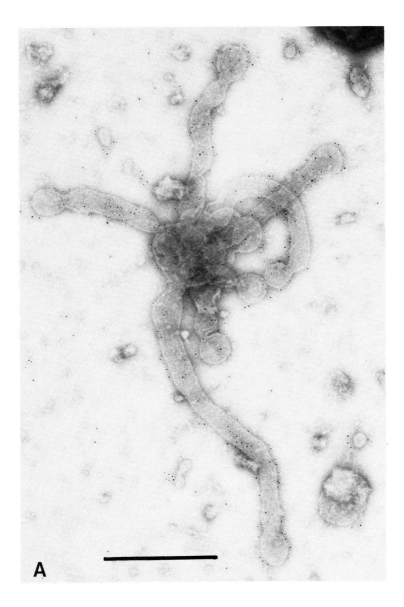

FIGURE 2. Tomato big bud MLO (mycoplasma-like organism) trapped from crude sap by ISEM using the (Fab')$_2$ fraction of the homologous polyclonal IgG (6 μg/ml), and decorated with the intact IgG (20 μg/ml), then 5-nm gold-goat anti-rabbit IgG, as in Figure 1. (A) Negatively stained with 0.5% ammonium molybdate, pH 6.8; (B) unstained. Bar = 500 nm (MLO and primary antiserum courtesy of M. F. Clark;[164,165] preparation and photo: Cristina Vera and Robert G. Milne.[163])

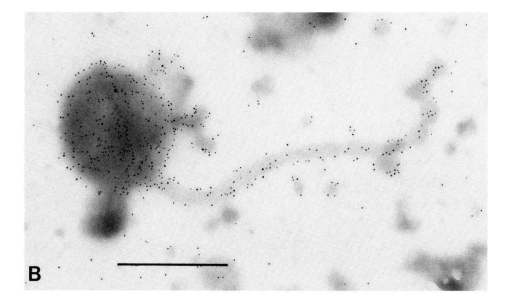

FIGURE 2B.

D. A STANDARD PROTOCOL

A standard protocol for gold labeling of virus particles simply adsorbed to grids might be as follows (modified from References 20 and 130), all operations being done at room temperature.

1. Adsorb the virus preparation to the grid (see Section IV), and rinse the grid carefully with the chosen buffer (for example, 0.1 *M* phosphate, pH 7). If the preparation is simply a sap extract, host components in it may provide adequate blocking, and use of BSA or NGS may give no further lowering of the background label. However, if the preparation is purified, blocking will be advisable. Experiments should be run with and without Tween 20, because it sometimes improves the preparation by lowering background labeling, but it sometimes wipes the virus off the grid, as was our experience with tospoviruses (Figure 1).
2. Block the preparation (but see above), for 1 h at room temperature with 1% BSA in 0.1 *M* phosphate, pH 7 (BSA-PB).
3. Decorate the preparation for 15 min with primary antibody diluted in BSA-PB. (Optionally, dilute the antibody a little more, and incubate for 1 h.)
4a. Rinse with BSA-PB; incubate with the appropriate second antibody-gold, diluted in BSA-PB, for 30 min to 1 h.
4b. Or, rinse with PB; incubate with PA-gold for 1 h.
5. Rinse with PB followed by water, then 1% UA or 0.5% ammonium molybdate, pH 6.5. Alternatively, leave the preparation unstained.

E. DOUBLE LABELING

Labeling two or more antigens in one preparation using gold of different sizes is possible, and the techniques for doing it have been discussed extensively.[5,20,147,157-159] As with single labeling, the gold can be attached to PA or to a secondary antibody. Namork[158] recommends the second-antibody system, whereas van Lent and Verduin[20] suggest that the PA system is preferable; the authors do not explain their preferences and it seems that either system will perform well if optimized.

A sophisticated alternative not open to everyone is to use monoclonal antibodies of different immunoglobulin classes such as IgM and IgG, followed by gold probes of different sizes attached to goat anti-mouse IgM or IgG.[158] Another possibility is to use, for example, rabbit IgG to detect one antigen and mouse IgG to detect the other; the gold of different sizes is then conjugated to goat anti-rabbit and goat anti-mouse IgG, but these secondary IgGs must be cross-absorbed, as rabbit and mouse IgGs do have some epitopes in common.

Where the primary immunoglobulins are of different classes or from different animals, double labeling can be done in one step, but if not, the two labelings must be done sequentially. In this situation, the smaller size of gold is used first, to minimize the chances that the second probe will be sterically blocked.

Double labeling of purified virus particles adsorbed to grids is likely to present fewer problems than double or multiple labeling of while cells or thin sections; such problems, some not directly relevant here, are discussed in References 157 to 161.

The simple and apparently effective protocol of van Lent and Verduin[20] is given below.

1. Adsorb the sample to the grid; all incubations are then done at room temperature.
2. Block the preparation with phosphate-buffered saline (PBS) containing 1% BSA (PBS-BSA) for 1 h.
3. Incubate the grids with the first antibody, diluted in PBS-BSA, for 1 h.
4. Rinse in PBS.
5. Incubate with 5- to 7 nm protein A-gold, 1 h.
6. Rinse in PBS.
7. Repeat steps 3 to 6, using the second antibody and 15- to 16-nm protein A-gold.
8. Rinse in water and stain in UA.

F. GOLD LABELING OF ISEM PREPARATIONS

When ISEM plus decoration is carried out without gold labeling and with the coating antibody used at relatively high dilution (Section VII.D), this background antibody does not interfere with assessment of decoration levels. However, when using gold labeling we are looking at a system whose sensitivity can be 2 to 4 $\log_2$ steps greater,[11,130] and gold labeling of the background can be a problem.

Of the various solutions available, the best seems that proposed by Louro and Lesemann:[130] to use (Fab')$_2$ for coating, and the intact IgG for decoration. The decorating antibody can then be labeled either with PA-gold or with second-antibody gold; in each case the coating antibody is not recognized since it lacks the Fc portion. Figure 2 illustrates the use of (Fab')$_2$ trapping and second-antibody-gold decoration of a mycoplasma-like organism (MLO). the protocol used[163] is given below. Longer incubation times could optionally be employed.

1. Use nickel grids covered with Formvar-backed carbon support film. Operate at room temperature. Where 0.1 M phosphate buffer, pH 7 (PB) is indicted, 0.1 M maleate buffer, pH 6.8, works equally well.
2. Prepare an MLO extract using young stems of infected *Catharanthus roseus* or another host plant, diluted 1/5 (w/v) in 0.3 M glycine buffer, pH 8.0, containing 20 mM MgCl$_2$. The stem material is passed through the rollers of a Pollähne-type press (Section IV.A.2) with dropwise addition of the buffer.
3. Coat the grids for 15 min with rabbit anti-MLO (Fab')$_2$ diluted in PB.
4. Rinse with PB.
5. Float grids for 3 h on the MLO extract.
6. Rinse with PB.
7. Float grids for 15 min on 0.1 to 1.0% GA diluted in PB.
8. Rinse with PB.
9. Incubate for 15 min with 2 to 20 μg/ml rabbit anti-MLO IgG diluted in PB.
10. Rinse with PB.
11. Incubate for 15 min with goat anti-rabbit 5-nm (or 15-nm) gold conjugate.
12. Rinse with PB.
13. Rinse with water.
14. Allow some grids to dry, unstained; negative-stain others with 0.5% ammonium molybdate, pH 6.5.

Some points to note about this protocol are the following: first, no blocking with BSA or other agent is used; crude sap apparently provides adequate blocking in itself, and additional blocking produces no improvement. Second, Tween 20 is omitted as it can prevent trapping of MLOs. Third, fixation of the trapped MLOs with GA is necessary, otherwise they wash off. Fourth, no quenching of the GA with, for example, glycine or ammonium chloride (Section IV.C) appears necessary in this situation.

A protocol very similar to the above should also be effective for virus preparations. As always, modifications will be necessary to optimize the system for the virus, the host plant, and the exact requirements of the experiment. ISEM using a polyclonal antiserum, then decoration with a monoclonal and gold labeling, could, for example, be an interesting option.

ACKNOWLEDGMENT

I thank Enrico Luisoni for critical discussion of the manuscript.

REFERENCES

1. **Langeveld, S. A., Dore, J. M., Memelink, J., Derks, A. F. L. M., van der Vlugt, C. I. M., Asjes, C. J., and Bol, J. F.,** Identification of potyviruses using the polymerase chain reaction with degenerate primers, *J. Gen. Virol.,* 72, 1531, 1991.
2. **Baker, K. K., Ramsdell, D. C., and Gillett, J. M.,** Electron microscopy: current applications to plant virology, *Plant Dis.,* 69, 85, 1985.
3. **Hampton, R., Ball, E., and De Boer, S., Eds.,** *Serological Methods for Detection and Identification of Viral and Bacterial Plant Pathogens, A Laboratory Manual,* APS Press, St. Paul, 1990.
4. **Katz, K. and Kohn, A.,** Immunosorbent electron microscopy for detection of viruses, *Adv. Virus Res.,* 29, 169, 1984.
5. **Kjeldsberg, E.,** Immunogold labeling of viruses in suspension, in *Colloidal Gold: Principles, Methods, and Applications,* Vol. 1, Hayat, M. A., Ed., Academic Press, San Diego, 1989, 433.
6. **Koenig, R.,** Serology and Immunochemistry, in *The Plant Viruses,* Vol. 4, *The Filamentous Plant Viruses,* Milne, R. G., Ed., Plenum, New York, 1988, 111.
7. **Lesemann, D. E.,** Advances in virus identification using immunosorbent electron microscopy, *Acta Hortic.,* 127, 159, 1983.
8. **Milne, R. G.,** Electron microscopy for the identification of plant viruses in *in vitro* preparations, *Meth. Virol.,* 7, 87, 1984.
9. **Milne, R. G.,** New developments in electron microscope serology and their possible applications, in *Developments and Applications of Virus Testing,* Jones, R. A. C. and Torrance, L., Eds., Association of Applied Biologists, Warwick, U.K., 1986, 179.
10. **Milne, R. G.,** Immunoelectron microscopy for virus identification, in *Electron Microscopy of Plant Pathogens,* Mendgen, K. and Lesemann, D. E., Eds., Springer-Verlag, Berlin, 1991, 87.
11. **Milne, R. G.,** Solid-phase immune electron microscopy of virus preparations, in *Immune Electron Microscopy for Virus Diagnosis,* Hyatt, A. D. and Eaton, B. T., Eds., CRC Press, Boca Raton, FL, 1992, chap 2.
12. **Milne, R. G.,** Immunoelectron microscopy of plant viruses and mycoplasmas, in *Advances in Disease Vector Research,* Vol. 9, Harris, K. F., Ed., Springer-Verlag, New York, 1992, 283.
13. **Milne, R. G. and Lesemann, D. E.,** Immunosorbent electron microscopy in plant virus studies, *Meth. Virol.,* 8, 85, 1984.
14. **Paliwal, Y. C.,** Immunoelectron microscopy of plant viruses and mycoplasmas, in *Current Topics in Vector Research,* Vol. 3, Harris, K. F., Ed., Springer-Verlag, New York, 1987, 217.

15. **Pares, R. D. and Whitecross, M. I.,** An evaluation of some factors important for maximizing sensitivity of plant virus detection by immuno-electron microscopy, *J. Virol. Meth.,* 11, 339, 1985.

16. **Plumb, R. T.,** Detecting plant viruses in their vectors, in *Advances in Disease Vector Research,* Vol. 6, Harris, K. F., Ed., Springer-Verlag, New York, 1990, 191.

17. **Roberts, I. M.,** Immunoelectron microscopy of extracts of virus-infected plants, in *Electron Microscopy of Proteins,* Vol. 5, *Viral Structure,* Harris, J. R. and Horne, R. W., Eds., Academic Press, London, 1986, 293.

18. **Roberts, I. M.,** Practical aspects of handling, preparing and staining plant virus particles for electron microscopy, in *Developments and Applications in Virus Testing,* Jones, R. A. C. and Torrance, L., Eds., Association of Applied Biologists, Warwick, U.K., 1986, 213.

19. **Van Lent, J. W. M. and Verduin, B. J. M..,** Specific gold-labelling of antibodies bound to plant viruses in mixed suspensions, *Neth. J. Plant Pathol.,* 91, 205, 1985.

20. **Van Lent, J. W. M. and Verduin, B. J. M.,** Immunolabeling of viral antigens in infected cells, in *Electron Microscopy of Plant Pathogens,* Mendgen, K. and Lesemann, D. E., Eds., Springer-Verlag, Berlin, 1991, 119.

21. **Williams, R. C. and Fisher, H. W.,** Electron microscopy of tobacco mosaic virus under conditions of minimal beam exposure, *J. Mol. Biol.,* 52, 121, 1970.

22. **Nixon, H. L. and Woods, R. D.,** The structure of tobacco mosaic virus protein, *Virology,* 10, 157, 1960.

23. **Hayat, M. A. and Miller, S. E.,** *Negative Staining,* McGraw-Hill, New York, 1990.

24. **Boccardo, G., Milne, R. G., Luisoni, E., Lisa, V., and Accotto, G. P.,** Three seed-borne cryptic viruses containing double-stranded RNA isolated from white clover, *Virology,* 147, 29, 1985.

25. **Luisoni, E. and Milne, R. G.,** White clover cryptic virus 2, *AAB Descriptions of Plant Viruses* No. 332, Association of Applied Biologists, Wellesbourne, U.K., 1988.

26. **Lisa, V., Milne, R. G., Accotto, G. P., Boccardo, G., Caciagli, P., and Parvizy, R.,** Ourmia melon virus, a virus from Iran with novel properties, *Ann. Appl. Biol.,* 112, 291, 1988.

27. **Conti, M., Milne, R. G., Luisoni, E., and Boccardo, G.,** A closterovirus from a stem-pitting diseased grapevine, *Phytopathology,* 70, 394, 1980.

28. **Stobbs, L. W.,** Effect of rotation in a magnetic field on virus adsorption in immunosorbent electron microscopy, *Phytopathology,* 74, 1132, 1984.

29. **Baumeister, W. and Hahn, M.,** Specimen supports, in *Principles and Techniques of Electron Microscopy: Biological Applications,* Vol. 8, Hayat, M. A., Ed., Van Nostrand Reinhold, New York, 1978, chap. 1.

30. **Evenson, D. P.,** Electron microscopy of viral nucleic acids, *Meth. Virol.,* 6, 219, 1977.

31. **Hayat, M. A.,** *Basic Techniques for Transmission Electron Microscopy,* Academic Press, Orlando, 1986.

32. **Keen, D. R.,** A method for producing dust-, streak- and hole-free Formvar films in laboratories having high atmospheric activity, *Stain Tech.,* 59, 56, 1984.

33. **Nermut, N. V.,** Advanced methods in electron microscopy of viruses, in *New Developments in Practical Virology,* Howard, C. R., Ed., Alan R. Liss, New York, 1982, chap. 1.

34. **Davison, E. and Colquhoun, W.,** Ultrathin Formvar support films for transmission electron microscopy, *J. Electron Microsc. Tech.,* 2, 35, 1985.

35. **Aebi, U. and Pollard, T. D.,** A glow discharge unit to render electron microscope grids and other surfaces hydrophilic, *J. Electron Microsc. Tech.,* 7, 29, 1987.

36. **Dubochet, J., Groom, M., and Mueller-Neuteboom, S.,** The mounting of macromolecules for electron microscopy with particular reference to surface phenomena and the treatment of support films by glow discharge, *Adv. Opt. Electron Microsc.,* 8, 107, 1985.

37. **Gregory, D. W. and Pirie, B. J. S.,** Wetting agents for biological electron microscopy. I. General considerations and negative staining, *J. Microsc.,* 99, 261, 1973.

38. **Hayat, M. A.,** *Principles and Techniques of Electron Microscopy: Biological Applications,* 3rd ed., MacMillan, Basingstoke, 1989.

39. **Pegg-Feige, K. and Doane, F. W.,** Effect of specimen support film in solid phase immunoelectron microscopy, *J. Virol. Meth.,* 7, 315, 1983.

40. **van Balen, E.,** The effect of pretreatments of carbon-coated Formvar films on the trapping of potato leafroll virus particles using immunosorbent electron microscopy, *Neth. J. Plant Pathol.,* 88, 33, 1982.

41. **Milne, R. G.,** unpublished observations, 1980.

42. **Barth, O. M.,** The use of polylysine during negative staining of viral suspensions, *J. Virol. Meth.,* 11, 23, 1985.

43. **Paliwal, Y. C.,** Detection of barley yellow dwarf virus in aphids by serologically specific electron microscopy, *Can. J. Bot.,* 60, 179, 1982.

44. **Brandes, J.,** Eine elektronenmikroskopische Schnellmethode zum nachweis faden- und stäbchenförmiger Viren, insbesondere in Kartoffeldunkelkeimen, *Nachrichtenbl. Dtsch. Pflanzenschutzdienstes (Braunschweig),* 9, 151, 1957.

45. **Milne, R. G.,** Electron microscopy of viruses, in *Principles and Techniques in Plant Virology,* Kado, C. I. and Agrawal, H. O., Eds., Van Nostrand Reinhold, New York, 1972, chap. 4.

46. **Roberts, I. M., Robinson, D. J., and Harrison, B. D.,** Serological relationships and genome homologies among geminiviruses, *J. Gen. Virol.,* 65, 1723, 1984.

47. **Srivastava, M. and Milne, R. G.,** unpublished observations, 1988.

48. **Accotto, G. P.,** Immunosorbent electron microscopy for detection of fanleaf virus in grapevine, *Phytopathol. Mediterr.,* 21, 75, 1982.

49. **Brunt, A. A.,** Purification of filamentous viruses and virus-induced noncapsid proteins, in *The Plant Viruses* Vol. 4, *The Filamentous Plant Viruses,* Milne, R. G., Ed., Plenum Press, New York, 1988, chap. 3.

50. **Dietzgen, R. G. and Francki, R. I. B.,** Reducing agents interfere with the detection of lettuce necrotic yellows virus in infected plants by immunoblotting with monoclonal antibodies, *J. Virol. Meth.,* 28, 199, 1990.

51. **Koenig, R., Tremaine, J. H., and Shepard, J. F.,** *In situ* degradation of the protein chain of potato virus X at the N- and C-termini, *J. Gen. Virol.,* 38, 329, 1978.

52. **Shukla, D. D., Strike, P. M., Tracy, S. L., Gough, K. H., and Ward, C. W.,** The N and C termini of the coat proteins of potyviruses are surface-located and the N terminus contains the major virus-specific epitopes, *J. Gen. Virol.,* 69, 1497, 1988.

53. **James, G. T.,** Inactivation of the protease inhibitor phenylmethylsulfonyl fluoride in buffers, *Anal. Biochem.,* 86, 574, 1978.

54. **Hayat, M. A.,** *Fixation for Electron Microscopy,* Academic Press, New York, 1981, chap. 3.

55. **Boccardo, G. and Milne, R. G.,** Enhancement of the immunogenicity of the maize rough dwarf virus outer shell with the cross-linking reagent dithiobis(succinimidyl) propionate, *J. Virol. Meth.,* 3, 109, 1981.

56. **Matthopoulos, D. P. and Tzaphlidou, M.,** Tissue culture fixation with diimidoesters. IV. A comparison between adipimidate and aldehydes of their effects on cell size and morphology, *Micron Microsc. Acta,* 21, 69, 1990.

57. **Wetz, K.,** Cross-linking of poliovirus with bifunctional reagents: biochemical and immunological identification of protein neighbourhoods, *J. Virol. Meth.,* 18, 143, 1987.

58. **Francki, R. I. B., Milne, R. G., and Hatta, T.,** *Atlas of Plant Viruses,* Vol. 1, CRC Press, Boca Raton, FL, 1985, chap. 1.

59. **Harris, J. R. and Horne, R. W.,** Eds., *Electron Microscopy of Proteins,* Vol. 5. *Viral Structure,* Academic Press, London, 1986.

60. **Horne, R. W.,** The development and application of negative staining to the study of isolated virus particles and their components: a personal account, *Micron Microsc. Acta,* 17, 149, 1986.

61. **Palmer, E. L. and Martin, M. L., Eds.,** *Electron Microscopy in Viral Diagnosis,* CRC Press, Boca Raton, FL, 1988.

62. **Spiess, E., Zimmermann, H. P., and Lünsdorf, H.,** Negative staining of protein molecules and filaments, in *Electron Microscopy in Molecular Biology,* Somerville, J. and Scheer, U., Eds., IRL Press, Oxford, 1987, 147.

63. **Harris, J. R. and Horne, R. W.,** Negative staining, in *Electron Microscopy in Biology,* Harris, J. R., Ed., IRL Press, Oxford, 203, 1991.

64. **Horne, R. W. and Cockayne, D. J. H., Eds.,** Special Issue: Negative Staining, *Micron Microsc. Acta,* 22(No. 4), 321, 1991.

65. **Gelderblom, H. R., Renz, R., and Özel, M.,** Negative staining in diagnostic virology, *Micron Microsc. Acta,* 22, 435, 1991.

66. **Horne, R. W.,** Early developments in negative staining technique for electron microscopy, *Micron Microsc. Acta,* 22, 321, 1991.

67. **Mahoney, J. B. and Chernesky, M. A.,** Negative staining in the detection of viruses in clinical specimens, *Micron Microsc. Acta,* 22, 449, 1991.

68. **Nermut, M. V.,** Unorthodox methods of negative staining, *Micron Microsc. Acta,* 22, 327, 1991.

69. **Steinbrecht, R. A. and Zierold, K., Eds.,** *Cryotechniques in Biological Electron Microscopy,* Springer-Verlag, Berlin, 1987.

70. **Hewat, E., Booth, T. F., Wade, R. H., and Roy, P.,** 3-D reconstruction of bluetongue virus tubules using cryoelectron microscopy, *J. Struct. Biol.,* 108, 35, 1992.

71. **Milne, R. G. and Masenga, V.,** Phosphotungstate negative staining of labile viruses at low pH, *Third Int. Cong. Plant Pathol., München, August 1978, Abstracts,* p. 18.

72. **Wrigley, N. G., Brown, E. B., and Skehel, J. J.,** Electron microscopy of infuenza virus, in *Electron Microscopy of Proteins,* Vol. 5, *Viral Structure,* Harris, J. R. and Horne, R. W., Eds., Academic Press, London, 1986, chap. 4.

73. **Wrigley, N. G., Brown, E. B., Daniels, R. S., Douglas, A. R., Skehel, J. J., and Wiley, D. C.,** Electron microscopy of influenza haemagglutinin-monoclonal antibody complexes, *Virology,* 131, 308, 1983.

74. **Masenga, V. and Milne, R. G.,** unpublished results, 1978.

75. **Milne, R. G. and Masenga, V.,** unpublished results, 1992.

76. **Pares, R. D. and Whitecross, M. I.,** A critical examination of the utilization of serum coated grids to increase particle numbers for length determination of rod-shaped plant viruses, *J. Virol. Meth.,* 7, 241, 1983.

77. **Casper, R., Meyer, S., Lesemann, D.-E., Reddy, D. V. R., Rajeshwari, R., Misari, S. M., and Subbarayudu, S. S.,** Detection of a luteovirus in groundnut rosette diseased groundnuts (*Arachis hypogaea*) by enzyme-linked immunosorbent assay and immunoelectron microscopy, *Phytopathol. Z.,* 108, 12, 1983.

78. **Forde, S. M. D.,** Strain differentiation of barley yellow dwarf virus isolates using specific monoclonal antibodies in immunosorbent electron microscopy, *J. Virol. Meth.,* 23, 313, 1989.

79. **Fribourg, C. E., Koenig, R., and Lesemann, D.-E.,** A new tobamovirus from *Passiflora edulis* in Peru, *Phytopathology,* 77, 486, 1987.

80. **Harrison, B. D., Muniyappa, V., Swanson, M. M., Roberts, I. M., and Robinson, D. J.,** Recognition and differentiation of seven whitefly-transmitted geminiviruses from India, and their relationships to African cassava mosaic and Thailand mung bean yellow mosaic viruses, *Ann. Appl. Biol.,* 118, 299, 1991.

81. **Makkouk, K. M., Koenig, R., and Lesemann, D.-E.,** Characterization of a tombusvirus isolated from eggplant, *Phytopathology,* 71, 572, 1981.

82. **Milne, R. G. and Lesemann, D.-E.,** An immunoelectron microscopic investigation of oat sterile dwarf and related viruses, *Virology,* 90, 299, 1978.

83. **Roberts, I. M., Tamada, T., and Harrison, B. D.,** Relationship of potato leafroll virus to luteoviruses: evidence from electron microscope serological tests, *J. Gen. Virol.,* 47, 209, 1980.

84. **Shukla, D. D. and Gough, K. H.,** Serological relationships among four Australian strains of sugarcane mosaic virus as determined by immune electron microscopy, *Plant Dis.,* 68, 204, 1984.

85. **van Regenmortel, M. H. V., Nicolaieff, A., and Burckard, J.,** Detection of a wide spectrum of virus strains by indirect ELISA and serological trapping electron microscopy (STREM), *Acta Hortic.,* 110, 107, 1980.

86. **Cohen, J., Loebenstein, G., and Milne, R. G.,** Effect of pH and other conditions on immunosorbent electron microscopy of several plant viruses, *J. Virol. Meth.,* 4, 323, 1982.

87. **Adam, G., Chagas, C. M., and Lesemann, D.-E.,** Comparison of three plant rhabdovirus isolates by two different serological techniques, *J. Phytopathol.,* 120, 31, 1987.

88. **Derrick, K. S.,** Quantitative assay for plant viruses using serologically specific electron microscopy, *Virology,* 56, 652, 1973.

89. **Derrick, K. S. and Brlansky, R. H.,** Assay for viruses and mycoplasmas using serologically specific electron microscopy, *Phytopathology,* 66, 815, 1976.

90. **Hamilton, R. I. and Nichols, C.,** Serological methods for detection of pea seedborne mosaic virus in leaves and seeds of *Pisum sativum, Phytopathology,* 68, 539, 1978.

91. **Kerlan, C., Mille, B., and Dunez, J.,** Immunosorbent electron microscopy for detecting apple chlorotic leaf spot and plum pox viruses, *Phytopathology,* 71, 400, 1981.

92. **Lesemann, D.-E., Bozarth, R. F., and Koenig, R.,** The trapping of tymovirus particles on electron microscope grids by adsorption and serological binding, *J. Gen. Virol.,* 48, 257, 1980.

93. **Luisoni, E., Milne, R. G., and Roggero, P.,** Diagnosis of rice ragged stunt virus by enzyme-linked immunosorbent assay and immunosorbent electron microscopy, *Plant Dis.,* 66, 929, 1982.

94. **Milne, R. G.,** Some observations and experiments on immunosorbent electron microscopy of plant viruses, *Acta Hortic.,* 110, 129, 1980.

95. **Milne, R. G. and Luisoni, E.,** Rapid immune electron microscopy of virus preparations, *Meth. Virol.,* 8, 85, 1977.

96. **Nicolaieff, A. and van Regenmortel, M. H. V.,** Specificity of trapping of plant viruses on antibody-coated electron microscope grids, *Ann. Virol. (Inst. Pasteur),* 131E, 95, 1980.

97. **Roberts, I. M. and Harrison, B. D.,** Detection of potato leafroll and potato mop-top viruses by immunosorbent electron microscopy, *Ann. Appl. Biol.,* 93, 289, 1979.

98. **van Regenmortel, M. H. V.,** *Serology and Immunochemistry of Plant Viruses,* Academic Press, New York, 1982.

99. **Bovey, R., Brugger, J. J., and Gugerli, P.,** Detection of fanleaf virus in grapevine tissue extracts by enzyme-linked immunosorbent assay (ELISA) and immune electron microscopy (IEM), *Proc. 7th Meeting Int. Council for the Study of Viruses and Virus-Like Diseases of the Grapevine,* Niagara Falls, 1980, McGinnis, A. J., Ed., 1982, 259.

100. **van Regenmortel, M. H. V. and Neurath, A. R., Eds.,** *Immunochemistry of Viruses: The Basis for Serodiagnosis and Vaccines,* Elsevier, Amsterdam, 1985.

101. **Goding, J. W.,** Use of staphylococcal protein A as an immunological reagent, *J. Immunol. Meth.,* 20, 241, 1978.

102. **Richman, D. D., Cleveland, P. H., Oxman, M. N., and Johnson, K. M.,** The binding of staphylococcal protein A by the sera of different animal species, *J. Immunol.,* 128, 2300, 1982.

103. **Bendayan, M. and Garzon, S.,** Protein G-gold complex: comparative evaluation with protein A-gold for high resolution immunocytochemistry, *J. Histochem. Cytochem.,* 36, 597, 1988.

104. **Timmons, T. S. and Dunbar, B. S.,** Protein blotting and immunodetection, in *Methods in Enzymology,* Vol. 182, Deutscher, M. P., Ed., Academic Press, New York, 1990, 679.

105. **Lesemann, D.-E. and Paul, H. L.,** Conditions for the use of protein A in combination with the Derrick method of immunoelectron microscopy, *Acta Hortic.,* 110, 118, 1980.

106. **Gough, K. H. and Shukla, D. D.,** Further studies on the use of protein A in immune electron microscopy for detecting virus particles, *J. Gen. Virol.,* 51, 415, 1980.

107. **Nicolaieff, A., Katz, D., and van Regenmortel, M. H. V.,** Comparison of two methods of virus detection by immunosorbent electron microscopy (ISEM) using protein A, *J. Virol. Meth.,* 4, 155, 1982.

108. **Shukla, D. D. and Gough, K. H.,** The use of protein A from *Staphylococcus aureus* in immune electron microscopy for detecting plant virus particles, *J. Gen. Virol.,* 45, 533, 1979.

109. **Milne, R. G. and Luisoni, E.,** Rapid high-resolution immune electron microscopy of virus preparations, *Virology,* 68, 270, 1975.

110. **Luisoni, E., Milne, R. G., and Boccardo, G.,** The maize rough dwarf virion II. Serological analysis, *Virology,* 68, 86, 1975.

111. **Yanagida, M., and Ahmad-Zadeh, C.,** Determination of gene product position in bacteriophage T4 by specific antibody association, *J. Mol. Biol.,* 51, 411, 1970.

112. **Tosi, M. and Anderson, D. L.,** Antigenic properties of bacteriophage φ29 structural proteins, *J. Virol.,* 12, 1548, 1973.

113. **Beckendorf, S. K.,** Structure of the distal half of the bacteriophage T4 tail fiber, *J. Mol. Biol.,* 73, 37, 1973.

114. **Anderson, F. A., and Stanley, W. M.,** A study by means of the electron microscope of the reaction between tobacco mosaic virus and its antiserum, *J. Biol. Chem.,* 139, 339, 1941.

115. **von Ardenne, M., Friedrich-Freksa, H., and Schramm, G.,** Elektronenmikroskopische untersuchung der Präcipitinreaktion von Tabakmosaikvirus mit Kaninchenantiserum, *Arch. gesamte Virusforsch.,* 2, 80, 1941.

116. **Schramm, G. and Friedrich-Freksa, N.,** Die Präcipitinreaktion des Tabakmosaikvirus mit Kaninchen und Schweineantiserum, *Z. Physiol. Chem.,* 270, 233, 1941.

117. **Ball, E. M. and Brakke, M. K.,** Leaf-dip serology for electron microscopic identification of plant viruses, *Virology,* 36, 152, 1968.

118. **Vetten, H. J., Lesemann, D. E., and Allen, D. J.,** Occurrence of potyvirus and potexvirus infections in black bryony (*Tamus communis*) in Devon, UK, *Plant Pathol.,* 36, 492, 1987.

119. **Stein, A., Salomon, R., Cohen, J., and Loebenstein, G.,** Detection and characterization of bean yellow mosaic virus in corms of *Gladiolus grandiflorus, Ann. Appl. Biol.,* 109, 147, 1986.

120. **Langenberg, W. G.,** Deterioration of several rod-shaped wheat viruses following antibody decoration, *Phytopathology,* 76, 339, 1986.

121. **Bourdin, D. and Lecoq, H.,** Evidence that heteroincapsidation between two potyviruses is involved in aphid transmission of a non-aphid-transmissible isolate from mixed infections, *Phytopathology,* 81, 1459, 1991.

122. **Milne, R. G., Conti, M., Lesemann, D.-E., Stellmach, G., Tanne, E., and Cohen, J.,** Closterovirus-like particles of two types associated with diseased grapevines, *Phytopathol. Z.,* 110, 360, 1984.

123. **Milne, R. G., Masenga, V., and Lovisolo, O.,** Viruses associated with white bryony (*Bryonia cretica* L.) mosaic in northern Italy, *Phytopathol. Mediterr.,* 19, 115, 1980.

124. **Murant, A. F., Raccah, B., and Pirone, T. P.,** Transmission by vectors, in *The Plant Viruses,* Vol. 4, *The Filamentous Plant Viruses,* Milne, R. G., Ed., Plenum, New York, 1988, chap. 7.

125. **Fukuda, M., Okada, Y., Otsuki, Y., and Takebe, I.,** The site of initiation of rod assembly on the RNA of a tomato and a cowpea strain of tobacco mosaic virus, *Virology,* 101, 493, 1980.

126. **Otsuki, Y., Takebe, I., Ohno, T., Fukuda, M., and Okada, Y.,** Reconstruction of tobacco mosaic virus rods occurs bidirectionally from an internal initiation region: demonstration by electron microscopic serology, *Proc. Natl. Acad. Sci. U.S.A.,* 74, 1913, 1977.

127. **Otsuki, Y. and Takebe, I.,** Production of mixedly coated particles in tobacco mesophyll protoplasts doubly infected by strains of tobacco mosaic virus, *Virology,* 84, 162, 1978.

128. **van Dijk, P., Verbeek, M., and Bos, L.,** Mite-borne virus isolates from cultivated *Allium* species, and their classification into two new rymoviruses in the family Potyviridae, *Neth. J. Plant Pathol.,* 97, 381, 1991.

129. **Conci, V., Nome, S. F., and Milne, R. G.,** Filamentous viruses of garlic in Argentina, *Plant Dis.,* 76, 594, 1992.

130. **Louro, D. and Lesemann, D.-E.,** Use of protein A-gold complex for specific labelling of antibodies bound to plant viruses. I. Viral antigens in suspensions, *J. Virol. Meth.,* 9, 107, 1984.

131. **Luisoni, E., Milne, R. G., Accotto, G. P., and Boccardo, G.,** Cryptic viruses in hop trefoil (*Medicago lupulina*) and their relationship to other cryptic viruses in legumes, *Intervirology,* 28, 144, 1987.

132. **Milne, R. G.,** Quantitative use of the electron microscope decoration technique for plant virus diagnostics, *Acta Hortic.,* 234, 321, 1988.

133. **Milne, R. G., Boccardo, G., Dal Bo, E., and Nome, F.,** Association of maize rough dwarf virus with Mal de Rio Cuarto in Argentina, *Phytopathology,* 73, 1290, 1983.

134. **Milne, R. G., Masenga, V., and Conti, M.,** Serological relationships between the nucelocapsids of some planthopper-borne rhabdoviruses of cereals, *Intervirology,* 25, 83, 1986.

135. **Natsuaki, T., Natsuaki, K. T., Okuda, S., Teranaka, M., Milne, R. G., Boccardo, G., and Luisoni, E.,** Relationships between the cryptic and temperate viruses of alfalfa, beet and white clover, *Intervirology,* 25, 69, 1985.

136. **Lundsgaard, T.,** Comparison of *Festuca* leaf streak virus antigens with those of three other rhabdoviruses infecting Gramineae, *Interviology,* 22, 50, 1984.

137. **Lundsgaard, T., Tien, P., and Toriyama, S.,** The antigens of wheat rosette stunt and northern cereal mosaic viruses are related, *Phytopathol. Z.,* 111, 232, 1984.

138. **Pares, R. D. and Whitecross, M. I.,** Gold-labelled antibody decoration (GLAD) in the diagnosis of plant viruses by immuno-electron microscopy, *J. Immunol. Meth.,* 51, 23, 1982.

139. **Lin, N.-S.,** Gold-IgG complexes improve the detection and identification of viruses in leaf dip preparations, *J. Virol. Meth.,* 8, 181, 1984.

140. **Giunchedi, L. and Langenberg, W. G.,** Efficacy of colloidal gold-labeled antibody as measured in a barley stripe mosaic virus-lectin-antilectin system, *Phytopathology,* 72, 645, 1982.

141. **Beesley, J. E. and Betts, M. P.,** Colloidal gold probes for the identification of virus particles: an appraisal, *Micron Micorsc. Acta,* 18, 299, 1987.

142. **Hayat, M. A., Ed.,** *Colloidal Gold: Principles, Methods, and Applications,* Vol. 1, Academic Press, San Diego, 1989.

143. **Hayat, M. A., Ed.,** *Colloidal Gold: Principles, Methods, and Applications,* Vol. 2, Academic Press, San Diego, 1990.

144. **Hayat, M. A., Ed.,** *Colloidal Gold: Principles, Methods, and Applications,* Vol. 3, Academic Press, San Diego, 1991.

145. **Baschong, W. and Wrigley, N. G.,** Small colloidal gold conjugated to Fab fragments or to immunoglobulin G as high-resolution labels for electron microscopy: a technical overview, *J. Electron Microsc. Tech.,* 14, 313, 1990.

146. **Carrascosa, J. L.,** Immunoelectron microscopical studies on viruses, *Electron Microsc. Rev.,* 1, 1, 1988.

147. **Beesley, J. E.,** Colloidal gold: immunonegative staining method, in *Colloidal Gold: Principles, Methods and Applications,* Vol. 2, Hayat, M. A., Ed., Academic Press, San Diego, chap. 12.

148. **Bingren Wu, Mahoney, J., and Chernesky, M.,** Comparison of three protein A-gold immune electron microscopy methods for detecting rotaviruses, *J. Virol. Meth.,* 25, 109, 1989.

149. **Dore, I., Weiss, E., Altschuh, D., and van Regenmortel, M. H. V.,** Visualization by electron microscopy of the location of tobacco mosaic virus epitopes reacting with monoclonal antibodies in enzyme immunoassay, *Virology,* 162, 279, 1988.

150. **Dore, I., Ruhlmann, C., Oudet, P., Cahoon, M., Caspar, D. L. D., and van Regenmortel, M. H. V.,** Polarity of binding of monoclonal antibodies to tobacco mosaic virus rods and stacked disks, *Virology,* 176, 25, 1990.

151. **Himmler, G., Brix, U., Steinkellner, H., Laimer, M., Mattanovich, D., and Katinger, H. W. D.,** Early screening for anti-plum pox virus monoclonal antibodies with different epitope specificities by means of gold-labelled immunosorbent electron microscopy, *J. Virol. Meth.,* 22, 351, 1988.

152. **Hu, J. S., Gonsalves, D., Boscia, D., and Namba, S.,** Use of monoclonal antibodies to characterize grapevine leafroll associated closteroviruses, *Phytopathology,* 80, 920, 1990.

153. **Lesemann, D.-E., Koenig, R., Torrance, L., Buxton, G., Boonekamp, P. M., Peters, D., and Schots, A.,** Electron microscopical demonstration of different binding sites for monoclonal antibodies on particles of beet necrotic yellow vein virus, *J. Gen. Virol.,* 71, 731, 1990.

154. **Zimmermann, D., Sommermeyer, G., Walter, B., and van Regenmortel, M. H. V.,** Production and characterization of monoclonal antibodies specific to closterovirus-like particles associated with grapevine leafroll disease, *J. Phytopathol.,* 130, 277, 1990.

155. **van Lent, J. W. M. and Verduin, B. J. M.,** Detection of viral protein and particles in thin sections of infected plant tissue using immunogold labelling, in *Developments and Applications in Virus Testing,* Jones, R. A. C. and Torrance, L., Eds., Association of Applied Biologists, Wellesbourne, U.K., 1986, 193.

156. **Handley, D. A.,** Methods for the synthesis of colloidal gold, in *Colloidal Gold: Principles, Methods, and Applications,* Vol. 1, Hayat, M. A., Ed., Academic Press, San Diego, 1989, chap. 2.

157. **Doerr-Schott, J.,** Colloidal gold for multiple labeling methods, in *Colloidal Gold: Principles, Methods, and Applications,* Vol. 1, Hayat, M. A., Ed., Academic Press, San Diego, 1989, 146.

158. **Namork, E.,** Double labeling of antigenic sites on cell surfaces imaged with backscattered electrons, in *Colloidal Gold: Principles, Methods, and Applications,* Vol. 3, Hayat, M. A., Ed., Academic Press, San Diego, 1991, 187.

159. **Wang, B.-L. and Larsson, L.-I.,** Simultaneous demonstration of multiple antigens by direct immunofluorescence or immunogold staining. Novel light and electron microscopical double- and triple-staining method employing primary antibodies from the same species, *Histochemistry,* 83, 47, 1985.

160. **Bastholm, L., Nielsen, M. H., and Larsson, L.-I.,** Simultaneous demonstration of two antigens in ultrathin sections by a novel application of an immunogold staining method using primary antibodies from the same species, *Histochemistry,* 87, 229, 1987.

161. **Nielsen, M. H., Bastholm, L., Chatterjee, S., Koga, J., and Norrild, B.,** Simultaneous triple-immunogold staining of virus and host cell antigens with monoclonal antibodies in ultrathin cryosections, *Histochemistry,* 92, 89, 1989.

162. **Gonsalves, D. and Trujillo, E. E.,** Tomato spotted wilt in papaya and detection of the virus by ELISA, *Plant Dis.,* 70, 501, 1986.

163. **Vera, C. and Milne, R. G.,** Manuscript in preparation.

164. **Clark, M. F., Morton, A., and Buss, S. L.,** Preparation of mycoplasma immunogens from plants and a comparison of polyclonal and monoclonal antibodies made against primula yellows MLO-associated antigens, *Ann. Appl. Biol.,* 114, 111, 1989.

165. **Clark, M. F.,** Immunodiagnostic techniques for plant mycoplasma-like organisms, in *Techniques for the Rapid Detection of Plant Pathogens,* Duncan, J. M. and Torrance, L., Eds., Blackwell Scientific, Oxford, 1992, 34.

Chapter 9

NUCLEIC ACID HYBRIDIZATION PROCEDURES

Roger Hull

TABLE OF CONTENTS

0-8493-4284-8/93/$0.00 + $.50

I. INTRODUCTION

Until recently, most virus-based diagnostic approaches focused on the viral coat protein. As was pointed out by Hull,[1] this uses only a small proportion of the viruses' genetic information and techniques based on properties of the viral nucleic acid are becoming more widespread. The explosion in the use of the technology associated with recombinant DNA research has led to the development of various techniques by which viruses can be diagnosed. Many of these are based upon the hybridization between the target viral nucleic acid, which in most cases now is immobilized onto a solid matrix, and a probe nucleic acid. The probe comprises the complementary nucleic acid to which is linked reporter molecules which "report" on successful hybridization to the target. Various other ways in which nucleic acids can be used in viral diagnosis are viewed by Chu et al.[2] and will not be discussed further in this chapter.

The application of nucleic acid probes and related techniques is becoming common in the detection of human and other animal viruses but has lagged somewhat in plant virology. There are many new approaches and techniques being developed in human virus detection which should be directly applicable to plant viruses and the reader is advised not to ignore the literature on the subject. The use of nucleic acid hybridization in plant virus detection has been reviewed in References 1 to 12.

In this chapter the theory of hybridization is described first, as this is most important in the understanding of this technology. There are several variables that can be used to maximize the information gained from this approach. The various elements of the most widely used formats are considered, followed by a discussion of some new systems which either might overcome some of the problems that have arisen or show great potential for the future. As many of the techniques are described in detail in the numerous laboratory manuals (see References 13 to 15) and in the literature available from molecular biology companies, they will not be repeated here.

II. HYBRIDIZATION THEORY

A. INTRODUCTION

The Watson and Crick model for the structure of double-stranded (ds) DNA showed that the two strands were held together by hydrogen bonds between specific (complementary) bases, namely adenine and thymidine, cytosine, and guanine. This interaction of base pairing is the basis of all molecular hybridization. The early studies revealed several important features of this process. The ability to use various physical and chemical procedures to disrupt base pairing and hence separate the strands (termed melting or denaturing the nucleic acid) and then to reinstate base pairing and thus renature the ds nucleic acid (termed hybridization) enabled the various factors controlling the stability of the duplex to be examined. Most of these basic studies were performed using both the target and probe DNAs in solution (liquid-liquid hybridization); there are also some data available for RNA:RNA and DNA:RNA interactions in solution. Although most of plant virus diagnosis now involves mixed-phase hybridization with the target immobilized on a solid matrix, the theory developed for liquid-liquid hybridization is still very relevant.

Below is a general account of some of the major factors that have to be considered in hybridization experiments. More details on the theory of hybridization can be found in References 16 to 18.

B. DENATURATION

Factors affecting denaturation include the following.

Temperature — Double-stranded nucleic acid will denature at high temperatures. Strand separation takes place over a relatively narrow range of temperatures, the temperature at which 50% of the sequences are denatured being called the melting temperature (Tm). The major factors affecting the Tm are the composition of the nucleic acid, the concentration of salt in, and the pH of, the solution, and the presence of materials that can disrupt hydrogen bonding, such as formamide.

Nucleic acid composition — Guanosine + cytosine (G + C) base pairs, which have three hydrogen bonds, are more stable than adenosine + thymidine (uridine) (A + T) base pairs, which have two. For perfectly base-paired DNA in 1 × SSC (SSC = 0.15 M NaCl, 0.015 M Na citrate) the Tm is related to the G + C content by:

$$Tm = 0.41 \ (\%G + C) + 69.3$$

Salt — The salt concentration has a marked effect on the Tm of a duplex. The Tm increases by almost 16°C for each tenfold increase in the concentration of monovalent cation over the range of 0.01 to 0.1 M; the effect is less at higher concentrations. Over the lower concentration range the Tm is given by:

$$Tm = (\%GC/2.44) + 81.5 + 16.6 \log M$$

where M is the molarity of the monovalent cation. Divalent cations have an even greater effect and should be removed by the use of chelators such as ethylene diaminetetraacetate (EDTA).

pH — The Tm is insensitive to pH in the range 5 to 9. Below pH 5 depurination will start and will become more rapid as the pH is lowered. This can be of use as a method for introducing nicks into DNA as apurinic acid is alkali labile. Above pH 9 denaturation of ds nucleic acid sets in, first in A + T-rich regions. Most duplex DNAs are fully denatured at pH 12. The phosphodiester bonds of RNA are degraded above pH 8; the higher the pH and temperature, the more rapid the degradation.

Organic solvents — Some organic solvents such as formamide, dimethyl formamide, and dimethylsulfoxide and also urea[19] lower the Tm. The Tm is reduced by 0.7°C for each percent of formamide.

Base-pair mismatch — Mismatched sequences are less stable than perfectly base-paired duplexes. In nucleic acids of more than 100 bp a mismatch of 1% reduces the Tm by about 1°C. Thus, mismatching can be assessed by varying the hybridization conditions (see stringency, Section II.D).

RNA:RNA and RNA:DNA duplexes — The Tm of dsRNA is significantly higher than that of dsDNA. A general value for the Tm of DNA in 1 × SSC is about 85°C and of RNA is close to 110°C. The Tm of RNA:DNA hybrids is about 4 to 5°C higher than that for DNA:DNA duplexes under the same conditions.

C. REASSOCIATION

The two main considerations relevant to reassociation are the rate at which it occurs and the stability of the products. The stability of the products is affected by the same factors which control denaturation and so will not be further discussed. The main factors affecting rate of reassociation are as follows.

Temperature — For a typical DNA:DNA reassociation reaction the graph relating the rate of formation of duplexes to temperature is a broad curve with the maximum rate at about 25°C below the Tm. This is taken to be the optimum temperature for reassociation. At temperatures well below the optimum, reassociation may effectively cease. Thus, it is possible to maintain the two strands of a duplex separated for considerable periods of time by melting at high temperature and then lowering the temperature rapidly (by plunging the tube into ice) to well below the Tm (termed quenching). If the Tm is lowered, say by the use of formamide, the optimum rate of reassociation is at the new Tm — 25°C. However, since the temperature is lower, the overall rate of hybridization will be slower than that in the absence of formamide.

Salt concentration — The concentration of monovalent cations affects the rate of reassociation markedly. Below 0.1 M NaCl, a twofold increase in salt concentration increases the rate by five- to more than tenfold. The rate continues to rise with increasing salt concentration, but becomes constant above 1.2 M NaCl. Divalent cations have very pronounced effects on the rate and should be removed by the use of chelators such as EDTA.

Base mismatch — The precision with which base-pairing occurs also affects the rate of reassociation. For each 10% mismatch the rate is halved when the reaction is under optimal conditions, i.e., Tm — 25°C.

Fragment length — If the complementary DNA strands are the same length, the rate of reassociation rises with the square root of the length. When the strands are of different lengths the rate depends upon which fragment is in excess and interpretation can be very complicated.

Concentration of nucleic acid — The time required for the formation of duplexes is directly proportional to the initial concentration of the interacting single-stranded (ss) molecules. Reactions are normally measured as the product of the initial concentration (C^0) of the nucleic acid in moles of nucleotide per liter and time (t) of reaction in seconds (C^0t expressed as moles $\times$ s $\times$ liter^{-1}). C^0t values are of great use in liquid-liquid hybridization in determining features such as amounts of unique sequence (complexity) and repeated sequences in nucleic acids.

Polymers — The anionic polymer, dextran sulfate, accelerates the rate of reassociation both in solution and on filters. A 10% solution of dextran sulfate (mol wt 500,000) increases the rate of hybridization in solution by about tenfold and of hybridization to immobilized nucleic acids by up to 100-fold. This increase is thought to be due to the exclusion of the nucleic acid from the volume occupied by the polymer, thus effectively increasing its concentration.

RNA:DNA hybridization — The factors affecting hybridization of RNA to complementary DNA are somewhat different from those affecting DNA:DNA interactions, probably because of the greater amount of strong secondary structure in the RNA. If RNA is in excess under moderate salt conditions (0.18 M NaCl) the rate is almost the same as with DNA:DNA hybridization. However, the rate does not rise as rapidly as does DNA:DNA hybridization with increasing salt concentration. If DNA is in excess, the rate is 4 to 5 times slower than that expected for DNA:DNA reassociation.

D. MIXED-PHASE HYBRIDIZATION

The kinetics of mixed-phase hybridization have been less well studied than those of liquid-liquid hybridization. It is generally assumed that the effects of the reaction conditions are qualitatively, and in most cases quantitatively, the same for filter hybridization as they are for liquid-liquid hybridization (for details see Reference 20). However, with the common practice in filter hybridization of using ds probes, the strandedness of the probe can have effects which may be important if kinetics are being studied.

Probe strandedness — When ds probes are used in filter hybridization two sets of hybridization take place, that between the probe and the immobilized target sequence and the self-annealing of the probe. The latter can have two effects, removal of the effective probe, which reduces sensitivity, and the formation of concatenates, which may then hybridize to the target nucleic acid and thus increase sensitivity but decrease specificity. To minimize these effects the sequences in the probe complementary to the target nucleic acid should be relatively short, the probe should be at low concentration in a small reaction volume, and the reaction should be at as high a temperature as possible. ss probes overcome many of these problems. However if ^{32}P-labeled ss probes are used at concentrations above 100 ng/ml, nonspecific binding to the membrane may occur.

Length of probe — For short oligonucleotide probes (less than 30 base pairs) the Tm can be estimated from:

$$Tm = 2(A + T) + 4(G + C)$$

where A, T, G, and C are the numbers of adenosine, thymidine, guanosine, and cytosine bases, respectively.[21] For filter hybridization the dissociation temperature (Td) is calculated:

$$Td = Tm - k$$

where the constant k has been determined experimentally to be 7.6°C.[22] The Tm of DNA longer than 50 bp can be calculated from:[23]

$$Tm = 81.5 + 16.6 \times \log[Na] + 0.41(\%GC) - 675/\text{length}$$
$$- 0.65(\% \text{ formamide})$$

Stringency — The term "stringency" (see References 20 and 24) is often imprecisely used. It relates to the effect of hybridization and/or wash conditions on the interaction between complementary nucleic acids which may be incompletely matched. The use of different stringencies is one of the more powerful tools of the hybridization technology. Filter hybridization can be used to determine the degree of relatedness between sequences. To achieve this, one has to be able to estimate approximately the proportion of mismatches in the hybrid. This is done by adjusting the reaction conditions so that the desired interaction can be examined. If close relationships are to be distinguished from distant ones more stringent conditions are used; if distant relationships are to be detected the conditions should be less stringent. The stringency can be varied at two stages in the procedure, at hybridization and at the post-hybridization wash. As a general rule, for distantly related sequences the hybridization conditions should be relaxed and the washes carried out under increasing stringency; for closely related sequences hybridization and washing should be under stringent conditions. Stringency can be varied by changing the temperature and/or salt concentration and/or formamide concentration.

Also to be considered in experiments to determine the relationship between nucleic acids are the relative concentrations of probe and target nucleic acid and the size of the probe. The probe nucleic acid should never be in excess, to ensure that it does not saturate the target nucleic acid. For probes larger than 100 bases the Tm of a DNA duplex is decreased by approximately 1°C for every 1% mismatch; for hybrids shorter than 20 bp the Tm decreases by about 5°C for each mismatched base pair.

As a general guideline for nucleic acids of more than 200 bp and 40 to 50% G + C content the following are conditions for various stringencies:

Low stringency	50°C	5 × SSC	Allowing approx. 25% mismatch
Stringent	65°C	2 × SSC	Allowing approx. 10% mismatch
High stringency	65°C	0.1 × SSC	Allowing <1% mismatch

III. TYPES OF HYBRIDIZATION FORMAT

Liquid-liquid hybridization has been used to examine the relationships between plant viruses. Essentially, the technique is to hybridize the target nucleic acid with the probe, which has been radioactively labeled (see Section V.B.1), then to digest away any unhybridized ss probe, precipitate the ds hybrid onto a filter membrane, and count the radioactivity. Some examples of the application of liquid-liquid hybridization to plant virology are found in References 25 and 26.

Most of the approaches to the use of nucleic acid hybridization for virus diagnosis now involve the use of mixed-phase hybridization. There are several basic forms of membranes, the most common being nitrocellulose and nylon. Nitrocellulose membranes are widely used but have two main disadvantages. As nucleic acids are attached to them by hydrophobic rather than covalent bonds, they can be released slowly during hybridization and washing. Nitrocellulose membranes also become brittle when dry and are frequently damaged if one wishes to strip a probe off and reprobe. Nylon membranes bind nucleic acids irreversibly and, as they are more flexible, can be used for stripping and reprobing experiments with less breakage problems. Nucleic acids can be immobilized onto nylon membranes in buffers of low ionic strength. The main disadvantage of nylon membranes is the tendency for high background hybridization especially with RNA probes and with some nonradioactive reporter systems (see Section V.B.2). There are two basic types of nylon membrane, unmodified and charge-modified. The latter, which has a positively charged surface, has a greater binding capacity for nucleic acids. There are numerous brands of nylon membranes available which differ in detail in their properties (see References 27 and 28). The manufacturers' instructions for specific brands should be followed closely to obtain the best results.

As noted above, nucleic acid has to be denatured to bind to membranes. This can be effected by alkali (but remember alkali treatment will degrade RNA), heating, and quenching the sample or by treating the sample with formamide, formaldehyde, or glyoxal. The latter two form complexes with nucleic acid which can inhibit hybridization and so the compound has to be removed by mild alkali treatment. Details of the various denaturation procedures are given in References 13 to 15.

After transfer to either nitrocellulose or nylon membranes the nucleic acid must be immobilized. This can be affected either by baking under vacuum for 0.5 to 2 h at 80°C or by exposing the side of the membrane carrying the nucleic acid to ultraviolet (UV) radiation (254 nm) (see References 27 and 29).

The simplest format is the dot blot, in which the target sample is spotted onto the membrane. This can be either directly or by using a template in which the samples are placed into holes or slots (for slot blots) in a piece of plastic placed over the membrane. In a more sophisticated version the template is incorporated in an apparatus which draws the sample through the membrane by suction; this enables dilute samples to be concentrated on the membrane.

The dot blot does not give any information on the size or number of species of the target nucleic acid. Such information is gained by electrophoresing the nucleic acid in a gel and then transferring it to the membrane. Details of protocols for transferring DNA (Southern blots) and RNA (Northern blots) are given in References 13 to 15.

Various other hybridization formats have been described. The one that shows most promise is sandwich hybridization.[30] In this procedure two sequences complementary to adjacent parts of the target nucleic acid are used. One, the capture sequence, is immobilized onto a solid matrix and is hybridized to the target nucleic acid to which the other complementary strand, the probe, is also being hybridized (liquid-liquid hybridization). In a modification of the technique[31] the capture strand is modified, e.g., by being sulfonated or by the attachment of a protein. Hybridization of the capture and probe strands to the target is performed in liquid and the hybrid is captured in the wells of a microtiter plate by an antiserum to the capture strand modifying molecule. Sandwich hybridization is relatively fast, as at least part of it involves liquid-liquid hybridization. It enables contaminating materials in crude sap extracts to be washed away and, in the modified form, can use enzyme-linked immunosorbent (ELISA) technology and equipment. It has been applied to the detection of plant viruses in at least a couple of cases.[10,32]

IV. TARGET SAMPLE PREPARATION

A. GENERAL CONSIDERATIONS

Using dot blot hybridization it is possible to identify specific nucleic acid sequences in preparations ranging from crude extracts to the purified RNA or DNA of interest. For qualitative detection of viral nucleic acid, crude homogenates can be tested, although interference from soluble and insoluble components in the samples may significantly reduce the limits of detection. For quantitation of viral DNA it appears that no additional sample preparation is necessary.[33] In contrast, the quantiation of viral RNA requires that various interfering compounds are removed before testing; this is usually effected by phenol extraction. For all the preparations of purified viruses so far tested the conditions required for DNA denaturation (alkali) or RNA binding to the nitrocellulose membrane (high salt) have been sufficient to disrupt virion structure, revealing the target nucleic acid and thus obviating the need for any pretreatment.

Experiments to examine relationships between viral sequences require known amounts of target sequences and, therefore, it is necessary to use either purified or cloned viral nucleic acids.

B. DOT BLOTTING CRUDE SAMPLES

The following is a protocol for the dot blotting of crude sap samples:

1. Cut samples of 0.5 to 1 g leaf tissue into small pieces, freeze in liquid N_2 in a prechilled mortar and grind to a fine powder.
2. Place powdered leaves in Eppendorf tubes and, as the powder thaws (turns from light to dark green), add 200 μl of cold TNE buffer (10 mM Tris-HCl, 100 mM NaCl, 1 mM EDTA, pH 8.0). Mix into a slurry and spin in an Eppendorf centrifuge for 1 min.
3. Estimate the volume of the supernatant and, if less than 200 μl, add more TNE, vortex and centrifuge again.

Separate procedures are followed for the binding of RNA and DNA to the membrane. ssDNA binds very efficiently at neutral or high pH; at neutrality binding is helped by a high concentration of salt. dsDNA will not bind to the membrane. ssRNA binds relatively inefficiently to nitrocellulose at low salt concentrations at neutrality and is degraded by high pH. For maximum binding of RNA, the presence of a high concentration of salt is essential.

4. For a DNA target sample take 100 μl extract, add 100 μl 1N NaOH and stand at room temperature for 10 min.

Handle the membrane with gloved hands.

5. Mark a piece of membrane (see Procedure 3) into 1.5 × 1.5 cm squares using a very soft pencil or a ball-point pen (making sure that the ink does not spread during the hybridization step) and label so that samples can be located. Wet the membrane in distilled H_2O), place in 20 × SSC and allow to equilibrate for 3 to 5 min. Transfer the membrane to a dry sheet of 3MM or other blotting paper, allow the surface liquid to drain off, transfer to a fresh sheet of 3MM paper and air-dry.
6. Spot 5 to 10 μl of the alkali-denatured sample into the center of each marked square according to your plan. If possible, include positive and negative (healthy plant) controls.
7. Neutralizing the filter: place 2 pieces of 3MM paper in each of four petri or similar dishes. Saturate the first two with 1 M Tris-HCl (pH 6.8), 0.6 M NaCl and the other two with 0.5 M Tris-HCl (pH 7.4), 1.5 M NaCl. Place the membrane, sample side uppermost, on the first saturated 3MM paper pad by lowering carefully from one corner, making sure no air bubbles are trapped; leave for 2 min. Transfer the membrane to the second pad for 2 min, the third and then the fourth for 5 min each.
8. Place the neutralized membrane on a sheet of dry 3MM paper, allow to air dry, and then bake under vacuum at 80°C for 1 h or more. If a vacuum oven is not available, dry the membrane at 37°C for 1 h and then bake it at 80°C for a further hour.

9.   For RNA target samples follow steps 5 and 6. If using nitrocellulose treat the membrane with 0.5 *M* Tris-HCL (pH 7.4), 1.5 *M* NaCl as described for the third and fourth pads in step 7, and immobilize the sample as in step 8. If using positively charged nylon there is no need for the Tris/NaCl treatment.

The filters can then be prehybridized and hybridized with probes as described in References 13 to 15.

V. PROBES

A. PROBE NUCLEIC ACID

The probe nucleic acid can be DNA or RNA, single- or double-stranded, so long as it has sequences complementary to the target nucleic acid. As over 75% of plant viruses have plus-strand RNA genomes,[34] the most common type of probe is cDNA made to the RNA genome by reverse transcription. This can either be made directly for each experiment or be cloned into a bacterial plasmid or phage vector. Details of generating cDNA probes directly are given by Palukaitis.[35] The cloning approach gives an unlimited supply of easily prepared, well-characterized, and uniform probe which can be designed for specific purposes (see Section VIII), like determining relationships between strains of a virus. Cloned DNA is double-stranded and thus, as noted above, during hybridization there will be some reannealing between the strands. By cloning into the bacteriophage M13 ssDNA probes can be made. These can prove difficult to label but one can use labeled ds replicative form of M13 to detect the hybrid. Methods for cDNA synthesis and cloning can be found in References 13 to 15.

RNA probes have the advantage of forming stronger duplexes with the target, especially if it is RNA, and thus being more sensitive. Although theoretically one could use dsRNA as a probe this is impractical because of preparation difficulties. A much better approach is to clone the cDNA into a transcription vector and synthesize complementary RNA *in vitro*.[13-15] This method has been used to prepare probes to plant viruses such as plum pox potyvirus[36] and potato leafroll luteovirus.[37] In both cases the cRNA probes were shown to be more sensitive than cDNA probes.

With the increasing knowledge of viral genome sequences, oligonucleotides which distinguish between targets, say of different virus strains, can easily be synthesized. Oligonucleotides can be made using a DNA synthesizer but for those laboratories without such a piece of automated equipment it is possible to make oligonucleotides manually.[38] As can be deduced from the discussion on hybridization theory (Section II) oligonucleotides can be used to discriminate between targets differing in just one nucleotide.[39]

B. REPORTER MOLECULES
1. Radioactive Reporter Molecules

Most of the hybridization techniques were developed in molecular biology laboratories using radioactively labeled nucleotides as reporter molecules. The most common labels are ^{32}P and ^{35}S and there are several basic techniques for incorporating these labeled nucleotides into probes. These include the synthesis of new nucleic acid strands in the presence of labeled nucleotides by nick translation or random-primed synthesis on a double-stranded template, by reverse transcription of RNA, or by transcription of RNA from a transcription vector. Nucleic acid can also be end-labeled with γ-labeled nucleotides by the use of polynucleotide kinase. Details of these procedures are found in References 13 to 15. There are obvious and well-documented problems associated with the use of radioactivity for routine diagnosis systems and recently much effort has gone into the development of nonradioactive reporter molecules. The main advantage that radioactive reporter molecules have over many of the nonradioactive systems is the ease of quantitation of hybrids for, say, determining closeness of relationships.

2. Nonradioactive Reporter Systems

There is an increasing number of nonradioactive reporting systems, several of which are commercially available. These systems fall into three basic types; those which directly modify bases in probe DNA, those which attach precursors to DNA or RNA, and those which incorporate labeled precursors into DNA or RNA. The first two involve modification of existing nucleic acid molecules and the third, synthesis of new nucleic acid molecules. Nonradioactive probing systems have been reviewed several times (See References 10 and 40).

a. Direct Modification of DNA

Various modified bases are specifically immunogenic and thus can be detected by antibodies to which an enzyme has been complexed. Examples of this form of reporter group are described in References 41 to 43.

b. Attachment of Precursors to DNA

One of the major approaches to nonradioactive reporters has been to cross-link compounds to DNA. These compounds can be either an enzyme which directly gives a color or luminescent reaction, or be a molecule which reacts to an antibody or other molecules carrying an enzyme. An example of the former is the cross-linking of horseradish peroxidase (HRP) or alkaline phosphatase (AP) to DNA,[44] both of which form the basis of commercially available products, e.g., the enhanced chemiluminescence (ECL) system[45] (see Section V.B.2.a). The disadvantage of HRP is that it is sensitive to higher temperatures and therefore hybridization has to be at 42°C in the presence of urea. Photobiotin is a photoreactive derivative of biotin which cross-links to ss- or dsDNA or RNA on irradiation with visible light.[46,47] Another biotin derivative, termed polybiotin, consists of an electronegative polymer to which several biotin molecules are attached.[48] The electronegative polymer + biotin is cross-linked to the DNA probe. The lengths of the cross-links between the DNA and electronegative polymer and between the electronegative polymer and biotin molecules are important in determining the sensitivity of the system.[48] This illustrates some of the considerations that have to be taken into account when designing a nonradioactive reporter system.

c. Incorporation of Labeled Precursors into DNA

This approach is essentially the same as that used for the radioactive labeling of nucleic acids described above. There are several forms of biotinylated nucleotides, for instance pyrimidine labeled at the 5 position[49,50] or adenine at the N6 position or cytidine at the N4 position.[51] The method used for incorporation of biotinylated dNTPs can influence the sensitivity of the probe.[52]

A commercially available kit is based upon the incorporation of digoxigenin-dUTP into the probe (see Reference 53). The digoxigenin is then detected by an antibody-enzyme complex.

d. Labeling of Oligonucleotides

The above methods can also be used to label oligonucleotide probes. For example, biotin and AP have been incorporated during the synthesis of oligonucleotides,[54,55] and the same molecules have been cross-linked to oligonucleotides.[56-59]

e. Detection of Reporter Molecules

The actual detection is usually by an enzyme which gives a colored product or a luminescent compound on reaction with a substrate. The colored product has to be insoluble (unlike the colored product of the ELISA reaction). The most commonly used enzymes are AP and HRP for which the detection systems are the same as in Western blots (see References 13 to 15, and 60).

Luminescent systems have several advantages over the color systems. They give very rapid results of which a permanent record can easily be made, as light released by the luminescence reaction can be detected by exposure of the membrane to photographic film. As it is easy to destory most luminescent compounds, it is possible to reprobe with several different probes. With the right equipment, one can quantitate the amount of hybridized probe and thus, as with radioactive probes, determine relationships. Luminescence systems based on both HRP and AP have been developed. In the ECL system, mentioned above, positive-charge modified, polymerized HRP, cross-linked to ssDNA or RNA, catalyzes the production of oxygen from H_2O_2 which then oxidizes luminol. This gives a chemiluminescent reaction, the light output being enhanced by the presence of various phenolic compounds.[45] This, and various other HRP-based systems, are reviewed by Durrant.[61] Among the AP-based systems is one in which the enzyme acts on D-luciferin-O-phosphate to liberate D-luciferin, which then emits light on oxidation by luciferase.[62] In another system AP deprotects phosphorylated phenyl dioxetane. Light output is enhanced by intermolecular energy transfer to a micelle-solubilized fluorophore.[63] In the digoxigenin system the substrate is 3-(2'-Spiroadamantane)-4-methoxy-4-(3''-phosphoryloxy)-phenyl-1,2-dioxetane (AMPPD) which, upon dephosphorylation at alkaline pH, releases light.[64]

Biotin can be detected by its close affinity for avidin (or streptavidin) which is one of the strongest associations between biological macromolecules;[65] the avidin usually has an enzyme conjugated to it, though gold-labeled avidin has been used. Biotin on probes can also be detected by an anti-biotin antibody conjugate. One conjugate of interest involves coating the antibody with colloidal gold, the presence of which can be further enhanced by silver treatment.[66]

Although biotin is widely used as a reporter molecule, it does have some disadvantages. Many sap samples contain significant amounts of endogenous biotin which can give false positives. Avidin frequently binds nonspecifically to nylon membranes giving high backgrounds, though this can be reduced by pretreating the membrane with a protease.[53] Both these problems are overcome with sandwich hybridization.[10]

VI. HYBRIDIZATION AND WASHING

Most of the considerations to be taken into account in hybridization and washing are discussed above (Sections II.C and D). Here are some additional points.

In the prehybridization step the sites on the membrane that might bind the probe nonspecifically have to be blocked. In the early states of the development of the technology a solution containing bovine serum albumin, polyvinyl pyridone, and Ficol[67] was used. This is now frequently replaced by a solution of nonfat dried milk[68] or even by one of a well-known alcoholic drink.[69] However, the use of milk as a blocking agent may inhibit the interactions between avidin (streptavidin) and biotin.[70]

Special care has to be taken to control the action of nucleases during hybridization and washing. DNases can be inhibited by including 10 mM EDTA in all solutions. To prevent RNase degradation of either RNA targets or probes the usual precautions for handling RNAs have to be observed. Hybridization in the presence of 50% formamide will inhibit RNases to a great extent and should be used especially when the target is in crude sap samples.

VII. THE POLYMERASE CHAIN REACTION

A recent development which is likely to have a major impact on the use of nucleic acids in viral diagnosis is the polymerase chain reaction (PCR).[71,72] This allows the amplification of specific pieces of nucleic acids which might be present in very low amounts. The technique involves the hybridization of synthetic complementary oligonucleotide primers to the target sequences and synthesis of multiple copies of complementary DNA of the sequence between the primers using heat-stable DNA polymerase. Since the primers are specific, this process can be performed on crude nucleic acid extracts. For full details of the process and methods the reader is referred to References 73 and 74.

As far as diagnostics are concerned the powerful tool of PCR has two basic uses. Specific fragments of DNA can be amplified from any virus infection provided enough sequence is known for synthesis of primers. These fragments can then be analyzed by gel electrophoresis for their presence and size and by restriction endonuclease digestion for heterogeneity (strain differences). Examples of the use of PCR for plant virus diagnosis can be found in References 75 to 80. The ability to make degenerate primers to regions conserved between the genomes of several or many different viruses[81] is a further feature of PCR which will allow the identification of a range of viruses and the recognition of new viruses.

PCR can also be used to generate and label probes. By using degenerate primers, specific probes to both known and unknown viruses can be synthesized. During the synthesis of primers, nucleotides containing nonradioactive reporter molecules can be incorporated.[82,83]

VIII. SOME APPLICATIONS TO PLANT VIRUSES

The most obvious application of hybridization and PCR technology to plant viruses is to virus diagnosis. However, the technology is such that it generates questions as to what the diagnostician actually wants. Is the need to determine if a plant just has a virus infection, or if the plant is infected with a specific virus, or if it is infected with a certain strain of a virus? Until recently the first question was unanswerable by hybridization technology as a defined probe was needed for each individual virus. With the development of PCR, and especially using degenerate primers to conserved sequences, it is possible to have more generalized virus detection systems. As noted in Sections II and V, by the use of defined probes and suitable conditions it is possible to have systems for the detection of either viruses or individual strains. Examples of the discrimination of plant virus strains are given in References 10 and 84–87. Nucleic acid hybridization has also been used to detect plant viruses in insect vectors (see References 88 and 89). Although it can be useful to know whether insects in a population are carrying a virus, it must be remembered that detection does not necessarily mean that the insects could transmit that virus.

One feature of a technique for virus detection that attracts much attention is the sensitivity. Many plant viruses occur in such high concentrations that this is irrelevant. However, for some viruses, and especially those which are phloem-limited or infect trees, it can be difficult to develop a reliable detection system. One of the reasons for this can be the irregular distribution of a virus in a tree, making it likely that there is nothing detectable by any means in some parts of the plant. Thus, it is important to develop a reliable sampling system. There are several comparisons of the sensitivities of various nucleic acid hybridization systems (see Reference 10). The recently developed chemiluminescent systems are claimed to be able to detect nucleic acids at the femtogram level[53,83] and PCR theoretically can detect a single molecule of target nucleic acid.

The other main use of mixed-phase hybridization is in determining virus relationships (see References 12 and 90). It was noted earlier (Sections V.B.1 and V.B.2) that either radioactive or chemiluminescent reporter groups have to be used for this. Various important considerations in the design of experiments to determine relationships are discussed by Koenig et al.[12]

IX. CONCLUSIONS

There are rapid advances being made in the molecular biology of plant viruses and in techniques by which they can be detected based on properties of their genomes. This is leading to an almost bewildering selection of diagnostic tools which, as noted above, face the diagnostician with questions as to what he actually wants from the diagnosis. Coupled with this are the pressures for reducing inputs into crops, such as insecticides, which increase the need for diagnostics, and the innate conservatism as to diagnostic techniques.[10] What hybridization technology will offer is a wide selection of tools for the diagnostician's toolkit.

REFERENCES

1. **Hull, R.,** The potential for using dot-blot hybridization in the detection of plant viruses, in *Developments and Applications in Virus Testing,* Jones, R. A. C. and Torrance, L., Eds., Association of Applied Biologists, Wellesbourne, 1986, 3.
2. **Chu, P. W. G., Waterhouse, P. M., Martin, R. R., and Gerlach, W. L.,** New approaches to the detection of microbial plant pathogens, *Biotechnol. Genet. Eng. Rev.,* 7, 45, 1989.
3. **Garger, S. J., Turpen, T., Carrington, J. C., Morris, T. J., Jordan, R. L., Dodds, J. A., and Grill, L. K.,** Rapid detection of plant RNA viruses by dot blot hybridization, *Plant Mol. Biol. Rep.,* 1, 21, 1983.
4. **Baulcombe, D. C., Boulton, R. E., Flavell, R. B., and Jellis, G. J.,** Recombinant DNA probes for detection of viruses in plants, *Pests Dis.,* p. 207, 1984.
5. **Hull, R.,** Rapid diagnosis of plant virus infections by spot hybridization, *Trends Biotechnol.,* 2, 88, 1984.
6. **Owens, R. A. and Diener, T. O.,** Spot hybridization for detection of viroids and viruses, *Meth. Virol.,* 7, 173, 1984.
7. **Hull, R.,** Detection of viruses and viroids in plants, *British Crop Protection Conference Monograph* No. 34, British Crop Protection Council, 1986, 123.

8. **Garger, S. J. and Turpen, T. H.,** Use of RNA probes to detect plant viruses, *Meth. Enzymol.,* 118, 717, 1986.

9. **Karjalainen, R., Rouhiainan, L., and Soderlund, H.,** Diagnosis of plant virus by nucleic acid hybridization, *J. Agric. Sci.,* 59, 171, 1987.

10. **Hull, R. and Al-Hakim, A.,** Nucleic acid hybridization in plant virus diagnosis and characterization, *Trends Biotechnol.,* 6, 213, 1988.

11. **Hopp, H. E., Giavedone, L., Mandel, M. A., Arese, A., Orman, B., Almonacid, F. B., Torres, H. N., and Mentaberry, A. N.,** Biotinylated nucleic acid hybridization probes for potato virus detection, *Arch. Virol.,* 103, 231, 1988.

12. **Koenig, R., An, D., and Burgermeister, W.,** The use of filter hybridization techniques for the identification and classification of plant viruses, *J. Virol. Meth.,* 19, 57, 1988.

13. **Sambrook, J., Fritsch, E. F., and Maniatis, T.,** *Molecular Cloning: a Laboratory Manual,* Vols. 1–3, 2nd ed., Cold Spring Harbor Laboratory Press, Cold Spring Harbor, NY, 1989.

14. **Perbal, B.,** *A Practical Guide to Molecular Cloning,* 2nd ed., John Wiley & Sons, New York, 1988.

15. **Ausubel, F. M., Brent, R., Kingston, R. E., Moore, D. D., Siedman, J. D., Smith, J. A., and Struhl, K., Eds.,** *Current Protocols in Molecular Biology,* Vols. 1 and 2, John Wiley & Sons, New York, 1989.

16. **Britten, R. J., Graham, D. E., and Neufeld, B. R.,** Analysis of repeating DNA sequences by reassociation, *Meth. Enzymol.,* 29, 363, 1974.

17. **Britten, R. J. and Davidson, E. H.,** Hybridization strategy, in *Nucleic Acid Hybridization: a Practical Approach,* Hames, B. D. and Higgins, S. J., Eds., IRL Press, Oxford, 1985, 3.

18. **Young, B. D. and Anderson, M. L. M.,** Quantitative analysis of solution hybridization, in *Nucleic Acid Hybridization: a Practical Approach,* Hames, B. D. and Higgins, S. J., Eds., IRL Press, Oxford, 1985, 47.

19. **Hutton, J. R.,** Renaturation kinetics and thermal stability of DNA in aqueous solutions of formamide and urea, *Nucleic Acids Res.,* 4, 3537, 1977.

20. **Anderson, M. L. M. and Young, B. D.,** Quantitative filter hybridization, in *Nucleic Acid Hybridization: a Practical Approach,* Hames, B. D. and Higgins, S. J., Eds., IRL Press, Oxford, 1985, 73.

21. **Suggs, S. V., Hirose, T., Miyake, E. H., Kawashima, M. J., Johnson, K. I., and Wallace, R. B.,** Synthetic oligonucleotides as hybridization probes for specific genes, *J. Supramol. Struct. Cell. Biochem. Suppl.,* 5, 429, 1981.

22. **Rychlik, W. and Rhoads, R. E.,** A computer program for choosing optimal oligonucleotides for filter hybridization, sequencing and *in vitro* amplification of DNA, *Nucleic Acids Res.,* 17, 8543, 1989.

23. **Baldino, F., Chesselet, M-F., and Lewis, M. E.,** High resolution *in situ* hybridization histochemistry, *Meth. Enzymol.,* 168, 761, 1989.

24. **Meinkoth, J. and Wahl, G.,** Hybridization of nucleic acids immobilized on solid supports, *Anal. Biochem.,* 138, 267, 1984.

25. **Gould, A. R. and Symons, R. H.,** A molecular biological approach to relationships among viruses, *Annu. Rev. Phytopathol.,* 21, 179, 1983.

26. **Blok, J., Gibbs, A., and Mackenzie, A.,** The classification of tymoviruses by cDNA-RNA hybridization and other measures of relatedness, *Arch. Virol.,* 96, 225, 1987.

27. **Khandjian, E. W.,** Optimized hybridization of DNA blotted and fixed to nitrocellulose and nylon membranes, *BioTechnology,* 5, 165, 1987.

28. **Twomey, T. A. and Krawetz, S. A.,** Parameters affecting hybridization of nucleic acids blotted onto nylon or nitrocellulose membranes, *BioTechniques,* 8, 478, 1990.

29. **Khandjian, E. W.,** UV crosslinking of RNA to nylon membrane enhances hybridization signals, *Mol. Biol. Rep.,* 11, 107, 1986.

30. **Ranki, M., Palva, A., Virtanen, M., Laaksonen, M., and Soderlund, H.,** Sandwich hybridization as a convenient method for detection of nucleic acids in crude samples, *Gene*, 21, 77, 1983.

31. **Syvanen, A.-C., Laaksonen, M., and Soderlund, H.,** Fast quantitation of nucleic acid hybrids by affinity-based hybrid collection, *Nucleic Acids Res.*, 14, 5037, 1986.

32. **Rouhiainen, L., Laaksonen, M., Karjalainen, R., and Soderlund, H.,** Rapid detection of a plant virus by solution hybridization using oligonucleotide probes, *J. Virol. Meth.*, 34, 81, 1991.

33. **Maule, A. J., Hull, R., and Donson, J.,** The application of spot hybridization to the detection of DNA and RNA viruses in plant tissues, *J. Virol. Meth.*, 6, 215, 1983.

34. **Zaitlin, M. and Hull, R.,** Plant virus-host interactions, *Annu. Rev. Plant Physiol.*, 38, 291, 1987.

35. **Palukaitis, P.,** Preparation and use of cDNA probes for detection of viral genomes, *Meth. Enzymol.*, 118, 723, 1986.

36. **Varveri, C., Candresse, T., Cugusi, M., Ravelonandro, M., and Dunez, J.,** Use of ^{32}P-labelled transcribed RNA probe for dot hybridization detection of plum pox virus, *Phytopathology*, 78, 1280, 1988.

37. **Robinson, D. J. and Romero, J.,** Sensitivity and specificity of nucleic acid probes for potato leafroll luteovirus detection, *J. Virol. Meth.*, 34, 209, 1991.

38. **Gait, M. J., Ed.,** *Oligonucleotide Synthesis, a Practical Approach,* IRL Press, Oxford, 1985.

39. **Alves, A.M., Holland, D., Edge, M. D., and Carr, F. J.,** Hybridization detection of single nucleotide changes with enzyme labelled oligonucleotides, *Nucleic Acids Res.*, 16, 8722, 1988.

40. **Gillam, I. G.,** Non-radioactive probes for specific DNA sequences, *Trends Biotechnol.*, 5, 332, 1987.

41. **Anon.,** User Manual Version 4.1, Chemiprobe Orgenic Ltd., 1987.

42. **Hopman, A. H. N., Weigant, J., Tesser, G. I., and van Duijn, P.,** Mercurated nucleic acid probes, a new principle for non-radioactive *in situ* hybridization, *Nucleic Acids Res.*, 14, 6471, 1986.

43. **Tchen, P., Fuchs, R. P. P., Sage, E., and Leng, M.,** Chemically modified nucleic acids as immunodetectable probes in hybridization experiments, *Proc. Natl. Acad. Sci. U.S.A.*, 81, 3466, 1984.

44. **Renz, M. and Kurz, C.,** A colorimetric method for DNA hybridization, *Nucleic Acids Res.*, 12, 3435, 1984.

45. **Durrant, I., Benge, L. C. A., Sturrock, C., Devenish, A. T., Howe, R., Roe, S., Moore, M., Scozzafava, G., Proudfoot, L. M. F., Richardson, T. C., and Mc-Farthing, K. G.,** The application of enhanced chemiluminescence to membrane-based nucleic acid detection, *BioTechniques*, 8, 564, 1990.

46. **Forster, A. C., McInnes, J. L., Skingle, D. C., and Symons, R. H.,** Non-radioactive hybridization probes prepared by the chemical labelling of DNA and RNA with a novel reagent, photobiotin, *Nucleic Acids Res.*, 13, 745, 1985.

47. **McInnes, J. L., Habili, N., and Symons, R. H.,** Non-radioactive, photobiotin-labelled DNA probes for the routine diagnosis of viroids in plant extracts, *J. Virol. Meth.*, 23, 299, 1989.

48. **Al-Hakim, A. H. and Hull, R.,** Studies towards the development of chemically synthesised non-radioactive biotinylated nucleic acid hybridization probes, *Nucleic Acids Res.*, 14, 9965, 1986.

49. **Langer, P. R., Waldrop, A. A., and Ward, D. C.,** Enzymatic synthesis of biotin-labelled polynucleotides: novel nucleic acid affinity probes, *Proc. Natl. Acad. Sci. U.S.A.*, 75, 6633, 1981.

50. **Leary, J. J., Briganti, D.J., and Ward, D. C.,** Rapid and sensitive colorimetric method for visualizing biotin-labelled DNA probes hybridized to DNA or RNA immobilized on nitrocellulose: Bioblots, *Proc. Natl. Acad. Sci. U.S.A.*, 80, 4045, 1983.

51. **Gebeyuhu, G., Rao, P. Y., Soo Chan, P., Simms, D. A., and Klevan, L.,** Novel biotinylated nucleotide-analogs for labelling and colormetric detection of DNA, *Nucleic Acids Res.,* 15, 4513, 1987.

52. **Eweida, M., Sit, T. L., Sira, S., and AbouHaidar, M. G.,** Highly sensitive and specific non-radioactive biotinylated probes for dot-blot, Southern and colony hybridization, *J. Virol. Meth.,* 26, 35, 1989.

53. **Martin, R., Hoover, C., Grimme, S., Grogan, C., Holtke, J., and Kessler, C.,** A highly sensitive, non-radioactive DNA labelling and detection system, *BioTechniques,* 9, 762, 1990.

54. **Kempe, T., Sundquist, W. I., Chow, F., and Hu, S.-L.,** Chemical and enzymatic biotin-labelling of oligodeoxyribonucleotides, *Nucleic Acids Res.,* 13, 45, 1985.

55. **Jablonski, E., Moomaw, E. W., Tullis, R. H., and Ruth, J. L.,** Preparation of oligonucleotide-alkaline phosphatase conjugates and their use as hybridization probes, *Nucleic Acids Res.,* 14, 6115, 1986.

56. **Murasugi, A. and Wallace, R. B.,** Biotin-labeled oligonucleotides: enzymatic synthesis and use as hybridization probes, *DNA,* 3, 269, 1984.

57. **Chollet, A. and Kawashima, E. H.,** Biotin-labeled synthetic oligodeoxyribonucleotides: chemical synthesis and uses as hybridization probes, *Nucleic Acids Res.,* 13, 1529, 1985.

58. **Riley, L. K., Marshall, M. E., and Coleman, M. S.,** A method for biotinylating oligonucleotide probes for use in molecular hybridizations, *DNA,* 5, 333, 1986.

59. **Li, P., Medon, P. P., Skingle, D. C., Lanser, J. A., and Symons, R. H.,** Enzyme-linked synthetic oligonucleotide probe: non-radioactive detection of enterotoxigenic *Escherischia coli* in faecal specimens, *Nucleic Acids Res.,* 15, 5275, 1987.

60. **Harlow, E. and Lane, D.,** *Antibodies: a Laboratory Manual,* Cold Spring Harbor Laboratory, Cold Spring Harbor, NY, 1988.

61. **Durrant, I.,** Light-based detection of biomolecules, *Nature,* 346, 297, 1990.

62. **Hauber, R. and Geiger, R.,** A sensitive, bioluminescence-enhanced detection method for DNA dot-hybridization, *Nucleic Acids Res.,* 16, 1213, 1988.

63. **Pollard-Knight, D., Simmonds, A. C., Schaap, A. P., Akhavan, H., and Brady, M. A. W.,** Nonradioactive DNA detection on Southern blots by enzymatically triggered chemiluminescence, *Anal. Biochem.,* 185, 353, 1990.

64. **Bronstein, I., Lazzari, K., Vary, C., and Voyta, J. C.,** Ultrasensitive chemiluminescent detection of genomic DNA using 1,2-dioxetane based alkaline phosphatase substrate; comparison with other detection methods, *Photochem. Photobiol.,* 49, 9, 1989.

65. **Hiller, Y., Gershoni, J. M., Bayer, E. A., and Wilcek, M.,** Biotin binding to avidin. Oligosaccharide side chain not required for ligand association, *Biochem. J.,* 248, 167, 1987.

66. **Tomlinson, S., Lyga, A., Huguenel, E., and Dattagupta, N.,** Detection of biotinylated nucleic acid hybrids by antibody-coated gold colloid, *Anal. Biochem.,* 171, 217, 1988.

67. **Denhardt, D. T.,** A membrane-filter technique for the detection of complementary DNA, *Biochem. Biophys. Res. Commun.,* 23, 641, 1966.

68. **Johnson, D. A., Gautsch, J. W., Sportsman, J. R., and Elder, J. H.,** Improved technique utilizing nonfat dry milk for analysis of proteins and nucleic acids transferred to nitrocellulose, *Gene Anal. Tech.,* 1, 3, 1984.

69. **Elbrecht, A., Rowley, D. R., and O'Malley, B. W.,** Irish cream liqueur as a blocking agent for DNA dot blots, *BMBiochemica,* p. 12, 1988.

70. **Hoffman, W. L. and Jump, A. A.,** Inhibition of the streptavidin-biotin interaction by milk, *Anal. Biochem.,* 181, 318, 1989.

71. **Saiki, R. K., Scharf, S., Faloona, F., Mullis, K. B., Horn, G. T., Erlich, H. A., and Arnheim, N.,** Enzymatic amplification of β-globin genomic sequences and restriction site analysis for diagnosis of sickle cell anemia, *Science,* 230, 1350, 1985.

72. **Saiki, R. K., Gelfand, D. H., Stoffel, S., Scharf, S. J., Higuchi, R., Horn, G. T., Mullis, K. B., and Erlich, H. A.,** Primer-directed enzymatic amplification of DNA with a thermostable DNA polymerase, *Science,* 239, 487, 1988.

73. **Innes, M. A., Gelfand, D. H., Sninsky, J. J., and White, T. J.,** *PCR Protocols; a Guide to Methods and Their Applications,* Academic Press, San Diego, 1990.

74. **Kumar, R.,** The technique of polymerase chain reaction, *Technique,* 1, 133, 1989.

75. **Wetzel, T., Candress, T., Ravelonandro, M., and Dunez, J.,** A polymerase chain reaction assay adapted to plum pox potyvirus detection, *J. Virol. Meth.,* 33, 355, 1991.

76. **Levy, L. and Hadidi, A.,** Development of a reverse transcription/polymerase chain reaction assay for the identification of plum pox potyvirus from microgram quantities of total nucleic acids, *Phytopathology,* 81, 1154, 1991.

77. **Robertson, N. L. and French, R.,** Amplification of mite- and fungus-transmitted potyviral 3'-terminal fragments by PCR, *Phytopathology,* 81, 1184, 1991.

78. **Kohnen, P. D., Dougherty, W. G., and Hampton, R. O.,** Pea seedborne mosaic virus (PSbMV) detection using the polymerase chain reaction (PCR), *Phytopathology,* 81, 1155, 1991.

79. **French, R. and Robertson, N. L.,** Detection of barley yellow dwarf and other luteo-viruses with group-specific primers and PCR, *Phytopathology,* 81, 1247, 1991.

80. **Geske, S. M., French, R., Robertson, N. L., and Carroll, T. W.,** Use of the polymerase chain reaction to distinguish and characterize the MT-RMV isolate of BYDV, *Phytopathology,* 81, 1247, 1991.

81. **Langeveld, S. A., Dore, J.-M., Memelink, J., Derks, A. F. L. M., van der Vlugt, C. I. M., Asjes, C. J., and Bol, J. F.,** Identification of potyviruses using the polymerase chain reaction with degenerate primers, *J. Gen. Virol.,* 72, 1531, 1991.

82. **Emanuel, J. R.,** Simple and efficient system for synthesis of non-radioactive nucleic acid hybridization probes using PCR, *Nucleic Acids Res.,* 19, 2790, 1991.

83. **Lanzillo, J. J.,** Chemiluminescent nucleic acid detection with digoxigenin-labelled probes: a model system with probes for angiotensin converting enzyme which detect less than one attomole of target DNA, *Anal. Biochem.,* 194, 45, 1991.

84. **Polston, J. E., Dodds, J. A., and Perring, T. M.,** Nucleic acid probes for detection and strain differentiation of cucurbit geminiviruses, *Phytopathology,* 79, 1123, 1989.

85. **Rosner, A. and Bar-Joseph, M.,** Diversity of citrus tristeza strains indicated by hy-bridization with cloned cDNA sequences, *Virology,* 139, 189, 1984.

86. **Rosner, A., Lee, R. F., and Bar-Joseph, M.,** Differential hybridization with cloned cDNA sequences for detecting a specific isolate of citrus tristeza virus, *Phytopathology,* 76, 820, 1986.

87. **Baulcombe, D. C. and Fernandez-Northcote, E. N.,** Detection of strains of potato virus X and of a broad spectrum of potato virus Y isolates by nucleic acid spot hybridization (NASH), *Plant Dis.,* 72, 307, 1988.

88. **Boulton, M. I. and Markham, P. G.,** The use of squash-blotting to detect plant patho-gens in insect vectors, in *Developments and Applications in Virus Testing,* Jones, R. A. C. and Torrance, L., Eds., Association of Applied Biologists, Wellesbourne, England, 1986, 55.

89. **Czonek, H., Ber, R., Navot, N., and Zamir, D.,** Detection of tomato yellow leaf curl virus in lysates of plants and insects by hybridization with a viral DNA probe, *Plant Dis.,* 72, 949, 1988.

90. **Gallitelli, D., Hull, R., and Koenig, R.,** Relationships among viruses in the tombusvirus group: nucleic acid hybridization studies, *J. Gen. Virol.,* 66, 1523, 1985.

Chapter 10

dsRNA IN DIAGNOSIS

J. A. Dodds

TABLE OF CONTENTS

I. INTRODUCTION

Detection and diagnosis of viruses in diseased plants relies on a battery of tools and the choice of which ones to use and in what order to use them is often a difficult one, especially when little is known about the disease in question.[1] When the suspected cause of a disease is from a fairly short list of expected viruses for a specific crop then appropriate serological tests will identify one or more of the viruses if they are present. However, as with all specific tests, other viruses not tested for will be overlooked, at least initially, and it may be much later before undetected agents are finally implicated as contributors or even the true cause of the disease problem. Most diagnostic laboratories recognize the value of less specific tests which can be used in addition to specific tests in the early testing stages, especially when the problems are new or unfamiliar.

One such nonspecific test is double-stranded (ds) RNA analysis. Many plant viruses have RNA genomes, either single-stranded (ss) or double-stranded.[2] In addition, many viruses or virus-like agents found in plant pathogenic fungi have dsRNA genomes and there is increasing interest in analyzing such fungi for dsRNAs.[3,4] RNA viral genomes are often segmented and the number and size of these segments is often quite diagnostic for a virus, group of viruses, or family of viruses.[2,5] As a general rule, for positive-sense ssRNA viruses each of these segments is represented in an infected plant either as a population of ssRNAs destined to act as a template for replication, as mRNAs destined to act in protein synthesis, as viral genomes to be encapsidated, or as equivalent dsRNAs, destined to be or to have been involved as intermediaries in replication.[2] In addition to these RNAs there are subgenomic segments of ssRNA which represent less than full-length copies of larger genomic RNA segments, and these are used as mRNAs for the translation of specific proteins that cannot be translated directly from the genomic RNAs because of the presence of upstream open reading frames.[2] dsRNAs corresponding to some of these subgenomic ssRNAs can also be detected in infected plants, though their origin may be uncertain.[6,7] The dsRNAs representing known genomic and subgenomic ssRNAs and additional virus-specific dsRNAs of even more uncertain origin are not present in equal amounts, and the most abundant ones seem to be those that correspond to the genomic ssRNAs. The presence, number, size, and abundance of major and minor dsRNAs for a specific virus form a collection of measurable properties which are of potential value for diagnosis. The practical value for attempting to detect and describe these dsRNAs lies in the ease with which they can be isolated and characterized, as described in this chapter.

The adaptation of dsRNA analysis to the detection and diagnosis of RNA viruses was first shown to be a fairly easy exercise by Morris, Dodds, and co-workers,[8,9] who simplified methods used earlier, including those for RNA virus replication studies.[10,11] After obtaining a nucleic acid extract from a plant, the method most commonly used for selective purification of dsRNA involves chromatographic adsorption and release from cellulose powder, and analysis of the purified and concentrated dsRNA by gel electrophoresis, though other methods are available.[12,13]

The remainder of this chapter will describe and comment on the extraction and analysis of dsRNA, give a detailed laboratory outline, catalog some examples where this technique has been used, describe a case history application, and discuss the strengths and weaknesses of the approach, including the ever-increasing reality that viral-like or nonviral dsRNAs are more common in plants than was once believed. Other reviews of this topic have addressed most of these issues to a greater or lesser extent.[14-19] The main objective of this chapter is to provide a detailed description of the basic methods for dsRNA analysis used in our laboratory and to document some of the newer literature. Many of the references cited were selected because they were published since the earlier reviews were written.

II. METHODS

A general scheme for purification and analysis of dsRNA based on CF-11 cellulose chromatography that can be copied for laboratory notebooks is provided in Section III. Some general comments on the method will be described in this section. Anyone who has not attempted dsRNA analysis should start with tobacco, *Nicotiana tabacum,* infected with tobacco mosaic *Tobamovirus* (TMV) and/or cucumber mosaic *Cucumovirus* (CMV), with sat RNA[20] if possible, in single or mixed infection. There are four reasons for this. These viruses are easily obtained, their dsRNAs are abundant in tobacco, their major RF dsRNAs provide a useful set of molecular weight markers for electrophoresis (TMV dsRNA $1 = 4.2 \times 10^6$, CMV dsRNA $1 = 2.3 \times 10^6$, CMV dsRNA $2 = 2.0 \times 10^6$, CMV dsRNA $3 = 1.4 \times 10^6$, CMV dsRNA $4 = 0.7 \times 10^6$, and CMV satellite dsRNA $= 0.2 \times 10^6$, see Figure 1), and unexpected dsRNAs of nonviral origin are not isolated from tobacco when using these methods. Citrus tristeza *Closterovirus* (CTV) RF dsRNA (13×10^6) is useful, if available (my laboratory can provide small amounts), since it is one of the largest known plant viral dsRNAs. Note that when mobility of dsRNA is plotted against log molecular weight a curvilinear relationship should be expected for this range of molecular weights[21] so estimates of molecular weight should never be made for dsRNAs that migrate outside of the range of the standards being used.

A. SELECTION AND PROCESSING OF TISSUES

Selection of tissue may not be particularly critical, since dsRNA is often as abundant in leaves that have been infected for long periods as it is in young new flush tissue. Indeed, knowledge of virion concentration or distribution in a plant may be a false guide for tissue selection for dsRNA analysis. We have observed host effects on dsRNA quantity and quality[22-24] as well as poor correlations between relative dsRNA levels and relative virus concentrations

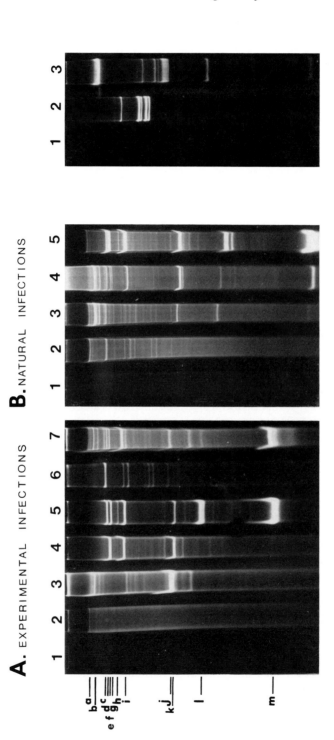

FIGURE 1A.

FIGURE 1. (appearing on page 276) dsRNAs purified from 0.5 g of leaf tissue or mycelial pad (one tenth of the original tissue) analyzed by 6.0% PAGE. Gels were stained with ethidium bromide and photographed. dsRNAs from plants include those purified from experimentally infected greenhouse-grown hosts (A) and naturally infected field plants (B). A1, Noninoculated tobacco; A2, tobacco infected with tobacco etch potyvirus (TEV); A3, tobacco infected with tobacco mild green mosaic tobamovirus (TMGMV) and satellite tobacco mosaic virus (STMV); A4 and A5, tobacco infected with cucumber mosaic cucumovirus — strain A (CMV-A) and strain B (CMV-B). CMV-B has an associated CMV satellite RNA. A6, *Nicotiana glauca* with dsRNA of a recently detected agent (unknown); A7, *N. glauca* mixedly infected with TEV, TMGMV, STMV, CMV, and CMV satellite RNA and the unknown agent. B1, a field grown *N. glauca* plant that seems to be free of dsRNA including the unknown agent. B2–B5 are from separate branches of a single field-grown *N. glauca* plant that is mixedly infected with TEV, TMGMV, STMV, CMV, CMV satellite RNA, and the unknown agent. B2, TEV, unknown agent; B3, TEV, TMGMV, STMV, CMV which is very weak, CMV-sat which is weak but definite, unknown agent; B4, TMV, CMV, STMV which forms a doublet with CMV dsRNA 4, CMV-sat, unknown agent; and B5, TEV, CMV, CMV-sat, unknown agent. dsRNAs from fungi include a dsRNA-negative (1) and dsRNA-positive (2) strain of *Trichoderma viride* and a dsRNA-positive strain of *Botryodiplodia theobromae* (3). Marker dsRNAs with molecular weights in parentheses are TEV RF dsRNA (a, 6.0×10^6); TMV RF dsRNA (b, 4.2×10^6; *N. glauca* unknown agent major dsRNA (c); CMV-A RF of dsRNA 1 (d, 2.3×10^6); CMV-B RF of dsRNA 1 (e); CMV-B RF of dsRNA 2, CMV-A RF of dsRNA 2 (g, 2.0×10^6); multimeric form of CMV satellite dsRNA (h); CMV RF of dsRNA 3 (i, 1.4×10^6); STMV RF dsRNA (j, 0.7×10^6); CMV RF of dsRNA 4 (k, $<0.7 \times 10^6$); dimeric form of CMV satellite dsRNA (l, 0.4×10^6); and monomeric form of CMV satellite dsRNA (m, 0.2×10^6). Northern blots with CMV satellite cDNA (not shown) have confirmed the relatedness of the three forms of CMV satellite dsRNA. Plant dsRNAs from experimentally infected plants and fungal dsRNAs were analyzed in the same electrophoretic analysis and plant dsRNAs from natural field infections in another, and a mobility difference is evident, especially for the lowest-molecular-weight dsRNAs such as the monomeric form of CMV satellite dsRNA (m).

for satellite tobacco mosaic virus (STMV)[22] and CTV,[23] which increased in dsRNA concentration before enzyme-linked immunosorbent assay (ELISA) values increased in early summer in sweet orange. dsRNAs of CMV can be present in readily detectable levels in leaves from which very little virus or viral ssRNA can be detected.[25] For viruses that are phloem-tissue limited, such as closteroviruses or luteoviruses, there is a definite advantage in using bark, stem, or midrib tissue rather than whole leaf tissue.[26] Trial and error is called for, and initial attempts should use both old and young tissues of leaf and stem and even roots.[27] Time of year should be considered when collecting field tissue. For example, little CTV dsRNA can be detected in the hottest months of the year.[23] Tissue can be stored for months or even years prior to extraction.[28,29]

Fungi should be grown in liquid culture and filtered to obtain mycelial pads which should be washed with either water or nucleic acid extraction buffer. Filtered and washed mycelia should be dried with absorbent paper to produce a leathery pad which can be used directly or frozen for later analysis.[30]

Good results are obtained when tissue is first ground to a fine powder after freezing it in liquid nitrogen in a well-cooled mortar and pestle. Frozen powder can be stored for months or years, or extraction can proceed immediately.

B. EXTRACTION OF NUCLEIC ACIDS

In order to begin dsRNA purification it is first necessary to obtain a total nucleic acid extract from the intended tissue source. Typical nucleic acid extraction buffers are normally adequate but special buffers and additives may be needed if the initial extract is unusually viscous.[31] dsRNAs are very stable molecules and so long as buffered extracts of tissue powder are solubilized and deproteinated, the resulting crude nucleic acid solution is adequate for subsequent steps. There is no need to add ribonuclease inhibitors.[28] We have found it sufficient to use sodium dodecyl sulfate (SDS) and phenol for extraction, and have also shown that there is no need for concern about how fresh the phenol solution should be.[28]

An alternate method that we have found useful is to extract dsRNA from the pellet obtained by centrifuging (8000 *g* for 30 min) a buffered extract of tissue. The buffer of choice is the one normally used for purification of the virus present in the tissue source. The pellet is resuspended in dsRNA extraction buffer and processed as if it were a normal tissue sample. By using this method it is possible to purify dsRNA (from the pellet) and virions (from the supernatant) from the same sample. We have found this to work well for TMV, STMV, CMV, and CTV and expect it to work for many viruses. The recovery of dsRNA from the pellet is presumed to be because dsRNA associates with membrane or other relatively large particulate fractions, as is suggested by ultrastructural studies.[32]

C. PURIFICATION OF dsRNA

The method that has worked consistently well for us is chromatography on chemically unmodified graded cellulose powder (e.g., Whatman CF11 powder), a method originally developed by Franklin.[10] dsRNA, but not other classes of nucleic acid, is bound to cellulose powder suspended in buffer in the presence of about 16% ethanol. The optimum ethanol concentration may vary depending on the batch of cellulose powder so it is wise to test this, and we are currently using 16.5% ethanol. When total nucleic acid adjusted to contain 16.5% ethanol is passed through such a column the dsRNA will be retained and other nucleic acids will pass through. After washing the column with buffered 16.5% ethanol, the retained dsRNAs can be eluted by passing ethanol-free buffer through the column. The dsRNA is collected and concentrated by ethanol precipitation. Two cycles of chromatography increase purity of dsRNA. dsRNA can be stored indefinitely under ethanol at $-20°C$[28] and can be shipped this way at ambient temperature to other laboratories, or even as an air-dried pellet, without any major loss of quality.

D. ANALYSIS OF dsRNA

The simplest type of analysis is gel electrophoresis of the concentrated dsRNA, followed by staining with ethidium bromide and photography[33] or silver staining.[34,35] We commonly start with 5.0 g (fresh weight) of tissue, finish up with 200 μl of dsRNA solution and analyze 20 μl (equivalent to

0.5 g of tissue) by gel electrophoresis. We prefer to use 6.0% polyacrylamide gel electrophoresis (PAGE) for most descriptive analyses because minor bands are well resolved by this method, and major bands with similar electrophoretic mobility are often separated. This is especially important when plants are infected with mixtures of strains of a virus such as CMV.[25] In addition, any contaminating ssRNA is normally too large to penetrate the polyacrylamide gel, which therefore serves as a final purification step, as well as the method for analysis. Analysis on 1.0% agarose gels is an easier alternative and Northern blotting is easier from agarose, but resolution is not as good as polyacrylamide gels, especially for minor bands, ssRNA contaminants enter the gel, and gels cannot be silver stained. Agarose gels are ideal when dsRNA analysis is being used to screen for specific major dsRNAs in multiple samples.

E. USE OF dsRNA AS A REAGENT OR TEMPLATE

All of the dsRNA purified from 5.0 g or more of tissue is rarely used up by a single electrophoretic analysis and it is a suitable reagent for techniques such as end labeling of RNA,[36-39] synthesis of cDNA from RNA for labeling[40,41] (also see Section III, step **19**), Northern or Southern blotting,[40-45] or for cDNA cloning,[46-50] including the use of the polymerase chain reaction (PCR).[51] Procedures are available besides electrophoresis from gel slices to recover specific dsRNAs from gels[52] for use with these methods. To use some of these procedures it is necessary to melt the dsRNA to ssRNA. This can be done by boiling in water and quick cooling on ice, or by using chemical treatments with methylmercuric hydroxide, glyoxal and dimethyl sulfoxide (DMSO), or formamide and formaldehyde.[53,54]

In addition to these uses, dsRNA denatured to ssRNA can also be translated *in vitro*.[55] Purified dsRNA is not infectious but becomes infectious after being melted by boiling and quick cooling.[56,57] dsRNA can be used as an immunogen, and this has resulted in the production of specific polyclonal and monoclonal antibodies that can be used to detect dsRNAs by ELISA and dot blot assays in some but not all plant extracts.[58,59] Use of the techniques mentioned in this section can be found in many more of the citations in the References, but those identified here are representative.

Some of these optional uses for purified dsRNA are especially attractive if encapsidated ssRNA is difficult to obtain, or if dsRNA is the only virus- or disease-specific molecule detected so far for a specific disorder. It would not be impossible to develop a "virus-tested" certification program using diagnostic probes created from disease-specific dsRNAs without ever purifying or characterizing the actual virus or viruses found in the plant being certified. Whether this would be wise is open to debate.

III. A LABORATORY OUTLINE

The following is a detailed outline of a basic method for dsRNA purification and analysis. More tissue can be used in Step 1 without changing the other steps, if larger amounts of dsRNA are needed, but this amount is usually more than enough for routine dsRNA detection. Scaling down to smaller tissue weights and smaller columns is a logical step once the amount of dsRNA obtained is known to be more than is needed for descriptive dsRNA analysis. Always run a noninoculated sample of the *specific cultivar* of every plant species being tested whenever possible. Eight is a good number of samples for a single run, since this will constitute a single low-speed centrifuge run, and is a good number for a single electrophoresis run, since it allows for additional dsRNA control samples and molecular weight markers to be analyzed along with the eight experimental samples.

1. Grind 5.0 to 7.0 g of tissue in a cold mortar and pestle with liquid nitrogen to make a fine POWDERED SAMPLE.
2. Transfer the powder with a brush to a 50-ml centrifuge tube, store frozen at $-20°C$ for later use or continue to step 3.
3. Add 12 ml of double-strength STE buffer ($2\times = 0.2\ M$ NaCl, 0.1 M Tris, 0.002 M ethylene diaminetetraacetate [EDTA], pH 6.8), 15 ml STE saturated phenol, and 1.5 ml 10% SDS.
4. Cap the tube well and shake at room temperature for 30 min.
5. Centrifuge at 8000 g for 20 min and recover the aqueous phase, which is the NUCLEIC ACID SOLUTION. Take the first 80%, but avoid the last 20% if this means transferring contaminants.
6. Allow the nucleic acid solution to come to room temperature and adjust the volume to 20 ml with single-strength STE buffer (or reduce to 20 ml if the tissue was a source of more than normal amounts of liquid) and add 4.2 ml of 95% ethanol. This is the COLUMN SAMPLE in STE-buffered 16.5% ethanol.
7. Make CELLULOSE COLUMNS while steps 5 and 6 are going on. Weigh 2.5 g of Whatman CF11 cellulose powder in a beaker, and make a slurry with 25 ml of STE-buffered 16.5% ethanol. Place a disk of miracloth or some glass wool in the bottom of a 30-ml disposable syringe barrel, and carefully transfer the cellulose powder slurry into the barrel with a plastic pasteur pipette made to have a wide mouth by cutting off the tip. Let the column bed drain and check for severe leakage of cellulose. If using several columns, hold them in a test tube rack placed on top of a second rack and place the racks in a pan which can collect unwanted eluates.
8. Carefully apply the column sample to the cellulose column, using a wide mouth plastic pasteur pipette, and allow the sample to percolate through the column to create a dsRNA CHARGED COLUMN. This is best done 1.0 ml at a time initially until the column bed is well settled with some solution above it. At this point the barrel can be filled with sample. The dsRNA in the sample will bind to the column and other nucleic acids will pass through. If you want host nucleic acids or viral ssRNAs "free" of dsRNA for other work, then collect the eluate;

otherwise, let it drip into the pan. If flow rate is slow this is often because of a crust on the surface of the column which can be disrupted by gentle stirring with a pasteur pipette tip. It may take between 30 min and 2 h for this step, depending on flow rate.

9. In order to be sure contaminating nucleic acids are kept to a minimum the charged column should be washed with at least 100 ml of STE-buffered 16.5% ethanol, in stages, to create a WASHED CHARGED COLUMN. Flow rates should speed up during this step.

10. The dsRNA is ready to be eluted from the column. Pass 15 ml of STE (3 × 5-ml aliquots) through the column. Discard the first 5 ml since it contains little or no dsRNA, but some contaminating nucleic acids are eluted. You should check that there is no loss of dsRNA for your system. Place a 50-ml collection tube beneath each column in the lower test tube rack. Collect the final 10 ml of eluate in the one tube. This eluate is the DILUTE dsRNA SAMPLE/ONE CYCLE.

11. Adjust the volume of the dilute dsRNA sample/one cycle to 20 ml and repeat Steps 6 to 10 if you are doing a SECOND ROUND OF CHRO-MATOGRAPHY, which is a recommended step that reduces levels of contaminating nucleic acids. This would result in DILUTE dsRNA SAMPLE/TWO CYCLES.

12. Add 2.5 volumes of 95% ethanol, and 1/20 volume of 3.0 M sodium acetate, pH 5.5, to the dilute dsRNA sample, store at $-20°C$ for 1 h or more then centrifuge at 8000 g for 30 min (cold storage is conventional, but we have found that no dsRNA is lost if the solution is centrifuged immediately[28]). Keep the pellet, which will contain dsRNA and usually some CF11 powder, which does not matter and actually helps identify where the pellet is in the tube.

13. Resuspend the pellet in 1.0 ml of STE (ENRICHED dsRNA) and transfer it to a small (4 to 5 ml) centrifuge tube. Add 2.5 volumes of 95% ethanol, and 1/20 volume of 3.0 M sodium acetate, pH 5.5, to the enriched dsRNA sample, store at $-20°C$ for 1 h or indefinitely.

14. Centrifuge at 3000 g for 30 min. Keep the pellet, which may be hard to see and will contain dsRNA, and resuspend it in 100 to 300 μl of water (if to be used for tests besides electrophoresis) or in electrophoresis sample buffer (electrophoresis buffer with 20% glycerol and an optional 0.05% bromophenol blue).

15. Analyze 5 to 20 μl of dsRNA samples by electrophoresis on 6.0% polyacrylamide (40:1 acrylamide:bis-acrylamide) gels cast in a small vertical gel apparatus (e.g., gel size of 8 cm × 7 cm × 1.5 mm) for 2 to 4 h at 110 V in 40 mM Tris, 20 mM sodium acetate, and 1 mM EDTA, pH 7.8, or on 1.0% agarose gels in a horizontal gel apparatus. Include marker nucleic acids, such as the dsRNAs of greenhouse isolates of CTV, TMV, CMV, and CMV satellite RNA or a *Hind*III or other restriction endonuclease digest of lambda DNA which can be obtained commercially (e.g., New England Biolabs) in all electrophoretic analyses.

16. Stain gels with ethidium bromide (10 ng/ml is adequate) for 15 min in electrophoresis buffer. Destain with 3×10-s rinses of water and photograph over an ultraviolet (260 to 300 nm) transilluminator. A pair of red (R-23-A, top) and yellow (y-9, bottom) filters are recommended. Be sure to try long exposures in order to avoid missing minor bands not visible by the protected eye. If gels destain they can be restained and can be held in a container for days or weeks in water or electrophoresis buffer. Polyacrylamide gels can also be silver-stained,[34,35] as an alternative to, or after staining with, ethidium bromide.

17. Nucleic acid contaminations and DNase or RNase digestions: the main contaminant observed within electrophoresed polyacrylamide gels is a DNA fraction which comigrates with, but is broader than, the TMV 4.2×10^6 major RF dsRNA. This DNA can be reduced to barely detectable levels by two cycles of chromatography or to nondetectable levels by DNAse treatment of the dsRNA sample after one or two cycles of chromatography. We avoid the routine use of nucleases until preliminary gel analysis of nontreated dsRNA is completed. DNase treatment is done as follows. Dilute dsRNA (in electrophoresis sample buffer) 1 to 5 with water, add 1/100 volume of $0.5 \, M \, MgCl_2$ and 0.5 to 1.0 unit of RNase-free DNase I (Promega), and incubate at 37°C for 1 h. DNAse that is not RNase free should be treated with proteinase K in the presence of $CaCl_2$ before use.[60] Extract with 1 volume of phenol and 1 volume of chloroform/isopentanol (24:1). Centrifuge 15 min at 3000 g, collect the aqueous phase, add 1/20 volume of $3.0 \, M$ sodium acetate, pH 5.5, and 2.5 volumes of 95% ethanol. Store at $-20°C$ prior to preparing a sample for electrophoresis (Step 14). dsRNAs also can be treated with ribonuclease[26,61-63] in salt concentrations that are low (in $0.01 \, M$ both ssRNA and dsRNA are digested) or high (in $0.15 \, M$ or higher, ssRNA is digested, but dsRNA is not) in order to remove contaminating ssRNAs or confirm the dsRNA nature of detected bands in gels. Removing contaminating ssRNA is not usually needed when analysis is by PAGE. Use either dsRNA in solution in electrophoresis buffer or an ethidium bromide-stained gel in 20 to 50 ml of electrophoresis buffer. Add 1/10 volume of $3.0 \, M$ NaCl, 1/100 volume of 50 μg/ml RNase 1A (final concentration of 0.5 μg/ml), and incubate 30 min at room temperature to digest ssRNA without digesting dsRNA. After treating a solution of dsRNA, extract it with phenol/chloroform then precipitate and concentrate the remaining dsRNA as described for DNase treatment. If it is a gel that was treated, rinse and possibly restain before observing. Note that you will need to keep the gel in 0.15 to 0.3 M NaCl, otherwise retained RNAse will begin to digest dsRNA bands as soon as the salt concentration around the band becomes sufficiently low for digestion to begin. NaCl can fog the gel and give somewhat undesirable background fluorescence with ethidium bromide, so in-gel ribonuclease digestion is best for observing no effect of ribonuclease on major dsRNA bands.

18. Northern blotting of dsRNA from polyacrylamide gels: this is an important method that most workers will want to use. It is our experience that it is difficult to obtain equivalent transfer of both the highest- and lowest-molecular-weight dsRNAs from a single gel. A selection of alternate methods for Northern blotting are to be found in many of the more recent citations found in the references section, and additional details can be found in laboratory manuals for molecular techniques.[53,54] A method we have found to work well is described here. Soak the gel in 50% deionized formamide, 6.5% formaldehyde, and 20 mM HEPES, pH 7.7, and incubate for 15 to 20 min at 70°C. Rinse the gels for 10 min in 10 mM Tris, pH 7.8, 5 mM sodium acetate, and 0.5 mM EDTA (TAE buffer). Place the gel in a blotting apparatus using a sheet of charged nylon (Zetaprobe, BioRad) as the solid support. Electroblot overnight at 30 V in TAE buffer, then for a final 6 h at 60 V with a buffer change every 1.5 h at 4°C. Remove and restain the gel to evaluate the relative efficiency of transfer of high- and low-molecular-weight dsRNAs. Rinse membranes briefly in TAE buffer then bake at 80°C for 2 h. Prehybridize the baked membrane in 0.75 M NaCl, 75 mM sodium citrate, 50 mM sodium phosphate, pH 6.5, containing 0.2% each of bovine serum albumin (BSA), Ficoll, and polyvinylpyrrolidone (PVP). An aliquot of cDNA or other probe is added to the prehybridization fluid to a final concentration of 1.5 × 10^6 cpm/ml and the blot is hybridized for 24 h at 42°C. Blots are removed from the hybridization fluid, washed in 0.3 M NaCl, 0.03 M sodium citrate, 0.025 M sodium phosphate, pH 6.5 (2 × SSCP), and 0.1% SDS for 0.5 h with two changes of buffer (low to moderate stringency), then washed in 1 × SSCP/0.1% SDS at room temperature for 1 h, then in 0.1 × SSCP/0.1% SDS at 55°C for 1 h with one change of buffer (high stringency). Blots are then air dried and placed in a Kodak X-Omatic cassette with intensifying screens next to a sheet of Kodak XAR-5 X-ray film. Exposures are for 12 to 72 h at −70°C.

19. Use of dsRNA to make randomly primed cDNA probes: labeled probes are needed for Northern blotting and cDNA probes can be made as follows. First treat dsRNA with DNAse and RNase (in high salt) as described previously. Dispense 5 to 10 μg of dsRNA into a small tube and adjust to 10 mM methylmercuric hydroxide (CH$_3$HgOH), and incubate for 10 min at room temperature to denature the dsRNA to ssRNA. Add 150 μg of random primer,[53] boil the mixture for 2 min then immediately chill on ice. Adjust to 75 mM 2-mercaptoethanol and add 25 units of RNasin (a ribonuclease inhibitor), 5× reverse transcriptase buffer (250 mM Tris, pH 8.3, 375 mM KCl, 50 mM dithiothreitol, and 15 mM MgCl$_2$, use 10 μl for a 50 μl final reaction volume) and 5 μg of BSA (final concentration of 100 μg/ml). Adjust the mixture to 0.3 mM dGTP, dATP, and dTTP and to 0.1 mM dCTP. Add 2000 units of Moloney-murine luekemia virus reverse transcriptase (BRL or

equivalent units from another manufacturer) (total volume is 50 μl) and add 100 μCi of deoxycytidine 5'-[α-^{32}P] triphosphate and incubate at 37°C for 1 h. Hydrolyze the RNA template by treating with 0.1 *M* Na$_2$EDTA (final concentration of 10 m*M*) and 1.0 *N* NaOH (final concentration of 0.35 *M*) for 1 h at 65°C. Neutralize with 1.0 *M* Tris-HCl (final concentration of 175 m*M*) and 1.0 *N* HCl (final concentration of 175 m*M*). Remove unwanted nucleotides with a spun column.[53]

IV. RECENT APPLICATIONS

dsRNA analysis has been used for the detection of many different plant or fungal viruses and most of the corresponding publications are listed in the following sections. This list is provided because there is no compendium of dsRNA profiles for plant viruses equivalent to the one available for plant virus inclusion bodies.[64] Additional examples can be found in previous review articles.[14-19]

A. DETECTION OF dsRNAs OF KNOWN VIRUSES INCLUDING SATELLITES

Many plant viruses have ssRNA genomes and the number and size of the ssRNA segments that make up the complete genome are characteristics of a virus and the group of which it is a member. Information on numbers and sizes of ssRNAs can be obtained from the *AAB Descriptions of Plant Viruses* published by the Association of Applied Biologists, National Vegetable Research Station, Wellsbourne, Warwick, U.K., and the 5th report of the International Committee on Taxonomy of Viruses.[5] Since each genomic ssRNA is normally represented by an equivalent dsRNA, the information obtained from dsRNA detection and analysis can be used in the identification of plant viruses.

A partial list of virus groups or individual viruses that have been detected recently by dsRNA analysis include tobamoviruses,[6,11,33] furoviruses,[43,65] potexviruses,[33,66-69] carlaviruses,[33,70] potyviruses,[33] tenuiviruses,[71] closteroviruses,[23,26,33,39,71-76] bromoviruses[6,77,78] cucumoviruses,[25] dianthoviruses,[7,56] tombusviruses,[8] luteoviruses,[34,63,79,80,81] sobemoviruses,[8,82] necroviruses,[8] tymoviruses,[8,15,33] reoviruses,[29] alfalfa mosaic virus, pea enation mosaic virus,[49] maize white line mosaic virus,[83] and strawberry mottle virus.[84] The dsRNAs of satellite viruses[22,40,85] and satellite RNAs[20,49,86] seem to be abundant, and for this reason dsRNA analysis is especially well suited to their detection, since a high abundance of a relatively low molecular weight RF dsRNA is both unexpected and hard to miss. Care should be taken to use Northern blotting to distinguish dsRNAs of satellites from the dsRNAs of defective interfering RNAs.[87]

The use of dsRNAs analysis to distinguish between strains of a single virus is well illustrated by work with CMV[25] and CTV.[23,24,39,88-91]

B. DETECTION OF dsRNAs IN POORLY CHARACTERIZED AGENTS

dsRNA has been detected in association with each of the following virus or virus-like diseases; lettuce big vein,[27] strawberry pallidosis,[92] strawberry june yellows,[31] rupestris stem pitting,[93,94] rose-rosette disease,[95] dodonaea yellows disease,[96] and latent virus-like agents in avocado,[62] cassava,[97] and small fruits.[98] A silvering disease of squash is an unusual example, since dsRNA may be associated with those populations of whiteflies that are associated with the disease.[99-101]

C. DETECTION OF dsRNAs IN PLANT PATHOGENIC AND EDIBLE FUNGI

This topic was described in some detail in previous reviews[3,4] and dsRNAs have been detected recently in several plant pathogenic fungi and cultivated edible mushrooms,[42,102-106] including rusts,[107-110] smuts,[111] *Cryphonectria (Endothia) parasitica*,[112] *Ophiostoma (Ceratocystis) ulmi*,[44] *Phytophthora infestans*,[113] *Rhizoctonia solani*,[114] *Leucostoma persoonii*,[115,116] *Septoria tritici*,[117] *Uncinula necator*,[118] and *Dreschlera* sp.[119]

D. DETECTION OF dsRNAs IN *N. glauca*: A CASE HISTORY

Tree tobacco, *Nicotiana glauca*, is a common solanaceous perennial in southern California, and is a reservoir for several plant viruses. Figure 1 shows dsRNAs isolated from four individual branches on a single plant, which was made up of many branches that made up a 3 m tall by 3 m wide bush. The dsRNAs detected are those of a tobacco etch *Potyvirus* (TEV), tobacco mild green mottle *Tobamovirus* (TMGMV or TMV U5) and its associated satellite tobacco mosaic virus (STMV), cucumber mosaic *Cucumovirus* (CMV), and an associated satellite RNA (CMV-satRNA[120]) and a putative virus that has been detected, but not yet diagnosed (unknown agent). The dsRNAs of each of the diagnosed viruses purified from experimentally infected tobacco or *N. glauca* is shown to the left of the dsRNAs from field grown plants of *N. glauca*. A Northern blot of the gel shows the specific detection of dsRNA of STMV (Figure 2). The figures illustrate the ability of dsRNA analysis to detect and diagnose viruses in field-collected leaf tissue, to detect multiple viruses in a single plant, to detect satellite viruses and satellite RNAs, and to describe the uneven distribution of viruses in a single plant. All this information was obtained from a single 2-day analysis of 5.0 g of tissue.

V. CONCLUSIONS AND COMMENTS

A. GENERAL APPLICABILITY

There are several advantages to dsRNA analysis. First, it is a nonspecific method much like inclusion body staining[64] since it detects any dsRNA that is present in the sample, regardless of the number of viruses present or if there are other possible sources of dsRNA besides virus infection. Because

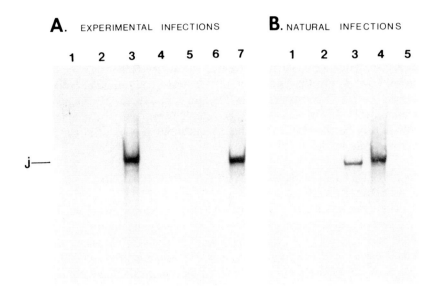

FIGURE 2. Northern blot analysis of the same plant dsRNAs shown in Figure 1. The gels which contained dsRNAs from experimentally infected (A) and naturally infected (B) plants were blotted and then probed with ^{32}P-labeled STMV minus-sense RNA transcript of a genomic cDNA clone. The distribution of STMV that was deduced from inspection of dsRNA profiles in Figure 1 was confirmed by this analysis.

of this, dsRNA analysis is very good for the detection of mixed virus infections, including the detection of satellite RNAs, satellite viruses, and defective interfering RNAs.

The sample that is brought in from the field can be tested directly,[90,121] as in Figure 1,which is a clear convenience and allows for very timely data collection when results are positive. Three levels of diagnosis can be anticipated from the use of dsRNA analysis. The first is at the level of presence or absence of an RNA virus, which is suggested by the detection of any dsRNA. The second is the possible group the virus does or does not belong to, which is determined by evaluating the number and size of the major dsRNAs detected, which correspond to the genomic ssRNAs of the virus.

The third level is the diagnosis of individual viruses or virus strains. This is determined only with considerable experience. We have commonly observed strain differences in the mobility of major dsRNAs corresponding to the genomes of specific viruses. This results in the detection of "doublet" bands on polyacrylamide gels when two strains are present in natural mixed infections or in artificially mixed samples. Doublets of dsRNA 1 and dsRNA 2 in mixtures of strains of CMV,[25] in the major RF dsRNAs of CTV in complex isolates,[122] and the major RFs of STMV isolates[40] have been detected. In addition to using major dsRNAs for strain differentiation, it is not unusual to be able to utilize minor dsRNAs of uncertain origin, which are constantly associated with infection with a specific virus or virus strain, for experiments on isolate or strain comparisons,[23,39,88-91] effects of host passage on isolates,[24] and on mixed infections.[122] Comparison of strains using polyacrylamide gels may well be an underused application of dsRNA analysis.

Only 10 to 20% of the purified dsRNA is normally used for gel electrophoresis and the remainder can be used as a reagent in other attempts to advance any virology project, as described in Sections II.E and III, steps **18** and **19**. An advantage of adding blotting techniques to gel analysis of dsRNAs is that this introduces specificity to the analysis. Genomic ssRNA is normally used for *in vitro* translation and cloning and sequencing when it is readily available. This may not be so for new viruses, or for viruses not easily purified, in which situation dsRNA may be the only choice. These techniques are described elsewhere and are only discussed here to point out potential advantages of dsRNA analysis.

B. LIMITATIONS

As with any method, there are limitations to dsRNA analysis, of which the most obvious is the impracticality of testing hundreds or thousands of samples. However, by scaling down the procedure by using less starting tissue, less cellulose powder, smaller columns, or even batch methods in beakers or small tubes,[15] it is surprising how many samples can be tested by a determined person.

Even when results are positive, skill and knowledge are needed if diagnosis is to be made with confidence, particularly if known viruses are unavailable for comparison. This need for knowledge can be seen as a limitation to the usefulness of dsRNA analysis. However, one advantage is the speed with which suspected viruses can be rejected from a diagnosis, based on the knowledge that they are normally readily detected by dsRNA analysis but their dsRNAs are absent. Timely dsRNA analysis, therefore, can be a major factor influencing the choice of subsequent tests aimed at a final correct diagnosis.

Not all virus groups are detected with the same ease, and even though most luteoviruses and potyviruses can be detected by this method, larger than normal tissue weights are usually needed. Note however, that a *Potyvirus* was detected readily in the *N. glauca* case history example in this chapter. Despite the apparent advantages of dsRNA analysis it will not be equally successful for all host/virus combinations and after an initial evaluation other approaches may have to be selected for specific projects.

All organisms can be infected with RNA viruses and this fact must be taken into consideration, since there may be nonplant organisms in or on the plant tissue being tested. This has in fact been shown to be a problem in grapes, where a fungal contaminant was the source of some dsRNAs, but not those that were disease specific,[94] and in strawberry, where spider mites were shown to be a source of dsRNA.[31] dsRNA analysis of plants infected with mite-transmitted viruses could be affected by this observation. Results with whitefly-associated dsRNAs were mentioned in Section IV.B. These observations reemphasize the need to attempt to approximate Koch's postulates when using dsRNA analysis to establish causal relationships, as was done for a study of lettuce big vein disease.[27]

DNA viruses are not detected by this method and there are no reports of detection of dsRNA in plants infected with negative-sense RNA viruses such as rhabdoviruses.

C. CRYPTIC VIRUSES AND NONVIRAL dsRNAs

Another major limitation of dsRNA analysis is the increasing frequency with which dsRNAs genomes of cryptic viruses[123] or nonviral dsRNAs[41,124-129] are being detected in plants as a result of the increased use of dsRNA analysis in plant virology. The nonviral dsRNAs are often unsegmented[41,50,124-126] (mol wt of about 4.0 to 6.0 $\times$ 10^6), are associated with organelles or other particulate fractions,[41,50,127] are not normally graft-transmissible,[41,125] but are seed-transmitted[41,50,125,129] and are found in some but not other cultivars.[41,50,124,125] They may[50] or may not[45] have some sequence similarity to host DNA. All users of dsRNA analysis should be aware of the existence of such dsRNAs. A practical consequence of this knowledge is the clear need to always run a noninoculated control of a specific cultivar for every plant species being tested. The necessity for this control cannot be over stressed.

This review is designed to allow nonusers to quickly add dsRNA analysis to their bag of tricks. It is an easy technique to learn and use and can give valuable results and reagents in a short time. It is a robust method and can be used at many stages of a plant virology project, including the initial collection of samples, testing and tracking a specific agent in infected symptomatic or asymptomatic host range plants, monitoring hosts for virus purification, testing purity of isolates, and comparing isolates and strains. We cannot help but note the large number of publications that have used dsRNA analysis, together with other methods, and look forward to seeing many more. Like any diagnostic method, it should always be used together with other methods, so that the different strengths each method has can complement each other.

ACKNOWLEDGMENTS

I would like to thank the technicians, undergraduate and graduate students, postdoctoral fellows, and visiting scientists who have run uncountable numbers of CF11 columns in my laboratory from 1977[56] to the present day. Dr. T. Jarupat holds the uncontested world record. Special thanks go to Deborah Mathews for the experimental work that allowed her to prepare the figures for this chapter.

REFERENCES

1. **Randles, J. W.**, Strategies for implicating virus-like pathogens as the cause of diseases of unknown etiology, in *Diagnosis of Plant Viruses Diseases,* Matthews, R. E. F., Ed., CRC Press, Boca Raton, FL, 1993, ch. 12.
2. **Matthews, R. E. F.**, *Plant Virology,* 3rd ed., Academic Press, San Diego, 1991, 650.
3. **Koltin, Y. and Liebowitz, M.**, *Viruses of Fungi and Simple Eukaryotes,* Marcel Dekker, New York, 1988, 434.
4. **Nuss, D. L. and Koltin, Y.**, Significance of dsRNA genetic elements in plant pathogenic fungi, *Annu. Rev. Phytopathol.,* 28, 37, 1990.
5. **Francki, R. I. B., Fauquet, C. M., Knudson, D. L., and Brown, F.**, Classification and Nomenclature of Viruses. Fifth Report of the International Committee on Taxonomy of Viruses, *Arch. Virol. Suppl. 2.,* Springer Verlag, New York, 1991, 450.
6. **Dawson, W. O. and Dodds, J. A.**, Characterization of sub-genomic double-stranded RNAs from virus-infected plants, *Biochem. Biophys. Res. Commun.,* 107, 1230, 1982.
7. **Osman, T. A. M. and Buck, K. W.**, Double-stranded RNAs isolated from plant tissue infected with red clover necrotic mosaic virus correspond to genomic and subgenomic single-stranded RNAs, *J. Gen. Virol.,* 71, 945, 1990.
8. **Morris, T. J. and Dodds, J. A.**, Isolation and analysis of double-stranded RNA from virus-infected plant and fungal tissue, *Phytopathology,* 69, 854, 1979.
9. **Morris, T. J., Dodds, J. A., Hillman, B., Jordan, R. L., Lommel, S. A., and Tamaki, S. J.**, Viral specific dsRNA: diagnostic value for plant virus disease identification, *Plant Mol. Biol. Rep.,* 1, 27, 1983.
10. **Franklin, R. M.**, Purification and properties of replicative intermediates of the RNA bacteriophage R17, *Proc. Natl. Acad. Sci. U.S.A.,* 55, 1504, 1966.
11. **Jackson, A. O., Mitchel, D. M., and Siegel, A.**, Replication of tobacco mosaic virus. I. Isolation and characterization of double-stranded forms of ribonucleic acid, *Virology,* 45, 182, 1971.
12. **Bar-Joseph, M., Rosner, A., Moskovitz, M., and Hull, R.**, A simple procedure for the extraction of double stranded RNA from viral infected plants, *J. Virol. Meth.,* 6, 1, 1983.
13. **Diaz-Ruiz, J. R. and Kaper, J. M.**, Isolation of viral double-stranded RNAs using LiCl fractionation procedure, *Prep. Biochem.,* 8, 1, 1978.
14. **Dodds, J. A., Morris, T. J., and Jordan, R. L.**, Plant viral double-stranded RNA, *Annu. Rev. Phytopathol.,* 22, 151, 1984.
15. **Jordan, R. L. and Dodds, J. A.**, Double-stranded RNA in detection of diseases of known and unproven viral etiology, *Acta Hortic.,* 164, 101, 1984.
16. **Dodds, J. A., Jordan, R. L., Heick, J. A., and Tamaki, S. J.**, Double-stranded RNA for the diagnosis of citrus and avocado viruses, Proc. 9th Conf. International Organization of Citrus Virology, 1985, 330.
17. **Dodds, J. A.**, The potential of using double-stranded RNAs as diagnostic probes for plant viruses, in *Developments and Applications in Virus Testing,* Jones, R. A. C. and Torrence, L., Eds., Association of Applied Biologists, Wellesbourne, England, 1986, 71.
18. **Dodds, J. A., Valverde, R. A., and Mathews, D. M.**, Detection and interpretation of dsRNA, in *Viruses of Fungi and Simple Eukaryotes,* Koltin, Y. and Leibowitz, M., Eds., Marcel Dekker, New York, 1987, 309.
19. **Valverde, R. A., Nameth, S. T., and Jordan, R. L.**, Analysis of double-stranded RNA for plant virus diagnosis, *Plant Dis.,* 74, 255, 1990.
20. **White, J. L. and Kaper, J. M.**, A simple method for detection of viral satellite RNAs in small plant tissue samples, *J. Virol. Meth.,* 23, 83, 1989.
21. **Bozarth, R. F. and Harley, E. H.**, The electrophoretic mobility of double-stranded RNA in polyacrylamide gels as a function of molecular weight, *Biochem. Biophys. Acta,* 432, 329, 1976.

22. **Valverde, R. A., Heick, J. A., and Dodds, J. A.,** Interactions between satellite tobacco mosaic virus, helper tobamoviruses and their hosts, *Phytopathology,* 81, 99, 1991.

23. **Dodds, J. A., Jarupat, T., Lee, J. G., and Roistacher, C. N.,** Effects of strain, host, time of harvest and virus concentration on double-stranded RNA analysis of citrus tristeza virus, *Phytopathology,* 77, 442, 1987.

24. **Dodds, J. A., Jarupat, T., and Roistacher, C. N.,** Effect of host passage on dsRNA profiles of selected strains of citrus tristeza virus, Proc. 10th Conf. International Organization of Citrus Virology, 1988, 39.

25. **Dodds, J. A., Lee, S. Q., and Tiffany, M.,** Cross protection between strains of cucumber mosaic virus: effects of host and type of inoculum on accumulation of virions and double-stranded RNA of the challenge strain, *Virology,* 144, 301, 1985.

26. **Dodds, J. A. and Bar-Joseph, M.,** Double-stranded RNA from plants infected with closteroviruses, *Phytopathology,.* 73, 419, 1983.

27. **Mirkov, T. E. and Dodds, J. A.,** Association of double-stranded RNA with lettuce big vein disease, *Phytopathology,* 75, 631, 1985.

28. **Jarupat, T., Lee, J. G., and Dobbs, J. A.,** Additional factors affecting dsRNA analysis of citrus tristeza virus, Proc. 11th Conf. International Organization of Citrus Virology, 1991, 137.

29. **Karan, M., Hicks, S., Harding, R. M., and Teakle, D. S.,** Stability and extractability of double-stranded RNA of pangola stunt and sugarcane fiji disease viruses in dried plant tissues, *J. Virol. Meth.,* 33, 211, 1991.

30. **Day, P. R., Dodds, J. A., Elliston, J. E., Jaynes, R. A., and Anagnostakis, S. L.,** Double-stranded RNA in *Endothia parasitica, Phytopathology,* 67, 1393, 1977.

31. **Watkins, C. A., Jones, A. T., Mayo, M. A., and Mitchell, M. J.,** Double-stranded RNA analysis of strawberry plants affected by june yellows, *Ann. Appl. Biol.,* 117, 73, 1990.

32. **Francki, R. I. B.,** Responses of plant cells to virus infection with special reference to the sites of replication, *UCLA Symp. New Ser.,* 54, 423, 1987.

33. **Valverde, R. A., Dodds, J. A., and Heick, J. A.,** Double-stranded ribonucleic acid from plants infected with viruses having elongated particles and undivided genomes, *Phytopathology,* 76, 459, 1986.

34. **Gildow, F. E., Ballinger, M. E., and Rochow, W. F.,** Identification of double-stranded RNAs associated with barley yellow dwarf virus infection in oats, *Phytopathology,* 73, 1570, 1983.

35. **Blum, H., Beier, H., and Gross, H. J.,** Improved silver staining of plant proteins, RNA and DNA in polyacrylamide gels, *Electrophoresis,* 8, 93, 1987.

36. **Rosner, A., Bar-Joseph, M., Moscovitz, M., and Mavarech, M.,** Diagnosis of specific viral RNA sequences in plant extracts by hybridization with polynucleotide kinase labelled double-stranded RNA probe, *Phytopathology,* 73, 699, 1983.

37. **Garger, S. J., Turpen, T., Carrington, J., Morris, T. J., Jordan, R. L., Dodds, J. A., and Grill, L. K.,** Rapid detection of plant viral RNA viruses by dot blot hybridization, *Plant Mol. Biol. Rep.,* 1, 21, 1983.

38. **Jordan, R. L. and Dodds, J. A.,** Hybridization of 5′ end-labelled RNA to plant viral (tobacco mosaic virus) RNA in agarose and acrylamide gels, *Plant Mol. Biol. Rep.,* 1, 31, 1983.

39. **Dodds, J. A., Jordan, R. L., Roistacher, C. N., and Jarupat, T.,** Diversity of citrus tristeza virus isolates indicated by dsRNA analysis, *Intervirology,* 27, 177, 1987.

40. **Valverde, R. A. and Dodds, J. A.,** Some properties of isometric virus particles which contain the satellite RNA of tobacco mosaic virus, *J. Gen. Virol.,* 68, 965, 1987.

41. **Valverde, R. A., Nameth, S., Abdalla, O., Al-Musa, A., Desjardins, P., and Dodds, A.,** Indigenous double-stranded RNA from pepper *(Capsicum annuum), Plant Sci.,* 67, 195, 1990.

42. **Romaine, C. P. and Schlagnhaufer, B.,** Hybridization analysis of the single-stranded RNA bacilliform virus associated with La France disease of *Agaricus bisporus, Phytopathology,* 81, 1336, 1991.

43. **Hutchinson, P. J., Henry, C. M., and Coutts, R. H. A.,** A comparison, using dsRNA analysis, between beet soil-borne virus and some other tubular viruses isolated from sugar beet, *J. Gen. Virol.,* 73, 1317, 1992.

44. **Rogers, H. J., Buck, K. W., and Brasier, C. M.,** A mitochondrial target for double-stranded RNA in diseased isolates of the fungus that causes dutch elm disease, *Nature,* 329, 558, 1987.

45. **Monroy, A. F., Gao, C., Zhang, M., and Brown, G. G.,** Double-stranded RNA molecules in *Brassica* are inherited biparentally and appear not to be associated with mitochondria, *Curr. Genet.,* 17, 427, 1990.

46. **Imai, M., Richardson, M. A., Ikegami, N., Shatkin, A. J., and Furuichi, Y.,** Molecular cloning of double-stranded RNA virus genomes, *Proc. Natl. Acad. Sci. U.S.A.,* 80, 373, 1983.

47. **Antinow, J. F., Linthorst, J. M., White, R. F., and Bol, J. F.,** Molecular cloning of double-stranded RNA of beet cryptic viruses, *J. Gen. Virol.,* 67, 2047, 1986.

48. **Jelkmann, W., Martin, R. R., and Maiss, E.,** Cloning of four plant viruses from small quantities of double-stranded RNA, *Phytopathology,* 79, 1250, 1989.

49. **Demier, S. A. and de Zoeten, G. A.,** Characterization of a satellite RNA associated with pea enation mosaic virus, *J. Gen. Virol.,* 70, 1075, 1989.

50. **Wakarchuk, D. A. and Hamilton, R. I.,** Partial nucleotide sequence from enigmatic dsRNAs in *Phaseolus vulgaris, Plant Mol. Biol.,* 14, 637, 1990.

51. **Wexler, A., Mawassi, M., Lachman, O., Amit, B., Wortzel, A., and Bar-Joseph, M.,** A procedure to amplify cDNA from dsRNA templates using the polymerase chain reaction, *Meth. Mol. Cell. Biol.,* 2, 273, 1991.

52. **Dulieu, P. and Bar-Joseph, M.,** Rapid isolation of double stranded RNA segments from disulphide crosslinked polyacrylamide gels, *J. Virol. Meth.,* 24, 77, 1989.

53. **Sambrook, J., Fritsch, E. F., and Maniatis, T.,** *Molecular Cloning, a Laboratory Manual,* 2nd ed., Cold Spring Harbor Laboratory Press, Cold Spring Harbor, NY, 1989.

54. **Ausubel, F. M., Brent, R., Kingston, R. E., Moore, D. D., Seidman, J. G., Smith, J. A., and Struhl, K.,** *Current Protocols in Molecular Biology,* Vols. 1 and 2, Wiley Interscience, New York, 1987.

55. **Dulieu, P. and Bar-Joseph, M.,** *In vitro* translation of the citrus tristeza virus coat protein from a 0.8 Kbp double stranded RNA segment, *J. Gen. Virol.,* 71, 443, 1990.

56. **Dodds, J. A., Tremaine, J. H., and Ronald, W. P.,** Some properties of carnation ringspot virus single- and double-stranded ribonucleic acid, *Virology,* 83, 322, 1977.

57. **Reddy, D. V. R., Murant, A. F., Raschke, J. H., Mayo, M. A., and Ansa, O. A.,** Properties and partial purification of infective material from plants containing groundnut rosette virus, *Ann. Appl. Biol.,* 107, 65, 1985.

58. **Powell, C. A.,** Preparation and characterization of monoclonal antibodies to double-stranded RNA, *Phytopathology,* 81, 184, 1991.

59. **Aramburu, J., Navascastillo, J., Moreno, P., and Cambra, M.,** Detection of double-stranded RNA by ELISA and dot immunobinding assay using an antiserum to synthetic polynucleotides, *J. Virol. Meth.,* 33, 1, 1991.

60. **Tullis, R. H. and Rubin, H.,** Calcium protects DNAse I from protease K, *Anal. Biochem.,* 107, 260, 1980.

61. **Dodds, J. A.,** Revised estimates of the molecular weights of dsRNA segments in hypovirulent strains of *Endothia parasitica, Phytopathology,* 70, 1217, 1980.

62. **Jordan, R. L., Dodds, J. A., and Ohr, H.,** Evidence for virus-like agents in avocado, *Phytopathology,* 73, 1130, 1983.

63. **Smith, O. P., Hunst, P. L., Hewings, A. D., Stone, A. L., Tolin, S. A., and Damsteegt, V. D.,** Identification of dsRNAs associated with soybean dwarf virus-infected soybean, *Phytopathology,* 81, 131, 1991.

64. **Christie, R. G. and Edwardson, J. R.,** Light and electron microscopy of plant virus inclusions, *Fla. Agric. Exp. Stn. Monogr. Ser.,* 9, 155, 1977.

65. **Kallender, H., Buck, K. W., and Brunt, A.,** Association of three RNA molecules with potato mop-top virus, *Neth. J. Plant Pathol.,* 96, 47, 1990.

66. **Mackie, G. A., Johnston, R., and Bancroft, J. B.,** Single- and double-stranded viral RNAs in plants infected with the potexviruses papaya mosaic virus and foxtail mosaic virus, *Intervirology,* 29, 170, 1988.

67. **Wakarchuk, D. A. and Hamilton, R. I.,** A potexvirus isolated from *Silene pratensis, Intervirology,* 30, 330, 1989.

68. **Jelkmann, W., Martin, R. R., Lesemann, D. E., Vetten, H. J., and Skelton, F.,** A new potexvirus associated with strawberry mild yellow edge disease, *J. Gen. Virol.,* 71, 1251, 1990.

69. **Jones, A. T., Mitchell, M. J., McGavin, W. J., and Roberts, I. M.,** Further properties of wineberry latent virus and evidence for its possible involvement in calico disease, *Ann. Appl. Biol.,* 117, 571, 1990.

70. **Meehan, B. M. and Mills, P. R.,** Cell-free translation of carnation latent virus RNA and analysis of virus-specific dsRNA, *Virus Genes,* 5, 175, 1991.

71. **Falk, B. W. and Tsai, J. H.,** Identification of single stranded RNA and double stranded RNA associated with maize stripe virus, *Phytopathology,* 74, 909, 1984.

72. **Mossop, D. W., Elliot, D. R., and Richards, K. D.,** Association of closterovirus-like particles and high molecular weight double stranded RNA with grapevines affected by leaf-roll disease, *N.Z. J. Agric. Res.,* 28, 419, 1985.

73. **Monette, P. L., James, D., and Godkin, S. E.,** Comparison of RNA extracts from *in vitro* shoot tip cultures of leafroll-affected and leafroll-free grapevine cultivars, *Vitis,* 28, 229, 1989.

74. **Rezaian, M. A., Krake, L. R., Cunying, Q., and Hazzalin, C. A.,** Detection of virus-associated dsRNA from leafroll infected grapevines, *J. Virol. Meth.,* 31, 325, 1991.

75. **Gunasinghe, U. B. and German, T. L.,** Purification and partial characterization of a virus from pineapple, *Phytopathology,* 79, 1337, 1989.

76. **Coffin, R. S. and Coutts, R. H. A.,** The occurrence of beet pseudo-yellows virus in England, *Plant Pathol.,* 39, 632, 1990.

77. **Haber, S. and Hamilton, R. I.,** Brome mosaic virus isolated from Manitoba, Canada, *Plant Dis.,* 73, 195, 1989.

78. **Valverde, R. A. and Glascock, C. B.,** Further examination of the RNA and coat protein of spring beauty latent virus, *Phytopathology,* 81, 401, 1991.

79. **Falk, B. W. and Duffus, J. E.,** Identification of small single- and double-stranded RNAs associated with severe symptoms in beet western yellows virus-infected *Capsella bursapastoris, Phytopathology,* 74, 1224, 1984.

80. **Dale, J. L., Phillips, D. A., and Parry, J. N.,** Double-stranded RNA in banana plants with bunch top disease, *J. Gen. Virol.,* 67, 371, 1986.

81. **Spiegel, S.,** Double-stranded RNA in strawberry plants infected with strawberry mild yellow edge virus, *Phytopathology,* 77, 1492, 1987.

82. **Chu, P. W. G., Francki, R. I. B., and Randles, J. W.,** Detection, isolation, and characterization of high molecular weight double stranded RNA in plants infected with velvet tobacco mottle virus, *Virology,* 126, 480, 1983.

83. **Zhang, L., Zitter, T. A., and Palukaitis, P.,** Maize white line mosaic virus double-stranded RNA, replication structure, and *in vitro* translation product analysis, *Phytopathology,* 81, 1253, 1991.

84. **Yoshikawa, N. and Converse, R. H.,** Purification and some properties of strawberry mottle virus, *Ann. Appl. Biol.,* 118, 565, 1991.

85. **Valverde, R. A. and Dodds, J. A.,** Evidence for a satellite RNA associated naturally with the U5 strain and experimentally with the U1 strain of tobacco mosaic virus, *J. Gen. Virol.,* 67, 1875, 1986.

86. **Kumar, I. K., Murant, A. F., and Robinson, D. J.,** A variant of the satellite RNA of groundnut rosette virus that induces brilliant yellow blotch mosaic symptoms in *Nicotiana benthamiana, Ann. Appl. Biol.,* 118, 555, 1991.

87. **Burgyan, J., Grieco, F., and Russo, M.,** A defective interfering RNA molecule in cymbidium ringspot virus infections, *J. Gen. Virol.,* 70, 235, 1989.

88. **Dodds, J. A., Jarupat, T., Roistacher, C. N., and Lee, J. G.,** Detection of strain specific double-stranded RNAs in *Citrus* species infected with citrus tristeza virus: a review, *Phytophylactica,* 19, 131, 1987.

89. **Moreno, P., Guerri, J., and Munoz, N.,** Identification of Spanish strains of citrus tristeza virus by analysis of double-stranded RNA, *Phytopathology,* 80, 477, 1990.

90. **Dodds, J. A. and Lee, J. G.,** An evaluation of types of citrus tristeza virus in selected sweet orange grove in southern California, Proc. 11th Conf. International Organization of Citrus Virology, 1991, 103.

91. **Guerri, J., Moreno, P., Munoz, N., and Martinez, M. E.,** Variability among Spanish citrus tristeza virus isolates revealed by double-stranded RNA analysis, *Plant Pathol.,* 40, 38, 1991.

92. **Yoshikawa, N. and Converse, R. H.,** Strawberry pallidosis disease: distinctive dsRNA species associated with latent infections in indicators and in diseased strawberry cultivars, *Phytopathology,* 80, 543, 1990.

93. **Monette, P. L., James, D., and Godkin, S. E.,** Double-stranded RNA from rupestris stem pitting-affected grapevine, *Vitis,* 28, 137, 1989.

94. **Azzam, A. I., Gonsalves, D., and Golino, D. A.,** Detection of dsRNA in grapevines showing symptoms of rupestris stem pitting disease and the variabilities encountered, *Plant Dis.,* 75, 960, 1991.

95. **Di, R., Hill, J. H., and Epstein, A. H.,** Double-stranded RNA associated with the rose rosette disease of multiflora rose, *Plant Dis.,* 74, 56, 1990.

96. **Borth, W. B., Gardner, D. E., and German, T. L.,** Association of double-stranded RNA and filamentous viruslike particles with dodonaea yellows disease, *Plant Dis.,* 74, 434, 1990.

97. **Gabriel, C. J., Walsh, R., and Nolt, B. L.,** Evidence for a latent viruslike agent in Cassava, *Phytopathology,* 77, 92, 1987.

98. **Jones, A. T., Abou El-Nasr, Mayo, M. A., and Mitchell, M. J.,** Association of dsRNA species with some virus-like diseases of small fruits, *Acta Hortic.,* 186, 63, 1986.

99. **Bharathan, N., Graves, W. R., Narayanan, H. R., Schuster, D. J., Bryan, H. H., and McMillan, R. T., Jr.,** Association of double-stranded RNA with whitefly-mediated silvering in squash, *Plant Pathol.,* 39, 530, 1990.

100. **Yakomi, R. K., Hoelmer, K. A., and Osborne, L. S.,** Relationships between the sweetpotato whitefly and the squash silverleaf disorder, *Phytopathology,* 80, 895, 1990.

101. **Bharathan, N., Narayanan, K. R., and McMillan, R. T., Jr.,** Characteristics of sweetpotato whitefly mediated silverleaf syndrome and associated double-stranded RNA in squash, *Phytopathology,* 82, 136, 1992.

102. **Koons, K., Schlagnhaufer, B., and Romaine, C. P.,** Double-stranded RNAs in mycelial cultures of *Agaricus bisporus* affected by La France disease, *Phytopathology,* 79, 1272, 1989.

103. **Romaine, C. P. and Schlagnhaufer, B.,** Prevalence of double-stranded RNAs in healthy and La France disease-affected basidiocarps of *Agaricus bisporus, Mycologia,* 81, 822, 1989.

104. **Morten, K. J. and Hicks, R. G. T.,** Changes in double-stranded RNA profiles in *Agaricus bisporus* during subculture, *FEMS Microbiol. Lett.,* 91, 159, 1992.

105. **Harmsen, M. C., Van Griensven, L. J. L. D., and Wessels, J. G. H.,** Molecular analysis of *Agaricus bisporus* double stranded RNA, *J. Gen. Virol.,* 70, 1613, 1989.

106. **Rytter, J. L., Royse, D. J., and Romaine, C. P.,** Incidence and diversity of double-stranded RNA in *Lentinula edodes, Mycologia,* 83, 506, 1991.

107. **Dickinson, M. J., Wellings, C. R., and Pryor, A.,** Variation in the double-stranded RNA phenotype between and within different rust species, *Can. J. Bot.,* 68, 599, 1990.

108. **Kim, W. K. and Klassen, G. R.,** Double-stranded RNAs in mitochondrial extracts of stem rusts and leaf rusts of cereals, *Curr. Genet.,* 15, 161, 1989.

109. **Dickinson, M. J. and Pryor, A. J.,** Isometric virus-like particles encapsidate the double-stranded RNA found in *Puccinia striiformis, Puccinia recondita* and *Puccinia sorghi, Can. J. Bot.,* 67, 3420, 1989.

110. **Dickinson, M. J. and Pryor, A. J.,** Encapsidated and unencapsidated double-stranded RNAs in flax rust, *Melampsora lini, Can. J. Bot.,* 67, 1137, 1989.

111. **Francki, M. G. and Kirby, G. C.,** Detection of dsRNA viruses in isolates of Australian smut fungi and their serological relationship to viruses found in *Ustilago maydis* from the USA, *Aust. J. Bot.,* 39, 59, 1991.

112. **Hillman, B. I., Tian, Y., Bedker, P. J., and Brown, M. P.,** A North American hypovirulent isolate of the chestnut blight fungus with European isolate-related dsRNA, *J. Gen. Virol.,* 73, 681, 1992.

113. **Newhouse, J. R., Tooley, P. W., Smith, O. P., and Fishel, R. A.,** Characterization of double-stranded RNA in isolates of *Phytophthora infestans* from Mexico, The Netherlands, and Peru, *Phytopathology,* 82, 164, 1992.

114. **Bharathan, N. and Tavantzis, S. M.,** Genetic diversity of double-stranded RNA from *Rhizoctonia solani, Phytopathology,* 80, 631, 1990.

115. **Snyder, B. A., Adams, G. C., and Fulbright, D. W.,** Association of a virus-like particle with a diseased isolate of *Leucostoma persoonii, Mycologia,* 81, 241, 1989.

116. **Hammar, S., Fulbright, D. W., and Adams, G. C.,** Association of double-stranded RNA with low virulence in an isolate of *Leucostoma persoonii, Phytopathology,* 79, 568, 1989.

117. **Zelikovitch, N., Eyal, Z., Ben-Zvi, B., and Koltin, Y.,** Double-stranded RNA mycoviruses in *Septoria tritici, Mycol. Res.,* 94, 590, 1990.

118. **Azzam, O. I. and Gonsalves, D.,** Detection of dsRNA from cleistothecia and conidia of the grape powdery mildew pathogen *Uncinula necator, Plant Dis.,* 75, 964, 1991.

119. **Shepherd, H. S.,** Characterization and localization of a virus-like particle in a *Dreschlera* species, *Exp. Mycol.,* 14, 294, 1990.

120. **Roossinck, M. J., Sleat, D., and Palukaitis, P.,** Satellite RNAs of plant viruses: structures and biological effects, *Microbiol. Rev.,* 56, 265, 1992.

121. **Dodds, J. A., Tamaki, S. J., and Roistacher, C. N.,** Indexing of citrus tristeza virus double-stranded RNA in field trees, Proc. 9th Conf. International Organization of Citrus Virology, 1985, 327.

122. **Jarupat, T. and Dodds, J. A.,** Interference of a non-seedling yellows by a seedling yellows strains of citrus tristeza virus in sweet orange, Proc. 11th Conf. International Organization of Citrus Virology, 1991, 31.

123. **Boccardo, G., Lisa, V., Luisoni, E., and Milne, R. G.,** Cryptic plant viruses, *Adv. Virus Res.,* 32, 171, 1987.

124. **Nameth, S. T. and Dodds, J. A.,** Double-stranded RNA detected in cucurbit varieties not inoculated with viruses, *Phytopathology,* 75, 1293, 1985.

125. **Valverde, R. A. and Fontenot, J. F.,** Variation in double-stranded ribonucleic acid among pepper cultivars, *J. Am. Soc. Hortic. Sci.,* 116, 903, 1991.

126. **Wang, B., Li, Y. N., Zhang, X. W., Hu, L., and Wang, J. Z.,** Double-stranded RNA in male sterility in rice, *Theor. Appl. Genet.,* 79, 556, 1990.

127. **Lefevre, A., Scalla, R., and Pfeiffer, P.,** The double-stranded RNA associated with '447' cytoplasmic male sterility in *Vicia faba* is packaged together with its replicase in cytoplasmic membranous vesicles, *Plant Mol. Biol.,* 14, 477, 1990.

128. **Harding, R. M., Teakle, D. S., and Dale, J. L.,** Double-stranded RNA in *Carica papaya* is not associated with dieback disease and is unlikely to be of viral origin, *Aust. J. Agric. Res.,* 42, 1179, 1991.

129. **Stace-Smith, R. and Martin, R. R.,** Occurrence of seed transmitted double stranded RNA in native red and black raspberry, *Acta Hortic.,* 236, 13, 1989.

Chapter 11

DIAGNOSTIC METHODS APPLICABLE TO VIROIDS

D. Hanold

TABLE OF CONTENTS

0-8493-4284-8/93/$0.00 + $.50

295

I. INTRODUCTION

Viroids are the smallest known pathogens and have only been found in plants. Unlike viruses, they do not have a protein coat and consist solely of a small, circular single-stranded infective RNA molecule that can both replicate in the host cell and be transmitted independently of any other microorganism. Viroids range in size from 246 to approximately 375[1] bases and have strong internal nucleotide sequence homologies, leading to substantial base pairing, which gives them a rod-like shape in their native state. They have a characteristic melting pattern with a transitional intermediate due to sequences in the conserved region[2] (Figure 1). Most viroids known so far have been found in cultivated plants and are transmitted mechanically by man's cultural practices. Some can spread naturally through insects, seed, and pollen,[3] or by still unknown means (coconut cadang-cadang viroid (CCCVd), coconut tinangaja viroid (CTiVd)[4]).

Not much is known yet about the physiology of pathogenicity and the replicative mechanisms of these pathogens, except that they appear to require RNA polymerase II for replication, but no DNA intermediates.[5] They have been classified, according to the presence of a central conserved region, into three major groups: the PSTVd, ASSVd, and ASBVd groups.[6]

With viroids thus defined by their molecular characteristics, diagnosis of a suspected pathogen as a member of this group must therefore rest on proof for a viroid-like molecule being present and responsible for the disease under investigation. Four important points should be borne in mind. First, most viroids such as CCCVd and CTiVd in palms, appear to be present in their hosts in very low concentrations. Also, the nature of the host tissue may be such that it is very difficult to extract the viroid RNA (e.g., palms, grapevine). It is therefore essential to first of all design an extraction method which is efficient and suitable for a potential viroid and which has also been optimized or modified for each different host species. Second, RNA is very susceptible to enzymatic RNAse degradation, and even more so when single-stranded; therefore great care must be taken with collection and storage of material as well as sample preparation, and appropriate controls of host plants of known disease status, both diseased and healthy, at all stages are crucial in order to check on efficiency of extraction and safe storage conditions. Third, due to the ubiquitous presence of small cellular nucleic acids, it is essential to prove every one of the characteristic properties of viroids — i.e., molecule consisting of RNA, correct size range, single-strandedness, circularity — in order to demonstrate the presence of a member of this group. Fourth, even if one or more viroids have been detected, association with disease needs to be established, for example by appropriate inoculation studies. This is because several "latent" viroids have been found accidentally which do not cause symptoms on some or all host plants,[7-10] and even closely related strains can vary in pathogenicity.

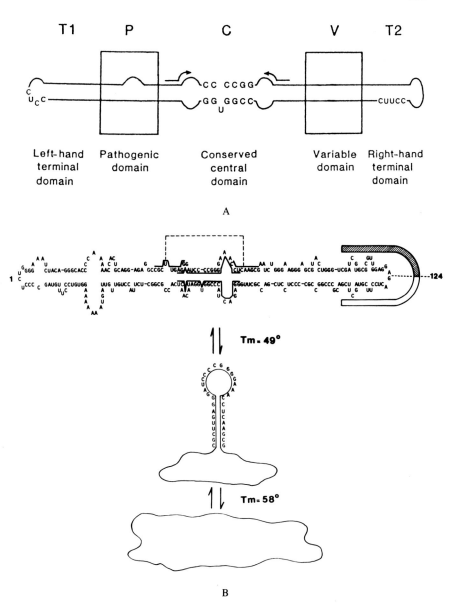

FIGURE 1. Viroid structure. (A) General model of viroid domains according to Keese and Symons.[1] Arrows indicate an inverted repeat sequence that can form a stem loop. (B) Primary sequence of coconut cadang-cadang viroid (CCCVd) showing probable base-paired and single-stranded regions.[28] The right-hand end enclosed by the horse-shoe bracket is reiterated to produce the slow forms of the viroid, an occurrence which is unique to CCCVd. The central conserved region is marked with a bold line inside. Complementary sequences to each side of the conserved region (connected by the dashed line) are able to base-pair just above the melting temperature (Tm) of 49°C to produce an intermediate in which the small single-stranded hairpin loop contains part of the central conserved region. This intermediate melts above 58°C.

Table 1. An Extraction Procedure for CCCVd-Like Viroids from Palms and Other Monocots

1. Chop 10–20 g of leaf and blend in 120 ml of 100 mM Na$_2$SO$_3$.
2. Strain through muslin and shake for 30 min at 4°C with 20 g/l polyvinyl polypyrrolidone (PVPP).
3. Mix vigorously for 5 min with 50 ml of chloroform; centrifuge at 10,000 g for 10 min.
4. To the aqueous supernatant add polyethylene glycol (PEG) 6000 at 80 g/l.
5. Precipitate for 2 h at 4°C; spin at 10,000 g for 10 min to collect the pellet.
6. Dissolve pellet in 2 ml of 10 g/l sodium dodecyl sulfate (SDS).
7. Add 2 ml of aqueous phenol (900 g/l) containing 1 g/l 8-hydroxy quinoline; shake vigorously for 1 h.
8. Collect aqueous supernatant by centrifugation, re-extract with 1 ml of phenol and 1 ml of chloroform for 5 min, spin at 10,000 g for 10 min.
9. Adjust NaCl in supernatant to 0.1 M, add cetyl trimethylammonium bromide (CTAB) to 3.3 g/l, and allow precipitation to occur for 30 min at 0°C.
10. Centrifuge at least 30 min at 10,000 g and wash pellet 3 times with 0.1 M Na-acetate in 75% ethanol.
11 Air-dry pellet and resuspend as needed.

From Hanold, D. and Randles, J. W., *Ann. Appl. Biol.*, 118, 139, 1991. With permission.

II. METHODOLOGY

A. PREPARATION OF SAMPLES
1. Collection and Storage of Material

Choose a range of young, but fully expanded, undamaged leaves with good symptoms (if present). Seal in plastic to prevent drying out, keep cool, and process immediately or store at −20°C. Discard brown, wilted, or bruised tissue, since this indicates cell breakdown with likely RNA degradation.

2. Viroid Purification

In order to limit contamination by RNAses, general sterile working practices must be adopted and gloves should be worn at all times. Sterilize all solutions, glassware, and other materials by appropriate means such as autoclaving, baking for 8 h at 180°C, or treatment with 2 M KOH in 90% ethanol. The work area should be cool and clean, i.e., free of drafts and dust, which might carry RNAse contamination.

An example of a protocol successfully used for the extraction of CCCVd, CTiVd, and related viroids from palms and herbaceous monocotyledons is given in Table 1. Dicotyledon tissue may allow simplified procedures. Antioxidants (SO$_3^{2-}$, β-mercaptoethanol, monothioglycerol) should be included in the extraction buffer to prevent oxidation of phenols and consequent browning. Work should proceed as quickly as practicable to minimize degradation by RNAses released during cell disruption. For CCCVd-related viroids, the addition of polyvinyl polypyrrolidone (PVPP) has proved advantageous during extraction, although its role is not exactly known. Note, however, that some viroids might stick to PVPP or bentonite (which is sometimes used to adsorb RNAses) and thus can be lost. Only CCCVd and CTiVd are known to be precipitated with polyethylene glycol (PEG).[11] Deproteinized viroids will not

Table 2. A Short Extraction Method for CCCVd Analysis in Coconut Tissue

1. Place approximately 1 g of leaf tissue in a strong plastic bag, with 1 vol of 100 mM Tris-HCl, pH 7.2, 100 mM Na-acetate, 10 mM ethylene diaminetetraacetic acid (EDTA), 0.5% thioglycerol.
2. Crush with a hammer, squeeze juice into a centrifuge tube, add 0.5 vol of 90% phenol, 0.5 vol of chloroform, mix.
3. Separate phases by low-speed centrifugation, precipitate nucleic acids with 75% ethanol.
4. Resuspend precipitate and assay for CCCVd.

From Randles, J. W., Hanold, D., Pacumbaba, E., and Rodriguez, J., *Plant Diseases of International Importance*, Vol. 4, Mukhopodhyay, A. N., Kumar, J., Chaube, H. S., and Singh, U. S., Eds., Prentice Hall, Englewood Cliffs, N.J., 1992, 227. With permission (See also Chapter 12.)

precipitate with PEG.[12] Efficient inhibitors of RNAse are 0.1 to 1% sodium dodecyl sulfate (SDS), as well as human placental RNAse inhibitor in the presence of at least 1 mM dithiothreitol (DTT), but restoration of RNAse activity will occur on removal of these inhibitors since the inhibition is a reversible process.

Other steps may be used in the purification of viroids. Treatment at 0°C for 16 h with 2 M LiCl can precipitate high-molecular-weight nucleic acids from the viroid extract. Extracts in TE buffer (10 mM Tris HCl pH 7.5, 0.1 mM ethylene diamino tetraacetic acid [EDTA]) can be purified either by adsorbtion to DE52 cellulose columns and elution with 1 M NaCl in TE, or fractionated on Quiagen[11] columns (Diagen G.m.b.H., Düsseldorf, Germany) with viriod-sized nucleic acids eluting at 0.4 to 0.7 M NaCl. Viroids are precipitated with 0.8 vol isopropanol or 3 vol ethanol and sedimented by centrifugation for at least 30 min at 10,000 g. Precipitation with cetyl tri-methylammonium bromide (CTAB) is efficient in removing polysaccharides and other contaminants and works well for CCCVd-like viroids, but its reliability needs to be checked for new viroids. Precipitate with 3.3 g/l CTAB in 0.1 M NaCl on ice for 30 min, spin for 30 min at 10,000 g, wash pellet three times by dispersion in 0.1 M Na-acetate in 75% ethanol. Density gradient centrifugation has been useful in obtaining pure viroid (e.g., PSTVd) when large quantities of material are available.

A widely used method for purification is preparative gel electrophoresis, usually with a combination of nondenaturing and denaturing gels, as described below.

Short extraction methods can sometimes be devised for routine screening of one particular host species after a viroid has been characterized, but good controls are crucial. Table 2 shows such a short method for CCCVd screening of only known cultivars of coconut in the Philippines. These methods should not be used for other viroids or on species or even host cultivars other than the ones they were designed for, without rigorous testing of their efficiency, since even small changes in the viroid molecule or the host physiology can render them unreliable.

B. GEL ELECTROPHORESIS

Due to their characteristic molecular properties, viroids migrate in most gel systems with a mobility less than that expected for a molecule of their molecular weight. Common single-stranded linear RNA size markers are therefore useless for estimation of their true molecular weight, and interpretation of gel patterns often difficult, even in comparison with known viroid markers. If linear viroid forms are present, their size can be estimated in denaturing gels by comparison to linear RNA markers. Because only a small percentage of total plant cell nucleic acids is represented by the infecting viroid, partially purified viroid preparations are usually required for analysis by gel electrophoresis.

1. Polyacrylamide Gel Electrophoresis (PAGE)

PAGE is a powerful and very flexible tool in viroid diagnosis.

Choice of gel matrix — By varying the polyacrylamide concentration from 5 to 20%, with a ratio of monomer to dimer from 99.96:0.4 to 97:3, pore size and gel strength can be adapted and finely tuned to suit particular applications. A 20% gel with 97:3 monomer:dimer ratio in a TBE[13] system (90 mM Tris, 90 mM boric acid, 3 mM EDTA) gives high resolution and sharp bands. This system is routinely used to separate the molecular forms of CCCVd, which differ in their secondary structure due to a single base addition, but it is also very susceptible to distortion from impurities in the sample and therefore requires highly purified preparations. In contrast, for example, in 5% PAGE with 98.8:1.2 monomer:dimer ratio, CCCVd forms differing by only one base are not resolved, but the sample tracks are not distorted by low levels of impurities (Figure 2).

Denaturing PAGE — Addition of 8 M urea to any polyacrylamide gel turns it into a "denaturing" gel, since under these conditions base pairing is ineffective and secondary structure of nucleic acids is abolished. Viroids in such systems are open circles — as opposed to their rod-like native structure — and thus show retardation of their migration in contrast to any linear nucleic acids, resulting in band patterns different from those obtained in nondenaturing systems. Denaturing conditions can also be achieved by high running temperatures or a change in gel pH, but in practice these parameters are experimentally more difficult to control and complete denaturation of the molecules therefore harder to obtain. It is essential to have 100% of all nucleic acid molecules in their denatured form, since otherwise interpretation of the resulting band patterns is impossible and may lead to wrong conclusions about the molecular species present. Diagnostic tests for viroids have often been based on PAGE under both nondenaturing and denaturing conditions and the characteristic change of behavior of these molecules when comparing one set of conditions to the other.

Consecutive PAGE — Samples are analyzed on a nondenaturing gel, bands under investigation are stained with either ethidium bromide or toluidine blue (without fixing) and cut out, or in the absence of visible bands, the region expected to contain possible viroids is cut out according to appropriate markers. The gel slices are then loaded on top of a second, denaturing gel containing 8 M urea of identical or different gel strength.[14] Viroid bands, due to their open circular configuration, should migrate more slowly than the now single-stranded linear nucleic acids that were in the same position in the first nondenaturing gel when the viroid had its rod-like native configuration.

Bidirectional PAGE — This method uses the same principle as consecutive PAGE, except that the gel is not cut; instead, after the first electrophoresis, the current is reversed and nucleic acids are denatured in the original gel by increasing either temperature or pH.[14] Since gel strength cannot be varied, good separation of viroid bands and linear nucleic acids is more difficult. Unless used by experienced investigators, this method also often fails to denature all nucleic acids completely for the return "denaturing" run and thus may not give a clear result. Small amounts of viroid can also remain trapped in a large excess of linear nucleic acids and thus escape detection.

Two-dimensional PAGE — In this system, a whole track of a nondenaturing gel is cut out and placed across a gel containing 8 M urea and then electrophoresed at a 90° angle to its first run.[15] This has the advantage that the complete molecular range can be screened for possible viroids, rather than having to cut out a certain area. This is especially advantageous when looking for some unknown pathogen. Gel strengths can be varied to get good separation in both dimensions. It is also possible to reverse the sequence of conditions, i.e., denaturing gel as the first, and nondenaturing as the second, dimension. This may result in less background if samples are not completely clean, and it will make staining of the second gel with ethidium bromide possible for preparative applications since the native molecules will stain much more strongly with this method than denatured viroids. Running the second gel at a 90° angle prevents trapping of small amounts of viroid by excess linear nucleic acids, since viroid molecules do not have to pass through a gel area contaminated with other molecular species. If used in connection with a sensitive detection method (silver stain, or molecular hybridization, see below), the two-dimensional PAGE system is a sensitive and definite test for the presence of small circular nucleic acids in a preparation, and thus a strong tool in the search for viroids (Figure 3).

Gradient electrophoresis — PAGE with temperature, urea, or pH gradients can be used to visualize the specific melting pattern characteristic for viroids.[2] A purified preparation is applied across a gel containing a gradient and electrophoresed in a direction perpendicular to the gradient. Characteristic molecular transitions occur at certain points of the gradient. These are visualized after staining, by a curve representing the altered electrophoretic mobilities of the native or partially melted molecular forms with sharp changes occurring at the respective melting points of distinct molecular structures. To determine the degree of relationship between viroids or strains, heteroduplex

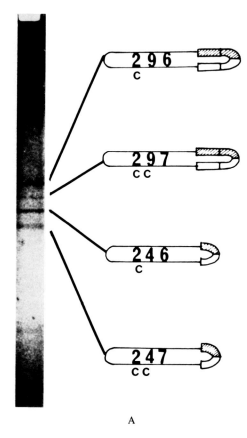

A

FIGURE 2.   Polyacrylamide gel electrophoresis (PAGE). (A) The four molecular forms of monomeric coconut cadang-cadang viroid (CCCVd) separated by 20% PAGE and visualized by silver stain.[4] The resolution of the 246 and 247 or 296 and 297 nucleotide bands is due to change in the secondary structure caused by the addition of an extra C residue. (B) CCCVd and oil palm orange spotting viroid analyzed by 5% PAGE;[4] 1,2, oil palm extracts containing viroid; s, slow (296/7 nucleotides); and f, fast (246/7) forms of CCCVd, both showing monomers (lower major bands) and dimers (upper minor bands).

molecules can be formed *in vitro* between the species under investigation by liquid hybridization. These can then be electrophoresed with the current parallel to a gradient.[16] Heteroduplexes will denature when they reach a specific point in a gradient, depending on their degree of homology, and will be slowed down there, resulting in different band positions even for molecules with only one base change.

2. Agarose

Agarose gels between approximately 1.2 and 2% can be used, either with a TBE[13] or TAE[13] buffer system and followed by staining with ethidium bromide, to analyze extracts. These systems are not too much affected by

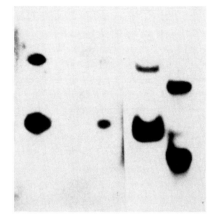

1   2 s f
Oil Palm CCCVd

FIGURE 2B.

impurities still present in the preparations, and are therefore suitable for screening partially purified samples. However, ethidium bromide staining is not as sensitive as the silver stain used for polyacrylamide gels, and bands are not as sharp as in PAGE, which results in a tendency for weak viroid bands to disappear in background fluorescence. Ethidium bromide stains denatured viroid molecules less efficiently than the native RNA, that is, at about 1/10 the sensitivity obtained for the native partially base-paired form. Agarose gels, with or without SDS, can be useful for preparative purposes.

3. Staining Methods

Silver — The most sensitive stain for polyacrylamide gels is silver staining.[17] In theory, detection of as little as 50 pg of viroid should be possible.[2] In practice, however, heavily staining background in the sample tracks often interferes with detection of bands. Depending on the host tissue, this background may be almost impossible to remove even after extensive purification (e.g., oil palm tissue extracted for CCCVd). With such materials, the alternatives are ethidium bromide staining (see below) or Northern blotting of gels with subsequent molecular hybridization assay (see below). The presence of bands derived from host cells is another frequently encountered problem. If for a particular host/viroid combination such bands are present in the viroid region, identification of the real viroid bands may be very difficult, even more so for a new viroid which is not yet available in purified form as a marker or a probe. Nucleic acids stained with silver are permanently fixed

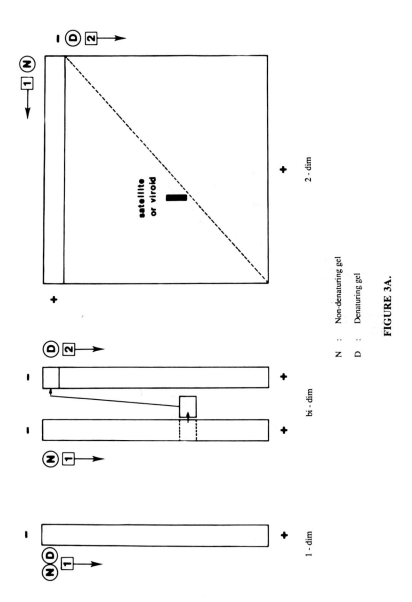

FIGURE 3A.

N : Non-denaturing gel

D : Denaturing gel

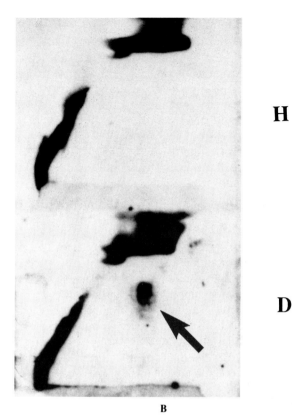

H

D

B

FIGURE 3. Two-dimensional polyacrylamide gel electrophoresis (2-D PAGE). (A) Principle of detection of circular nucleic acids by PAGE. (B) An example of 5% nondenaturing/denaturing 2-D PAGE followed by electroblot hybridization assay under low stringency using a ^{32}P-labeled cRNA probe;[9] healthy asymptomatic oil palm (top) and orange-spotted palm (bottom) containing a CCCVd-related viroid indicated by the arrow. Loading represents viroid contained in total nucleic acid extract from approximately 5 g of host tissue.

in the gel. This stain, therefore, cannot be used for consecutive or two-dimensional PAGE or for preparative gels.

Ethidium bromide — Ethidium bromide is an intercalating fluorescent dye which is reversibly bound to nucleic acids and can be used in all gel systems. In nondenaturing gels it is 10 to 50 times less sensitive than silver stain. It causes fewer problems with background staining, but denatured viroid molecules are inefficiently stained and for these, ethidium bromide cannot be recommended. It is ideal for preparative gels, since it can be incorporated in the gel during electrophoresis, thus minimizing risk of breakdown of viroid due to extensive manipulation of the gel. Ethidium bromide is strongly mutagenic and a suspected carcinogen, and appropriate safety precautions should be taken to avoid any skin contact.

Toluidine blue — Toluidine blue staining is at least ten times less sensitive than ethidium bromide, except for denatured viroid molecules, where it provides about the same level of detection. It can be used in conjunction with acetic acid to fix nucleic acids permanently in the gel, or as an aqueous stain for preparative gels. Since the procedure requires an extensive time of destaining, the aqueous stain usually results in partly diffused bands, and risk of RNA breakdown in the gel must be considered.

Sequential staining of gels is possible, but only in the order ethidium bromide — toluidine blue — silver stain.

4. Viroid Recovery from Gel Pieces

If electrophoresis is used for preparative purposes, all equipment should be RNAse-free. Great care must be taken not to contaminate the preparations with RNAses during recovery from the gel.

RNA can be eluted from gel pieces by incubation at 37°C in 10 to 20 vol of NH$_4$-acetate/SDS/EDTA[18] overnight, removing the liquid and adding the same amount of fresh buffer, incubation for another 1 to 2 h, and ethanol precipitation of the viroid from the combined eluates. The presence of SDS will effectively inhibit RNAses. Recovery is best from agarose, and less and less efficient from PAGE with higher and higher gel strength, but it is not more than approximately 50 to 70%. Alternatively, gel pieces can be inserted into dialysis tubes filled with 1 to 2 ml of electrophoresis buffer containing a few micrograms of carrier nucleic acid, RNAsein, and 1 mM DTT, and electroeluted for 1 to 2 h. Reverse the current for 1 min at the end of the run in order to free RNA adhering to the tubing. Remove buffer, rinse bag, and precipitate with ethanol. Recovery should be similar for all gel systems and somewhat higher than by the first method. The major source of material loss is RNA sticking to the inside of the bags. The risk of RNAse contamination may be slightly higher than by the diffusion method described above, due to more extensive handling of materials. DE52 (Whatman anion exchanger) can be used to concentrate RNA from large volumes.[13]

Nucleic acids can also be extracted from gel pieces by centrifugal[19] force. Inside a larger centrifuge tube (e.g., 1.5 ml Eppendorf tube) place a shorter one (e.g., 0.5 ml Eppendorf tube) with a rim diameter exceeding that of the inside of the large tube and with a fine hole at its lowest point. Place siliconized glass wool, then the gel slice (agarose or polyacrylamide) in the upper tube and centrifuge at 10,000 g for 10 to 15 min to remove the aqueous phase, including nucleic acid, from the gel piece. Discard top tube with gel, vacuum dry the liquid obtained from the gel piece or add salt and ethanol to precipitate viroid (depending on amount present and the required purity).

Viroid purified in low-melting-point agarose can be recovered by melting and phenol extraction.[13] However, partial denaturation of the RNA is likely to occur during the melting of the agarose.

C. ELECTRON MICROSCOPY

Electron microscopy can be used to prove circularity and estimate the size of purified molecules thought to be viroids when spread under denaturing

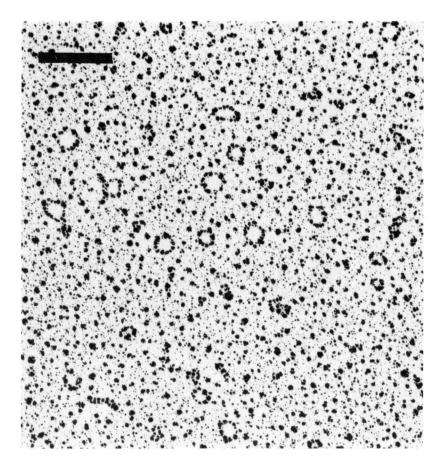

FIGURE 4. Electron microscopy of purified coconut cadang-cadang RNA spread under denaturing conditions, showing circularity of molecules.[29] Grids were stained with uranyl acetate and rotary shadowed. Bar represents 300 nm.

conditions[20] (Figure 4). It cannot be used for diagnostic purposes on tissue sections or crude extracts, since the small viroid rods or circles cannot be identified in such mixtures of unknown components.

D. TESTS FOR NUCLEIC ACIDS CHARACTERISTICS
1. Distinction Between DNA and RNA

For details see Chapter 12, Table 2. In summary, the methods applicable to viroids are the following:

1. *Alkali:* RNA is degraded by incubation in 0.5 M KOH at room temperature for 30 to 60 min. DNA is not affected.[21]

2. *RNAse:* Selective degradation of RNA by DNAse-free RNAse. To prepare DNAse-free RNAse, dissolve RNAse A in 10 m*M* Tris (pH 7.5), 15 m*M* NaCl at 10 mg/ml, heat to 100°C for 15 min, allow to cool slowly. Use at 10 μg/ml in TE buffer, incubate at room temperature or 37°C for 1 h.[21]

3. *DNAse:* Selective degradation of DNA by RNAse-free DNAse. Use RNAse-free DNAse I[13] always in the presence of RNAsein and DTT.[13]

2. Distinction Between Single/Double-Strandedness

1. *Enzymatic:* S1 nuclease digests only single-stranded nucleic acids. RNAse A degrades only single strands at high salt (300 m*M* NaCl) and both single and double strands at low salt (50 m*M* NaCl) conditions.[21]

2. *Acridine orange:* Acridine orange can be used to stain gels in a similar way to ethidium bromide, but it will show red fluorescence for single strands, and green fluorescence for double strands.[22] It has mutagenic and carcinogenic properties similar to ethidium bromide and should be handled with equal care.

3. *Electrophoretic:* Bands on gels show no difference in migration pattern under denaturing and nondenaturing conditions if the molecules are single-stranded. Bands of double strands change their patterns when the molecules are denatured to single strands. Denaturation can be achieved, e.g., by boiling the samples with 50% formamide, glyoxylation,[22] or by denaturing gels (see above).

For all these tests, problems can arise due to the internal base pairing which occurs in viroid RNAs, making them "rod-like" in their native state. This results in partial double-strandedness and strong secondary structure which can cause abnormal behavior compared to "true" single strands.

3. Distinction Between Linear/Circular Nucleic Acids

1. *Two-dimensional PAGE:* Two-dimensional PAGE (see above) is the method of choice to prove the presence of small circles in a nucleic acid mixture. Bidirectional or consecutive PAGE can also be used.

2. *Electron microscopy:* Electron microscopy of both denaturing and nondenaturing spreads of the purified nucleic acid under investigation,[20] using standard techniques, can show the presence of rods/circles in the preparation (Figure 4).

E. MOLECULAR HYBRIDIZATION ASSAYS

The following will describe only problems of specific concern when working with viroids. For general methodology and theoretical background, see Chapter 9.

Table 3. An Example of a Protocol for the Detection of CCCVd-Related Viroids with a Heterologous CCCVd-cRNA Probe

1. Prepare ^{32}P-cRNA probe by transcription from full-length CCCVd clone in pSP64 vector and purification through PAGE.[9]
2. Fractionate nucleic acids by PAGE, equilibrate gels in TAE[13] (9 mM Tris, 4 mM Na-acetate, 0.4 mM EDTA, pH 7.4) for 15 min, transfer RNA to Zeta Probe (BioRad) nylon membrane in TAE at 60 V, 0.6 A for 5 h, and bake membranes for 2 h at 80°C.
3. Prewash blots for 1 h, 67°C in 0.1× SSC (20× is 3 M NaCl, 0.3 M Na-citrate), 1 g/l SDS, then prehybridize for 15–20 h at 37°C in 0.75 M NaCl; 75 mM Na-citrate; 50 mM sodium phosphate at pH 6.5; 5 mM EDTA; 2 g/l each of bovine serum albumin (BSA), Ficoll 400 (Pharmacia), polyvinyl pyrrolidone of mol wt 40,000, SDS; 1 mg/ml denatured herring testis DNA; 50% deionized formamide.
4. Heat cRNA probe to 80°C for 1 min in 50% formamide, add to hybridization mix (i.e., prehybridization buffer with 110 g/l dextran sulfate) at about 10^6 cpm/ml final concentration, and hybridize filters at 37°C for 20–40 h in slowly rotating roller bottles.
5. Rinse filters at room temperature in 0.5× SSC, 1 g/l SDS, then wash for 1 h at 55°C with agitation in 1× SSC, 1 g/l SDS for low stringency. Wrap filters in plastic to keep moist, expose at −70°C using intensifying screens for approximately 2 days.
6. Rewash filters for 2 h in 0.1× SSC, 1 g/l SDS with agitation for high stringency. Re-expose for 7–10 days.

From Hanold, D. and Randles, J. W., *Ann. Appl. Biol.*, 118, 139, 1991. With permission.

1. Preparation of Blots

For both dot and gel blots, the positively charged Zeta Probe nylon membrane (BioRad) gives the best results due to its superior ability to bind and retain small RNA molecules and its ease of handling. It is therefore preferable to uncharged nylon or other kinds of supporting membranes such as nitrocellulose. We routinely prewash the blots for 1 h at 65°C in 0.1× SSC (20× is 3 M NaCl, 0.3 M Na-citrate), 0.5% SDS, and include carrier DNA at 1 mg/ml in the prehybridization/hybridization mixtures to minimize background signals[9] (see hybridization protocol, Table 3).

Both electroblotting and capillary blotting work well for viroids from all types of gels, including even 20% PAGE. Gels are not pretreated for capillary blots[23] and only briefly equilibrated in transfer buffer for electroblotting to minimize diffusion of bands and loss of the small amounts of viroid in the sample tracks. After transfer, blots are not rinsed, but immediately air-dried and baked.[9]

2. Preparation of Probe[13,24]

One of the following is necessary for the preparation of probe:

- Purified viroid suitable for reverse transcription
- Cloned viroid in a suitable vector (e.g., SP6 system) to make cDNA or cRNA
- Specific primers and polymerase chain reaction (PCR) amplification working in crude extracts
- Knowledge of the taxonomic group to which the viroid belongs, so that oligonucleotide sequences in the conserved region may be synthesized for use as probes, or so probes to related viroids can be used in low-stringency hybridization

• Knowledge of at least part of the sequence for oligonucleotide probe synthesis

The best approach will depend on whether the viroid under investigation is a new one or a variant of a previously known isolate.

cRNA probes are preferable to cDNA probes, because they bind more strongly to the target RNA, giving stronger signals, thus allowing conditions of higher stringency for hybridization and washing which result in less background and reduce the nonspecific binding of probe. Table 3 gives an example of a protocol for detection of CCCVd-like viroids with heterologous CCCVd-cRNA probe.

Since high sensitivity is desirable for diagnosis, and with viroid concentrations possibly low, probes with high specific labeling are needed. So far, none of the nonradioactive labeling methods available give satisfactory results in our experience, because they show either (1) low sensitivity, (2) nonspecific binding to plant extracts, or (3) nonspecific signals due to an endogeneous reaction between the sample and substrate, in the absence of conjugated probe. This occurs especially when alkaline phosphatase is used in the detection reaction, e.g., with a biotin-streptavidin system. Further disadvantages are that consecutive washes at increasing stringencies (see below), multiple probing, and variation of exposure times to increase sensitivity are not possible with most nonradioactive probes. Thus, radioactive labeling with ^{32}P is the method of choice in our laboratory. Nick translation and end-labeling do not result in high specific activity of the probe. Our best results, by far, are obtained with a monomeric ss cRNA probe of high purity from an SP6 transcription system. We consider the extra effort spent on preparation of this kind of probe well worthwhile.

3. Hybridization Parameters

Stringency of hybridization can be lowered by using 37°C instead of 42°C, if heterologous viroid isolates are to be detected.[25]

Variants of viroids with different levels of homology to the probe can be detected and their homology estimated by first washing blots at low stringency, exposing, then rewashing at higher stringency,[25] re-exposing, and continuing in this way over several steps of stringency, if desired. Ensure that the membrane is kept moist at all times if further washes are planned, since drying will fix the probe.

The rate of loss of probe compared to a homologous marker preparation can give an estimate of the degree of homology of the strains to the probe.

Probe can be removed by boiling the blots for 20 min in $0.2 \times$ SSC, 0.5% SDS, and they can then be reprobed following the original protocol. This can be repeated several times when using Zeta Probe nylon membrane without excessive loss of sample.

F. AMPLIFICATION AND NUCLEOTIDE SEQUENCE DETERMINATION

To obtain sufficient material for diagnosis, characterization, or determination of nucleotide sequence, it may be necessary, for some viroids, to use molecular amplification of the RNA. Amplification may be achieved either by cloning cDNA in bacteria,[26,27] or by *in vitro* amplification with the polymerase chain reaction.[13] Available methodology can be used, taking into consideration the special properties and problems of viroids. Nucleotide sequencing can be done directly on the purified RNA, or by conventional methods on cDNA.[1] However, the strong secondary structure of viroid RNA will require special approaches to overcome compressions on sequencing gels.

G. INOCULATION METHODS

As a final proof that an isolated viroid is the cause of a disease, Koch's postulate needs to be satisfied and reproduction of the original disease by a purified preparation demonstrated.

Many viroids show high infectivity in crude plant extracts. However, for some viroids or particular host species, crude extracts may not contain a high enough viroid concentration for successful inoculation. It is then preferable to use partly purified viroid preparations (e.g., a PEG precipitate in the case of CCCVd[17]). cDNA clones have also been shown to be infectious for some viroids.[5]

Inoculation methods can be abrasion, slashing, or grafting,[3] but in the case of CCCVd, for example, only high-pressure injection close to the meristem of young seedlings gives sufficiently high rates of infection in coconut[17] (Figure 5).

III. CONCLUSIONS

A. IMPLICATION OF A VIROID PATHOGEN

If a nucleic acid has been implicated as disease-associated (see Chapter 12), a distinction has to be made between a viral or a viroid pathogen.

To achieve this, the following sequence of tests is suggested:

1. Is the nucleic acid RNA?
2. Does it have a circular form?
3. Is it single-stranded?
4. Is its size in the range of 246 to 380 nucleotides?
5. Is it infectious when purified?

The properties listed as points 1 to 5 need to be demonstrated to show the viroid nature of a pathogen. Infection may be latent in particular species, and molecular methods are generally necessary to establish host species. For example, HLVd,[10] ASBVd,[7] CLVd,[8] and CCCVd in herbaceous monocots[4,9] are latent in some or all known hosts; or symptoms are so mild or nonspecific

FIGURE 5. Inoculation of coconut seedling with CCCVd using high-pressure injection (Panjet) close to the meristem.

that unequivocal diagnosis by symptoms is difficult; or symptoms take too long to develop to be of practical use (CCCVd in palms[17]).

If steps 1 to 5 demonstrate the viroid nature of the agent, then two further steps can be undertaken:

6. Is its nucleotide sequence characteristic of viroids, and in which taxonomic group does it fit?
7. Develop a diagnostic test which is specific, sensitive, rapid.

B. CLASSIFICATION AND IDENTIFICATION

When the nucleotide sequence of a new viroid is known, classification can be attempted using the sequence of the central conserved region.[6] This leads to the possibility of designing group-, viroid-, or strain-specific probes which at different levels of stringency can be used to establish degrees of relatedness or to look for new group members and strains, and for diagnostic purposes. Even single base mutations can be detected using probe heteroduplexes in temperature gradient gel electrophoresis.[16]

C. LATENCY

Viroids are probably much more widespread than we are currently aware.[9] Besides the ones that have been identified because they are causing serious diseases in crop plants, there is likely to be a large number that does not cause obvious symptoms. However, these so-called "latent" viroids may still cause yield reduction or some growth inhibition, and "latent" hosts may also act as reservoirs for a viroid that can cause disease under changed environmental conditions or in other economically relevant species. Apparent freedom from symptoms can by no means be regarded as freedom[4,9] from viroid pathogens.

REFERENCES

1. **Keese, P. and Symons, R.,** Physical-chemical properties: molecular structure (primary and secondary), in *The Viroids,* Diener, T. O., Ed., Plenum Press, New York, 1987, 37.
2. **Riesner, D.,** Physical-chemical properties: structure formation, in *The Viroids,* Diener, T. O., Ed., Plenum Press, New York, 1987, 63.
3. **Diener, T. O.,** Biological properties, in *The Viroids,* Diener, T. O., Ed., Plenum Press, New York, 1987, 9.
4. **Hanold, D. and Randles, J. W.,** Coconut cadang-cadang disease and its viroid agent, *Plant Dis.,* 75, 330, 1991.
5. **Sänger, H.,** Viroid function in viroid replication, in *The Viroids,* Diener, T. O., Ed., Plenum Press, New York, 1987, 117.
6. **Koltunow, A. M. and Rezaian, M. A.,** A scheme for viroid classification, *Intervirology,* 30, 194, 1989.
7. **Desjardins, P.,** Avocado sunblotch, in *The Viroids,* Diener, T. O., Ed., Plenum Press, New York, 1987, 299.
8. **Diener, T. O.,** Columnea latent, in *The Viroids,* Diener, T. O., Ed., Plenum Press, New York, 1987, 297.
9. **Hanold, D. and Randles, J. W.,** Detection of coconut cadang-cadang viroid-like sequences in oil and coconut palm and other monocotyledons in the south-west Pacific, *Ann. Appl. Biol.,* 118, 139, 1991.
10. **Puchta, H., Ramm, K., and Sänger, H.,** The molecular structure of hop latent viroid (HLVd), a new viroid occurring worldwide in hop, *Nucleic Acids Res.,* 16, 1988, 4197.
11. **Randles, J. W.,** personal communication, 1991.
12. **Randles, J. W., Rillo, E., and Diener, T.,** The viroid-like structure and cellular location of anomalous RNA associated with the cadang-cadang disease, *Virology,* 74, 128, 1976.
13. **Sambrook, J., Fritsch, E., and Maniatis, T.,** *Molecular Cloning: a Laboratory Manual,* 2nd ed., Cold Spring Harbor Laboratory, Cold Spring Harbor, NY, 1989.

14. **Schumacher, J., Meyer, N., Weidemann, H. L., and Riesner, D.,** Diagnostic procedure for detection of viroids and viruses with circular RNAs by "return" gel electrophoresis, *J. Phytopathol.*, 115, 332, 1986.

15. **Schumacher, J., Randles, J. W., and Riesner, D.,** A two-dimensional electrophoretic technique for the detection of circular viroids and virusoids, *Ann. Biochem.*, 135, 288, 1983.

16. **Wartell, R., Hossein, S., and Moran, C.,** Detecting base pair substitutions in DNA fragments by temperature-gradient gel electrophoresis, *Nucleic Acids Res.*, 18, 2699, 1990.

17. **Imperial, J., Bautista, R., and Randles, J. W.,** Transmission of the coconut cadang-cadang viroid to six species of palm by inoculation with nucleic acid extracts, *Plant Pathol.*, 34, 391, 1985.

18. **Maxam, A. M. and Gilbert, W.,** Sequencing end-labelled DNA with base-specific chemical cleavages, in *Methods in Enzymology,* Vol. 65, Grossman, L. and Moldave, K., Eds., Academic Press, New York, 1980, 499.

19. **Heery, D. M., Gaunan, F., and Powell, R.,** A simple method for subcloning DNA fragments from gel slices, *Trends Genet.,* 6, 173, 1990.

20. **Randles, J. W., Hanold, D., and Julia, J. F.,** Small circular single-stranded DNA associated with foliar decay disease of coconut palm in Vanuatu, *J. Gen. Virol.*, 68, 273, 1987.

21. **Randles, J. W., Julia, J. F., Calvez, C., and Dollet, M.,** Association of single-stranded DNA with the foliar decay disease of coconut palm in Vanuatu, *Phytopathology,* 76, 889, 1986.

22. **McMasters, G. and Carmichael, G.,** Analysis of single- and double-stranded nucleic acids on polyacrylamide and agarose gels by using glyoxal and acridine orange, *Proc. Natl. Acad. Sci. U.S.A.,* 17, 4835, 1977.

23. **Thomas, P.,** Hybridisation of denatured RNA and small DNA fragments transferred to nitrocellulose, *Proc. Natl. Acad. Sci. U.S.A.,* 77, 5201, 1980.

24. **Arrand, J.,** Preparation of nucleic acid probes, in *Nucleic Acid Hybridsation, A Practical Approach,* Hames, B. and Higgins, S., Eds., IRL Press, Oxford, 1985, 17.

25. **Anderson, M. and Young, B.,** Quantitative filter hybridisation, in *Nucleic Acid Hybridsation, A Practical Approach,* Hames, B. and Higgins, S., Eds., IRL Press, Oxford, 1985, 73.

26. **Vos, P.,** cDNA cloning of plant RNA viruses and viroids, in *Plant Gene Research: Plant DNA infectious Agents,* Hohn, T. and Schell, J., Eds., Springer-Verlag, New York, 1987, 53.

27. **Puchta, H. and Sänger, H.,** An improved procedure for the rapid one-step-cloning of full-length viroid cDNA, *Arch. Virol.,* 101, 137, 1988.

28. **Randles, J. W., Hanold, D., Pacumbaba, E., and Rodriguez, J.,** Cadang-cadang disease of coconut palm, in *Plant Diseases of International Importance,* Vol. 4, Mukhopadhyay, A. N., Kumar, J., Chaube, H. S., and Singh, U.S., Eds., Prentice Hall, Englewood Cliffs, N.J., 1992, 277.

29. **Randles, J. W. and Hatta, T.,** Circularity of the ribonucleic acids associated with cadang-cadang disease, *Virology,* 96, 47, 1979.

Chapter 12

STRATEGIES FOR IMPLICATING VIRUS-LIKE PATHOGENS AS THE CAUSE OF DISEASES OF UNKNOWN ETIOLOGY

J. W. Randles

TABLE OF CONTENTS

I. INTRODUCTION

Virus diseases are plant disorders caused by genetically distinct transmissible intracellular nucleocapsids which replicate and produce associated cytopathic changes. The success of modern plant virology in ascribing viruses, or virus-like agents such as viroids, to diseases, is largely due to the use of Koch's postulates, in a modified form. Thus, a virologist implicates a virus by:

1. Recognizing a disease syndrome or disorder
2. Transmitting the syndrome
3. Biologically isolating putative causal virus(es) on primary or alternate hosts
4. Inoculating putative agents alone or together to reproduce disease
5. Purifying, characterizing, and describing the agents
6. Demonstrating by inoculation to appropriate hosts that the purified agent can reproduce the disease

These Steps 1 to 6 constitute the standard procedure for establishing that a recognized disease in the field is caused by a particular virus, or viroid; and the procedure has frequently been successful. However, a number of diseases are now known, some of them economically very important, where the straightforward application of Steps 1 to 6 has not been effective in identifying the causal pathogen.

The scope of this chapter is to outline and discuss procedures that can be used when this classical approach for establishing the cause of a virus-like disease is neither successful nor appropriate. In particular, the chapter describes an overall plan for gaining sufficient information on nucleic acids found only in diseased plants, to provide an unambiguous preliminary classification by type, size, and circularity; and thus to lead to identification of the agent as a particular type of virus or viroid.

II. THE PROBLEM OF DISEASES OF UNKNOWN ETIOLOGY (DUEs)

The path towards determining the cause of plant diseases is well trodden, and virologists are well aware of the sequence of events which commonly leads them to being asked to study a disease. In simple terms, a virologist is asked to intervene when no cellular pathogen can be found to be associated with a disorder. In the absence of a detectable virus, the virologist is left with a choice of viroids, environment, nutrition, herbicides, or genotype as possible causes of symptoms or poor plant performance. Correction of unsuitable environmental and other factors may eliminate the disease, but if not, it becomes known as a disease of unknown etiology (DUE). DUEs are the subject of this chapter. They have a number of characteristics which make their study difficult and which may occur singly or together. The most important of these are summarized in this section.

A. LACK OF A MEANS OF EXPERIMENTAL TRANSMISSION
Ability to spread is one of the main characteristics of infectious diseases, and the mapping of disease distribution and movement may be used to provide evidence of transmission in the field. Experimental work with viruses depends on an ability to transfer symptoms from the diseased plant to others of the same, or other, species. Grafting and dodder transmission provide the most reliable means of transmission for virus-like agents. However, monocotyledonous hosts cannot be grafted, so that virus-like agents not transmissible in this group of hosts by any other means cannot be transferred experimentally. Failure to transmit by grafting does not necessarily preclude the involvement of a virus, as the cryptic viruses are not graft-transmissible.[1]

B. INABILITY TO INCREASE AMOUNTS OF DISEASED TISSUE
Some plants, including many species of palm, cannot be cloned vegetatively to allow an increase in the amount of diseased material. Under these conditions, a number of separate plants with similar symptoms may have to be used to scale up amounts of diseased tissue, with attendant complications arising from the possible mixing of pathogens and variants from a population of plants.

C. VECTORS ARE UNKNOWN
Vectors may be unknown, or may be difficult to manipulate for experimental transmission studies. The unequivocal demonstration of transmission by vectors normally requires their culture, and many have unknown or complex conditions for such culture.

D. SEED TRANSMISSION
If seed transmission is the only mode of transmission suspected, several generations need to be followed through to assess whether the pattern of

transmission through successive generations follows Mendelian segregation (and is thus probably genetic) or resembles a pattern expected for virus transmission. With low rates of seed transmission, large-scale trials are required to determine the actual rates of transmission, and these trials need to be done under conditions where other possible modes of transmission cannot operate.

E. INCUBATION TIMES ARE LONG

Intervals between inoculation and disease development may be measured in years. As experiments with viruses require numerous infectivity assays to evaluate their biological and physical properties, and to develop protocols for purification, the rate at which experiments can be done may be too prolonged for satisfactory results to be obtained within a defined time frame.

F. NATURAL SPREAD IS IMPERCEPTIBLE OR SLOW

Where natural spread is occurring at rates of below about 10% per plant generation, not only will field experimentation on epidemiology be impractical, but the disease will probably have escaped the interest and concern of crop agronomists. Diseases of perennial crops commonly fall into this category. For example, a striate mosaic of unknown etiology in sugar cane in central Queensland is regarded as unimportant because it does not appear to be spreading outwards from its present boundaries.[32]

G. DISEASE PATTERNS ARE NOT CONSTANT

Pattern analysis in space and time is a technique for obtaining clues about the epidemiology and nature of a suspected pathogen.[2] If patterns of increase and distribution vary in different regions, conflicts arise in understanding epidemiology. These patterns may also indicate that more than one pathogen could be involved in a disease syndrome.

H. NO DISEASE IS APPARENT

The level of severity at which a disease becomes apparent usually depends on the observer. In addition, environmental influences and variability in host genotype may overshadow effects due to pathogens and obscure a typical syndrome. Some pathogens may induce subtle effects which may be defined only as poor performance, but this may in turn lead to a cascade of events that may overshadow the original minor change. For example, the less vigorous plant may be less competitive with neighbors, or have increased susceptibility to secondary organisms.

III. EXAMPLES OF DUEs

Perennial plants which are not able to be reproduced by vegetative cloning are good candidates for harboring pathogens causing diseases of unknown etiology. Two of the examples given here are of DUEs in the coconut palm,

Cocos nucifera. One is now known to be inducible by a viroid, coconut cadang-cadang viroid,[3] but the approach that led to the detection of the viroid was an approach designed to detect a putative virus agent. The other example from palms describes the identification of coconut foliar decay virus.[4,5]

A. COCONUT CADANG-CADANG (CCC)

Coconut cadang-cadang (CCC) was first reported as a disease in 1937, following an epidemic in a coconut plantation on an offshore island in the Philippines. The disease progressed through the island, reaching an incidence of 25% in some areas and ultimately only 100 or so survivors remained from the original roughly 250,000 palms.[6] This epidemic was thought to signal the point of origin of CCC, but mapping and estimates of rates of spread by Zelazny and co-workers[6] led to the conclusion that the disease was endemic in a much wider area at the time of the first report.

CCC had not previously been remarked upon in the scientific literature probably because of a combination of characteristics, such as the subtle nature of the symptoms, the slow rate of disease development, the low incidence of disease, its random distribution, and the mixture of coconut palm ages and genotypes in the areas of incidence. This would have led to a lack of uniformity in the crop and hence an inability to recognize the poor performance of individual palms as a distinct syndrome.

Following the discovery of the viroid agent of CCC (CCCVd) using the technique described in Table 3 a further chance discovery was made of CCCVd-related sequences in both coconut and oil palm in the Pacific region where typical CCC has never been observed. Despite the absence of CCC in this region, a significant proportion of poorly yielding coconut palms occurs in all areas in the region. Absence of a recognizable disease syndrome could be due to any of a number of factors, such as variation in palm genotype, variation in the CCCVd-related sequences, lack of uniformity in the growth conditions in plantations, or environmental effects on symptom expression.

CCC is a good example of the need for alternative strategies for implicating pathogens in disease because (1) the host cannot be grafted or cloned routinely; (2) no vector of CCC is known; (3) spread of CCC is slow, at 0.1 to 1% per year; (4) symptoms are subtle, especially at the early stages; (5) first symptoms develop at about 2 years postinoculation, and unmistakable symptoms are evident 4 to 14 years postinoculation; and (6) no alternate (nonpalm) host species of the viroid agent which would be suitable for experimental work are known.

B. COCONUT FOLIAR DECAY (CFD)

Coconut foliar decay (CFD) was first reported as a lethal disease of exotic coconut palm selections in Vanuatu in 1964. No cellular organisms were found to be constantly associated with the disease, and the lack of effect of tetracyline treatment had virtually precluded mycoplasmas as a possible cause.[7]

As with CCC, difficulties arose in determining the etiology of CFD because the characteristics of the host prevented experimental transmission to primary or secondary hosts. Unlike CCC, however, the plant hopper vector was known at the commencement of attempts to find the agent, sensitive cultivars were known, symptoms appeared 6 to 10 months postinoculation, incidence was high, and epidemics were well described and mapped. The limiting factor for determining etiology again proved to be the intractable nature of the host for *in vitro* isolation of potential agents. The strategy described in Table 3 gave a lead which eventually identified a novel virus (CFDV)[5] as the probable causal agent of CFD.

C. CRYPTIC VIRUSES

The cryptic viruses frequently cause no symptoms, and are not experimentally transmitted except by seed and pollen.[1] Most known representatives of this group occur in dicotyledonous hosts, so difficulties specific to monocotyledons are not relevant. However, the absence of infectivity assays has led to reliance on the detection of dsRNA, virus particles, or coat protein antigen for demonstrating their involvement in circumstances where they cause disease.

IV. A SOLUTION TO THE PROBLEM OF DUEs: COMPONENT ANALYSIS

The steps involved in implicating a virus-like particle (VLP) in a disease include: (1) demonstration that the VLP can induce the disease after it has been purified and inoculated to a healthy indicator; and (2) characterization of the *in vitro* properties of the VLP or similar agent.

Component analysis is a strategy that makes use of current knowledge about the composition of viruses and the relevance of such knowledge to taxonomy. The concept of component analysis utilizes the fact that all viruses have components different from those of the host. The use of component analysis depends on the availability of technology to distinguish viral components from those of the host. Once a nonhost component has been found and shown to be correlated with the presence of a specific host disorder, methods for its detection can be optimized for diagnosis, and a search can be made for the agent from which it was derived. The two major components of viruses are considered here for their relative value in application to component analysis.

A. PROTEIN

Structural proteins make up most of a virus particle. Coat protein molecular weights can be used to allocate viruses to provisional taxonomic groups.[8] Isoelectric focusing and molecular filtration by gel electrophoresis can be used alone or together to separate proteins, and to identify those that are disease-associated.

Table 1. The Sizes of Major Nucleic Acid Classes

Source		Mol wt × 10⁶	Nucleotides
Plant	DNA	>300	>10^6 bp
	rRNA		
	25S	1.3	3800
	23S	1.1	3200
	18S	0.7	2000
	16S	0.56	1600
	5s RNA	0.05	140
	tRNA	0.03	80
Plasmid	Circular DNA	1–200	1500–3 × 10^5 bp
Virus	Genomic ssRNA	1.3–6.5	4000–20000
	Genomic dsRNA	0.5–3.0	850–5000 bp
	Genomic ssDNA	0.4–0.8	1000–2800
	Genomic dsDNA	5	8000 bp
	Satellite ssRNA	0.1–1.0	300–3000
Viroid	ssRNA	0.1	246–375

A number of potential disadvantages in using virus structural protein for component analysis should be noted: (1) a protein has to be detected against a background of over 1000 host polypeptides; (2) isolation of unique proteins at a high level of purity is often difficult to achieve; (3) amplification is not possible; (4) virus structural proteins do not have unique physical properties, such as size, structure, or stability, compared with those in the host organisms in which they occur; (5) sensitive detection methods, such as those based on Ag^+ staining, immobilize the protein; (6) proof of uniqueness is dependent on immunological or amino acid sequencing techniques which require microgram amounts of protein at high purity.

B. NUCLEIC ACID

Total nucleic acid extracts of leaf of almost all plants show the same pattern when fractionated under nondenaturing conditions by polyacrylamide gel electrophoresis (PAGE). DNA has the lowest electrophoretic mobility, with the 25S, 23S, 18S, 16S, and 5S ribosomal RNAs, and 4S tRNA separating ahead of the DNA. Table 1 shows their approximate sizes. In principle, virus nucleic acids, most of which have specific sizes distinct from the host plant RNAs, would be detectable under the appropriate conditions as additional species against the background of host plant nucleic acids.

Viral genomic nucleic acid detection and characterization as an aid to identifying virus-like agents is favored over viral proteins for the following reasons:

1. A single dimension of gel electrophoresis normally suffices to separate additional nucleic acid species, and even when molecular weights are very close to those of the host, modifications may be made to the technique of electrophoresis (variation of medium, gel concentration, inclusion of denaturants, length of gel) to allow their separation

2. Nucleic acids are extractable by rigorous procedures which allow efficient recovery
3. Simple tests of physical and chemical properties (susceptibility to enzymes and alkali, effects of denaturation) can be tested in conjunction with gel electrophoresis
4. Nucleic acids can be isolated, further purified, then cloned[9] for applications such as synthesis of probes, sequencing, and testing the infectivity of DNA and RNA transcripts
5. Reference to comparative taxonomic data[8] allows rapid allocation of the putative virus to a possible group or family, or can indicate its possible uniqueness
6. In contrast to strategies which implicate viruses by detecting intermediate double-stranded (ds) RNA forms, isolation of viral genomic nucleic acids is applicable to pathogens without dsRNA forms, such as DNA viruses
7. This form of analysis may be applicable to nonviral endoparasites such as phytomonads, mycoplasmas, fastidious xylem-inhabiting bacteria, and fungi

Thus, a possible virus etiology may sometimes be excluded, and the technique may be extended to identify unique sequences associated with a nonviral parasite.

1. Methods for Distinguishing the Groups of Nucleic Acids

Table 2 summarizes the commonly used methods that can be used to distinguish the kinds of nucleic acids found in cells, in viruses and viroids, and in tissues in which these agents are replicating.

Nucleic acids found in virus- or viroid-infected tissue fall into single-stranded (ss) or double-stranded (ds) RNA or DNA groups. Each of the groups may also have linear or circular forms. Identification of any of these nucleic acid structures can assist greatly in a preliminary classification of a putative virus-like agent. For example, if a small circular ssRNA form is present, it could originate from either a viroid or a circular satellite RNA. Presence of a helper virus linear ssRNA genome would confirm the nature of the circular RNA as a satellite.[10] A small circular dsRNA would be expected to be the replicative form of a viroid or satellite. Detection of both circular and linear ss forms of about the same size would probably be due to the presence of both covalently closed circles and linear forms arising from "nicked" circles of the same sequence. The known genomic ssDNAs are circular, and are frequently accompanied by a linear form and dsDNA.[11] The known sizes range from about 1 kb for the new group of tentatively named circoviruses,[12,13] to about 2.8 kb for the geminiviruses. The dsDNA genomes of plant viruses are also circular[14,15] (caulimoviruses and the nonenveloped bacilliform plant viruses) with some linear DNA likely to be associated. The characteristic

Table 2. A Scheme for Distinguishing the Types and Forms of Nucleic Acids

Kind of nucleic acid	Presence in:			Labeling of:		Sensitivity to:						Dye Binding	Buoyant density (g/cm³)		Thermal transition		Electrophoretic mobility	
	Pro- and Eukaryote	Virus	Viroid and Satellite	5' end[a]	3' end[b]	Alkali	RNAse A	RNAse T1	DNase[c]	S1 nuclease	Mung-bean nuclease	Acridine orange[d]	In CsCl	In Cs_2SO_4	Sharp transition at T_m	T_m in 10 mM NaCl (°C)	Mobility proportional to size in PAGE	Detection by 2D PAGE
ss linear RNA	+	+	+[e]	+	+	+	+	+	–	+	–	Red	>1.85[f]	1.61–1.69	–		Yes	No
ss circular RNA	–	+	+[e]	–	–	+	+	+	–	+	–	Red	>1.85[f]	1.61–1.69	+[g]	ca. 50	No	Yes
ds linear RNA	+[h]	+	+[e]	+	+	+	+(–)[i]	+(–)[i]	–	–	–	Green	>1.85[f]	1.60–1.64	+	ca. 85	Yes	Nt
ds circular RNA	–	–	+[e]	–	–	+	+(–)[i]	+(–)[i]	–	–	–	Green	>1.85[f]	1.60–1.64	+	ca. 85	Nt	Nt
ss linear DNA	–	+[e]	–	+	+	–	–	–	+	+	+	Red	1.72–1.73	1.45	–		Yes	Nt
ss circular DNA	–	+	–	–	–	–	–	–	+	+	+	Red	1.72–1.73	1.45	–		No	Yes[j]
ds linear DNA	+	+[e]	–	+	+	–	–	–	+	–	–	Green	1.69–1.74	1.42–1.46	+	ca. 80	Yes	Nt
ds circular DNA	+[k]	+	–	–	–	–	–	–	+	–	–	Green	1.69–1.74	1.42–1.46	+	ca. 80	Nt	Nt

Nt, not tested.

a Uncapped 5' termini labeled following dephosphorylation and reaction with γ-^{32}P ATP and polynucleotide kinase.[9]
b 3' termini of ss or ds DNA labeled with terminal transferase.[9]
c Mg^{2+} dependent.
d Used with glyoxal denaturation and UV excitation.[17]
e May be intermediate or cleavage product of circular form.[14]
f See Szybalski.[18]
g Fully or partially internally base-paired molecules such as viroids and circular satellite RNAs.[19,20]
h See Wakarchuk and Hamilton.[31]
i Tolerance to low RNase activity at Na^+ concentrations >0.3 M.[16]
j CFDV DNA detectable.[12]
k Plasmid.

retardation of both circular ssRNA and circular ssDNA in denaturing gels, allowing them to be detected in two-dimensional PAGE, is a valuable adjunct to electron microscopy and to end-labeling techniques for identifying conformation. The applicability of two-dimensional PAGE to the large circular virus dsDNA genomes or to detection of plasmids has yet to be tested. One of the constraints would appear to be the availability of a gel medium suitable for larger nucleic acid circles which is also amenable to the use of denaturants and to the detection of components by staining or probing.

If nucleic acids are found to be associated with the disease under study, chemical and physical methods can be used for further characterization.

a. Chemical Methods

The chemical methods indicated in Table 2 allow RNA to be distinguished from DNA by the sensitivity of the former to alkali (e.g., 0.25 N KOH, 25°C, 60 min) or ribonuclease (RNase), and the sensitivity of DNA to deoxyribonuclease (DNase). dsRNA is distinguishable from ssRNA by its salt-dependent (above about 0.3 M NaCl) resistance to RNase.[16] ds forms of both RNA and DNA are distinguished by their resistance to the single-strand-specific S1 nuclease, and their green fluorescence in ultraviolet light when stained with acridine orange.[17] Mung bean nuclease is considered to be specific for ssDNA, but not dsDNA.[9]

b. Physical Methods

Isopycnic gradients in CsCl allow a clear separation of RNA and DNA on the basis of buoyant density; RNA is normally found at the bottom of gradients set up to have a midpoint density of about 1.7 g/ml, whereas DNA bands about midway. Cs_2SO_4 gradients allow both RNA and DNA to band on the same gradient, with RNA heavier than DNA.[18] Completely base-paired dsRNA and dsDNA have a thermal denaturation point (Tm) in 20 mM NaCl of between 80 and 90°C, with a sharp transition from a lower to a higher A_{260} value around the Tm.[19] Partially base-paired molecules such as viroids or circular satellite RNAs also show a sharp transition, but at a temperature in low salt of about 50°C.[19,20] Single-stranded nucleic acids show hyperchromicity over a wide temperature range with no specific Tm.[18]

Circular RNA and DNA show anomalous electrophoretic mobilities in denaturing gels, in that their mobility is not proportional to their molecular weight when linear molecular weight markers are used. This is illustrated by viroids in which the linear form migrates ahead of the circular form in denaturing gels. Two-dimensional PAGE[12,21] makes use of this phenomenon to identify circular ssRNA and ssDNA. It is easier to interpret than bidirectional and return gel electrophoresis. The principle of the two-dimensional PAGE method (see Chapter 11) is that total nucleic acid extracts are run in the first dimension under nondenaturing conditions of electrophoresis. The gel strips are then excised and embedded in a gel containing 7 to 8 M urea as a denaturing agent, and the nucleic acids are run transversely. Under these conditions, the linear RNA components denature and migrate in proportion to their size, forming a diagonal front. Circular forms dissociate to open circles with consequent greatly retarded mobility, and are detectable, usually by staining with Ag^+, as a component trailing behind the diagonal front of linear RNAs.

Electron microscopy (not included in Table 2) is a valuable technique for identifying circular nucleic acids in purified preparations.[12] It involves denaturation, spreading on protein films, picking up on grids, and rotary shadowing. Length, coiling, and the presence of internal base pairing can be evaluated by electron microscopy.

V. BASIC CONSIDERATIONS IN ESTABLISHING A PROTOCOL

Because of the relatively low amounts of virus-associated nucleic acids usually present in tissue from plants with a DUE, methods must be developed for concentrating the putative virus (and its nucleic acid) relative to host nucleic acids, and reducing the level of other compounds which may interfere with detection of the target nucleic acid.

This section describes the aspects which should be considered, and attempts to outline a basic system from which a more appropriate system may be developed for individual diseases. The system is based essentially on methods used for partial purification of viruses,[22] and many hints are derived from the author's and his colleagues' experience with palm tissue.

A. PLANT TISSUE

Early symptomatic tissue, if available, has the highest probability of containing a viral pathogen at its maximum concentration, and it should be trimmed of any nonsymptomatic tissue. If symptomatic tissue is not available or is impractical to select, young, differentiated tissue (i.e., leaves or root tissue close to the meristems) would be most likely to support the replication of a putative virus. A range of different tissues or organs can be tested simultaneously. Control tissue should come from a number of disease-free plants, including plants of the same species from other areas free of the disease under consideration.

B. EXTRACTION METHOD

The extraction procedure must maximize cell and organelle breakage, and may range from blending or juice extraction for soft tissues, to mechanical pulverization[23] or blending in liquid nitrogen for highly fibrous tissue where the parenchyma and vascular bundles may be protected by structural fibers of thickened sclerenchyma.

C. EXTRACTION MEDIUM

Extraction media must perform a number of functions. pH must be maintained in a desirable range, depending on whether the putative virus or viroid is to be kept in suspension or allowed to precipitate at a pH near its isoelectric point. Phosphate buffers should be avoided in some situations because the salts coprecipitate with nucleic acids in ethanol and interfere with PAGE. Tris, acetate-, or borate-based buffers are used instead if appropriate.

Polyphenoloxidase (PPO) activity leads to the rapid browning of extracts and can be inhibited by chilling the buffers, maintaining pH above 8, excluding the PPO cofactor Cu^{2+} by the use of chelating agents, such as ethylene diaminetetraacetate (EDTA) and disodium diethyldithiocarbamate (DIECA), and including a reducing agent usually with thiol or sulfite groups. Buffers may be complex[24] (e.g., for chrysanthemum) or as simple as Na_2SO_3.[25] This salt, at 0.1 M, holds the pH of extracts at about 8, and maintains the greenness of perennial plant extracts for several hours if kept cold. It is particularly useful if economies in cost, labor, or facilities are necessary.

The effects of buffer components and concentration on the structural integrity of viruses in general need to be considered when choosing a buffer. The use of enzymes during extraction has been developed for extracting viruses confined to the phloem. For example, an incubation in cellulase improves the extractability of luteoviruses.[26] Direct alkali extraction of coconut leaf tissue (0.5 N KOH, 1 to 2 h, 25°C) enhances the extractability of the DNA of CFDV for subsequent assay by molecular hybridization,[33] but this is not appropriate for isolating intact particles.

D. CLARIFICATION

Clarification may be achieved by the use of organic solvents, detergents, thermal denaturation of plant proteins, adsorbents, filtering agents, and low-speed centrifugation.[22] A range of methods needs to be tested because of the possibility of denaturation, cosedimentation with denatured plant components, adsorption, or aggregation of putative virus components. The following can be recommended as possible starting points: emulsification with chloroform or carbon tetrachloride added at 0.5 to 1 volumes, 9% *n*-butanol, 1% Triton X100, bentonite (1% w/v), activated charcoal (0.05 g/ml), followed by filtration through a celite pad[27] and centrifugation at 10,000 to 20,000 g for 10 min.

Some tissue extracts do not need clarifying agents. Others, particularly, those in which polysaccharides, gums, or starch are prevalent, are difficult to handle and may need to be avoided by the selection of more tractable plant parts (e.g., the roots instead of leaves of *Malvaceae*).

E. CONCENTRATION

Polyethylene glycol (mol wt 6000 to 8000) can be titrated in the range 4 to 12% w/v to find the optimum concentration to precipitate a putative virus (or viroid in the case of CCCVd[28]) while maximizing the amount of plant material retained in suspension. A good starting point is 8%.

High-speed centrifugation designed to sediment all particles down to 80S would cover the known range of virus particles, while leaving viroids in the supernatant. Modifications to centrifugal methods to improve separation of virus particles from other components include differential, rate zonal, and isopycnic density gradient centrifugation in either cesium salts or Nycodenz®[29] and sedimentation through cushions of sucrose or cesium salts.

F. ASSAY OF THE COMPONENTS

Aliquots representing all stages of the concentration procedure, from both symptomatic and control material, need to be assayed for their nucleic acid content. Standard reliable methods include protease digestion (with concomitant dialysis if appropriate) and sodium dodecyl sulfate (SDS)-phenol extraction, followed by precipitation with ethanol (2.5 vol), isopropanol (0.8 vol), or 0.3% cetyl trimethylammonium bromide (CTAB) in the presence of 0.1 M NaCl. The latter two precipitants minimize coprecipitation of polysaccharides with the nucleic acid. CTAB precipitates are washed with sodium acetate in 75% ethanol before redissolving in appropriate buffers for analysis.

Precipitated nucleic acids may need an extended period of solubilization, and it is useful to resuspend pellets in electrophoresis buffer containing 50% glycerol and store at $-20°C$ before analysis after storage. Preparations are vortexed, aliquots are removed and incubated at 37 to 60°C for 15 min to equilibrate nucleic acid secondary structure, separated from undissolved material by centrifugation at 10,000 g for 10 min, then loaded on appropriate gels with molecular weight markers. A range of loadings is applied and run, and gels can be stained sequentially with the first stain comprising aqueous 1 ppm ethidium bromide (with observation at 254 to 310 nm), the second stain comprising toluidine blue (0.01% in 5% acetic acid followed by destaining in water), and the final stain containing 0.2% AgNO$_3$ to detect bands containing down to about 1 ng.

Any bands associated with the disease being studied can be characterized by the methods outlined in Table 2, and separation procedures then modified to optimize their isolation and detection. Nucleic acids can be recovered from gels by either electroelution, direct elution,[9] or centrifugation procedures.

VI. NUCLEIC ACIDS AS SIGNPOSTS

Reference to taxonomic texts such as the reports of the International Committee on Taxonomy of Viruses (ICTV),[8] allows the type, size, and structure of any newly discovered disease-associated nucleic acid to be compared with those already described for the groups of plant viruses and viroids.

If the properties fall outside those known for previously described viruses, the virus from which the nucleic acid is derived could be expected to be a new type and a search for a potentially different type of particle can commence.

The search for nucleic acid-associated particles uses the nucleic acid assay procedure in place of the traditional bioassay for the establishment of a virus purification procedure. Thus, the efficacy of clarification, sedimentation, and various fractionation procedures can be determined. When a fraction contains essentially pure disease-associated nucleic acid, an associated particle will also be approaching a high degree of purity if it is still intact, and its characterization by electron microscopy, protein properties, and production of antibodies can commence.

If it is difficult to achieve relatively high levels of purity of particles associated with the nucleic acid, the nucleic acid may be used directly for further steps in characterizing the virus and for developing appropriate diagnostic tests. Thus, the nucleic acid can be copied and/or amplified, cloned, sequenced, open reading frames (ORFs) identified, and their amino acid sequences determined.[9,14] Examination of potential ORF products can indicate by specific motifs the possible functions of the ORF products[14] (e.g., nucleotide triphosphate [NTP] binding sites, zinc fingers, and other amino acid sequences indicative of replicase or coat protein function). The arrangement of the ORFs shows the genome organization and may allow classification of the virus at group, family, or superfamily level.[14]

A further advantage of the nucleic acid isolation methodology is the ability to clone selected ORFs in expression vectors, express and purify the protein,[9] and use it to immunize rabbits. Resultant antibodies can be used for *in situ* localization of virus and its precursors in the plant, and those that recognize coat protein may be used diagnostically.

VII. METHODS APPLICABLE TO SOME SPECIFIC HOSTS

Section V outlined basic considerations for establishing a general protocol suitable for many host plant species. Any worker considering the nucleic acid isolation approach to the determination of etiology needs to develop a basic starting procedure, and needs then to define, on the basis of a range of comparative tests, an improved system. This requires not only an evaluation of the most favorable conditions for original tissue extraction, but a knowledge of, and familiarity with, nucleic acid extraction and assay procedures. The first methods tested should be simple to facilitate trouble-shooting; and the use, for example, of a clarification agent such as bentonite should not be included without first testing samples with and without the agent.

Table 3 outlines two methods which were used in the first successful trials for isolating CCCVd and CFDV from coconut palm. They are similar and a single common method for identifying either of these nucleic acids could be developed if needed for a specific purpose such as a survey. Alternatively, each can be refined, and, in fact, the methods shown in Table 3 have been further modified to optimize the sensitivity and specificity of detection of each of the disease-associated nucleic acids.

Table 3. Examples of Procedures with Two Tree Hosts for Detecting Disease-Associated Nucleic Acids

| | Coconut | | Spruce (*Picea* spp.) |
	CCCVd[23,28]	CFDV[12]	DUE[30]
Tissue	Leaflets	Leaflets	Needles
Medium	Buffer, chelating agent, thiol compounds	Buffer, chelating agent, thiol compounds	Buffer, chelating agent, thiol, phenol, SDS
Extraction	Blend 1 g/4–6 ml	Blend 1 g/4–6 ml	Blend 1 g/13 ml
Clarification	Low-speed centrifugation	Triton X100 + bentonite	—
Concentration	5% PEG 6000	8% PEG 6000	—
Nucleic acid isolation	Protease; phenol-SDS; ethanol precipitation	Phenol chloroform	Phenol chloroform 2-phase system with PEG 6000, CTAB precipitation
Analysis	3.3–20% PAGE, various stains	5% PAGE + urea, Ag+ stain	2-Dimensional 5% PAGE, Ag+ stain

Table 3 also includes, for comparison, a procedure described for isolating and examining the small RNAs of healthy and diseased *Picea* spp.,[30] a host expected to present difficulties for nucleic acid isolation. The method is more complex than for coconut, but the product gives a high degree of resolution when subjected to the subsequent fractionation procedure.

VIII. CONCLUSIONS AND POINTS TO CONSIDER IN INTERPRETATION

Diagnosticians need to be resourceful in finding alternative methods for identifying the causal agents of DUEs, and a range of strategies needs to be devised from the techniques available. No recipe for success can be formally described, but it is helpful to read widely, remain in contact with colleagues, and to remain active in the laboratory so that new techniques can be developed, modified, and evaluated.

The rationale of using nucleic acids as signposts to the agents of DUEs is not new, but it has been used with quite narrow research goals in mind, in the past. With the widening data-base of virus properties, it should be possible to develop a relatively simple range of techniques which could be used to find any nucleic acids associated with a viral or viroid pathogen, and to use this information to indicate likely taxonomic associations. Table 2 summarizes these techniques in the form of a key. It is important to remember that a combination of techniques should be used to support a conclusion about the structure of a nucleic acid. For example, even though circular nucleic acid molecules may be seen in preparations by electron microscopy, the technique is not quantitative. To confirm that circular nucleic acid is a major component in the nucleic acid preparation, a technique such as two-dimensional PAGE should be used in conjunction with electron microscopy.[12]

Other points to consider are that an unusual nucleic acid may be a subgenomic component, that it may represent only a part of the genome of a virus with a multipartite genome, or that it may be a satellite, or derive from a mixed virus infection. Moreover, the detection of any unusual disease-associated nonhost nucleic acid should not lead automatically to the conclusion that it derives from a viroid or virus agent. It could be derived from an endoparasitic prokaryote or eukaryote. It could also be a bacterial plasmid or be derived from a bacteriophage. It is advisable to remain sceptical about the significance of a disease-associated nucleic acid until further experiments can be done to link it to a specific causal agent.

REFERENCES

1. **Boccardo, G., Lisa, V., Luisoni, E., and Milne, R. G.,** Cryptic plant viruses, *Adv. Virus Res.,* 32, 171, 1987.
2. **Thresh, J. M.,** Progress curves of plant virus disease, *Adv. Appl. Biol.,* 8, 1, 1983.
3. **Hanold, D. and Randles, J. W.,** Coconut cadang-cadang disease and its viroid agent, *Plant Dis.,* 75, 330, 1991.
4. **Randles, J. W., Julia, J. F., Calvez, C., and Dollet, M.,** Association of single-stranded DNA with the foliar decay disease of coconut palm in Vanuatu, *Phytopathology,* 76, 889, 1986.
5. **Randles, J. W. and Hanold, D.,** Coconut foliar decay virus particles are 20-nm icosahedra, *Intervirology,* 30, 177, 1989.
6. **Zelazny, B., Randles, J. W., Boccardo, G., and Imperial, J. S.,** The viroid nature of the cadang-cadang disease of coconut palm, *Sci. Filipinas,* 2, 46, 1982.
7. **Julia, J. F., Dollet, M., Randles, J. W., and Calvez, C.,** Foliar decay of coconut by *Myndus taffini* (FDMT): new results, *Oleagineux,* 40, 25, 1985.
8. **Francki, R. I. B., Fauquet, C. M., Knudson, D. L., and Brown, F.,** Classification and nomenclature of viruses, *Fifth Report of the International Committee on Taxonomy of Viruses,* Springer-Verlag, Wien, 1991, 1–450.
9. **Sambrook, J., Fritsch, E. F., and Maniatis, F.,** *Molecular Cloning: A Laboratory Manual,* 2nd ed., Cold Spring Harbor Laboratory Press, Cold Spring Harbor, NY, 1989.
10. **Randles, J. W., Davies, C., Hatta, T., Gould, A. R., and Francki, R. I. B.,** Studies on encapsidated viroid-like RNA I. Characterization of velvet tobacco mottle virus, *Virology,* 108, 111, 1981.
11. **Honda, Y. and Ikegami, M.,** Mung bean yellow mosaic virus, *AAB Descriptions of Plant Viruses,* No. 323, Association of Applied Biologists, Wellesbourne, 1986.
12. **Randles, J. W., Hanold, D., and Julia, J. F.,** Small circular single-stranded DNA associated with foliar decay disease of coconut palm in Vanuatu, *J. Gen. Virol.,* 68, 273, 1987.
13. **Chu, P. W. G. and Helms, K.,** Novel virus-like particles containing circular single-stranded DNAs associated with subterranean clover stunt disease, *Virology,* 167, 38, 1988.
14. **Matthews, R. E. F.,** *Plant Virology,* 3rd ed., Academic Press, San Diego, 1991.
15. **Lockhart, B. E. L.,** Evidence for a double-stranded circular DNA genome in a second group of plant viruses, *Phytopathology,* 80, 127, 1990.
16. **Rohozinski, J., Francki, R. I. B., and Chu, P. W. G.,** The *in vitro* synthesis of velvet tobacco mottle virus-specific double-stranded RNA by a soluble fraction in extracts from infected *Nicotiana clevelandii* leaves, *Virology,* 155, 27, 1986.
17. **McMaster, G. K. and Carmichael, G. G.,** Analysis of single- and double-stranded nucleic acids on polyacrylamide and agarose gels by using glyoxal and acridine orange, *Proc. Natl. Acad. Sci. U.S.A.,* 74, 4835, 1977.
18. **Szybalski, W.,** Use of cesium sulfate for equilibrium density gradient centrifugation, in *Methods in Enzymology,* Vol. 12B, Grossman, L. and Moldave, K., Eds., Academic Press, New York, 1987, 63.
19. **Riesner, D.,** Physical-chemical properties: structure formation, in *The Viroids,* Diener, T. O., Ed., Plenum Press, New York, 1987, 63.
20. **Randles, J. W., Steger, G., and Riesner, G.,** Structure transitions in viroid-like RNAs associated with cadang-cadang disease, velvet tobacco mottle virus, and *Solanum nodiflorum* mottle virus, *Nucleic Acids Res.,* 10, 5569, 1982.
21. **Schumacher, J., Randles, J. W., and Riesner, D.,** A two-dimensional electrophoretic technique for the detection of circular viroids and virusoids, *Anal. Biochem.,* 135, 288, 1983.

22. **Francki, R. I. B.,** Purification of viruses, in *Principles and Techniques in Plant Virology,* Kado, C. I. and Agrawal, H. O., Eds., Van Nostrand Reinhold, New York, 1972, 295.

23. **Randles, J. W.,** Coconut cadang-cadang viroid: detection methods and their application, in *Proceedings International Seminar Viroids of Plants and their Detection,* Warsaw Agricultural University Press, Warsaw, 1988, 29.

24. **Palukaitis, P. and Symons, R. H.,** Purification and characterization of the circular and linear forms of chrysanthemum stunt viroid, *J. Gen. Virol.,* 46, 477, 1980.

25. **Hanold, D. and Randles, J. W.,** Detection of coconut cadang-cadang viroid-like sequences in oil and coconut palm and other monocotyledons in the south-west Pacific, *Ann. Appl. Biol.,* 118, 139, 1991.

26. **Takanami, Y. and Kubo, S.,** Enzyme-assisted purification of two phloem-limited plant viruses: tobacco necrotic dwarf and potato leafroll, *J. Gen. Virol.,* 44, 153, 1979.

27. **Francki, R. I. B.,** Plant rhabdoviruses, *Adv. Virus Res.,* 18, 257, 1973.

28. **Randles, J. W.,** Association of two ribonucleic acid species with cadang-cadang disease of coconut palm, *Phytopathology,* 65, 163, 1975.

29. **Gugerli, P.,** Isopycnic centrifugation of plant viruses in Nycodenz density gradients, *J. Virol. Meth.,* 9, 249, 1984.

30. **Beuther, E., Köster, S., Loss, P., Schumacher, J., and Riesner, D.,** Small RNAs originating from symptomless and damaged spruces (*Picea* spp.). I. Continuous observation of individual trees at three different locations in NRW, *J. Phytopathol.,* 121, 289, 1988.

31. **Wakarchuk, D. A. and Hamilton, R. I.,** Cellular double-stranded RNA in *Phaseolus vulgaris, Plant Mol. Biol.,* 5, 55, 1985.

32. **Mullins, R.,** personal communication.

33. **Randles, J. W.,** unpublished results.

Chapter 13

COMPLEXES OF TRANSMISSION-DEPENDENT AND HELPER VIRUSES

A. F. Murant

TABLE OF CONTENTS

I. INTRODUCTION

Most studies on transmission of plant viruses by vectors are concerned with single virus infections. Indeed care is usually taken to ensure, as far as possible, the clonal purity of the virus being studied so as to simplify interpretation of experimental results. In nature, however, mixed virus infections are common, and there are many examples of plant viruses or virus-like agents that interact with others in various ways and may even depend on the interaction for their survival. Almost every imaginable kind of interaction has been found to exist in nature, but prominent among them are those in which one of the viruses lacks some essential molecular function that the other provides. Thus, *satellites*[1-3] depend on an independent virus for th replicase function. Those that require only this function are called *satellite viruses,* the best known example being the tobacco necrosis virus satellite.[4,5] *Satellite RNA* molecules, such as those associated with cucumber mosaic *Cucumovirus*[6] and tomato black ring *Nepovirus,*[7] depend on a helper virus not only for replication but also for encapsidation — and hence for transmission by the specific vector of the helper virus, if it has one. There is another type of dependent virus which relies on a helper virus for transmission by a vector, but not for replication in the inoculated plant. Such viruses can therefore occur alone in plants, and cause disease symptoms, but are then no longer transmissible by the vector. An increasing number of complexes of transmission-dependent and helper viruses is now known, and form the subject of this chapter. The topic, in whole or in part, has been reviewed several times in recent years.[8-12] This chapter will be concerned primarily with discussing general features that may be of use in the recognition or diagnosis of such complexes.

II. COMPLEXES INVOLVING TRANSMISSION-DEFECTIVE VARIANTS

A. POTYVIRUSES

Viruses in the *Potyvirus* group have flexuous filamentous particles about 750 nm long, and are transmissible by aphids in the classical[13] nonpersistent manner. The viruses have single-stranded (ss) RNA genomes of about 9.5 kb and a single coat protein species of about 30 kDa. The first evidence for the existence of transmission-defective variants of some plant viruses came from the work of Bawden and Sheffield[14] with an aphid nontransmissible virus they called potato virus C, but which is now regarded as a strain of potato Y *Potyvirus* (PVY) and is called PVY[c]. Later work[15-17] showed that PVY[c], though not aphid-transmissible on its own, becomes so when the source plants are also infected with aphid-transmissible isolates of PVY or several other potyviruses. The helper factor in PVY infections was shown to be a nonparticle protein, the so-called helper component (HC).[18-20] All potyviruses tested have been found to produce HC proteins, and the serological and molecular properties of HC proteins from different potyviruses have been compared.[21-23] Transmission-defective variants have been reported for several other potyviruses besides PVY, including bean yellow mosaic virus (BYMV),[24] plum pox virus (PPV),[25] tobacco etch virus (TEV),[26,27] and zucchini yellow mosaic virus (ZYMV).[28,29]

In PVY[c] the defect appears to be, not a failure to produce HC, but a relatively minor change in the HC protein leading to loss of activity. A polypeptide that reacts with antiserum to PVY HC and comigrates with PVY HC can be detected by immunoblotting, and is produced in normal amounts in PVY[c]-infected plants; amino acid sequence comparisons revealed two amino acids that were conserved in the HC proteins of five aphid-transmissible potyviruses but differed in the HC of PVY[c].[30] In other instances, as with poorly aphid-transmissible or non-aphid-transmissible isolates of TEV[31] and ZYMV,[29] the defect appears to lie in the virus coat protein rather than in the HC, which is functional with other isolates. Attention has been drawn[32,33] to an asp-ala-gly motif which lies near to the N-terminus of the coat protein of several potyvirus isolates that are aphid-transmissible but is modified in isolates that are not.[25,32] At first glance it would seem that isolates in which the defect lies in the coat protein would fail to be aphid-transmissible even in the presence of active HC (and would therefore be strictly outside the scope of this chapter) but this is not necessarily so: the aphid nontransmissible isolate of TEV can in fact be transmitted by aphids that have acquired the HC of PVY,[31] which suggests that the defect in this instance is the inability of the particle protein to interact with its own HC.

The ability of the HC from one *Potyvirus* to assist aphid transmission of another is also widely reported,[20,34,35] although some specificity in the interaction is usually observed. Thus, the HC of BYMV assisted the aphid transmission of PVY and TVMV, whereas PVY HC assisted that of tobacco vein mottling *Potyvirus* (TVMV) but not BYMV, and TVMV HC assisted that of PVY but not BYMV.[20] There is even an example of the *coat protein* of one potyvirus being able to assist the aphid transmission of another: a non-aphid-transmissible isolate of ZYMV, which has a transmission-deficient coat protein, was not transmitted from singly infected plants by aphids already carrying the HC of papaya ringspot *Potyvirus* (PRSV), but was aphid-tranmissible from plants that were also infected with PRSV.[36] Evidence was presented to show that this was a result of the ZYMV RNA becoming coated with a mixture of its own coat protein and that of PRSV (a phenomenon termed ''phenotypic mixing,'' see below[8]).

Although many of these examples of transmission-defective variants and mutual assistance among potyviruses could perhaps be regarded as laboratory curiosities, it is a fair inference that many field infections of potyviruses consist of populations of variants, among which there are transmission-defective forms of the same or even of different potyviruses that rely on normal forms to spread and survive. These field populations may also include totally unrelated dependent viruses, as discussed in detail below.

B. CAULIMOVIRUSES

Caulimoviruses have about 50 nm diameter isometric particles containing a double-stranded (ds) DNA of about 8 kbp and a single particle protein of 42 kDa. They induce the formation in infected cells of characteristic "currant bun" inclusion bodies which consist of a proteinaceous matrix in which the virus particles are embedded. Although cauliflower mosaic *Caulimovirus* (CaMV) seems to be transmitted in a semipersistent[37] manner by aphids,[38] it depends for this, like the potyviruses, on a HC,[39-43] which is a virus-coded nonparticle protein. Aphid-non-transmissible isolates of CaMV can be transmitted by aphids that have already acquired HC from an aphid-transmissible isolate, and such aphids can also transmit CaMV after feeding though membranes on particle preparations.[39] The CaMV HC does not assist aphid transmission of PVY, nor do the HC proteins of PVY or TEV assist aphid transmission of CaMV.[39]

The CaMV HC is an 18-kDa protein (p18) encoded by open reading frame (ORF) II of the viral genome.[43] Both p18 and the helper activity are associated with the inclusion bodies.[42,44] Tests with two aphid nontransmissible isolates of CaMV showed that one of them (Campbell) induced the formation in plants of a protein that reacted with antiserum to p18 whereas the other (CM4–184), which has a large deletion in ORF II, did not.[45] However, in cells infected with the Campbell isolate, p18 was absent from the inclusion bodies.[44] The lack of aphid transmissibility of the Campbell isolate was attributed[46] to a small portion of ORF II which showed only two nucleotide sequence differences from that of the aphid-transmissible isolate. Only one of these changes resulted in an altered amino acid (from glycine to arginine). Thus, this single difference in the amino acid sequence probably causes loss of the helper function.

Two other caulimoviruses, carnation etched ring and figwort mosaic (FMV), which are distantly serologically related to CaMV, can act as helpers for aphid-non-transmissible isolates of CaMV,[47] indicating that they too induce the formation of HC proteins. Other evidence suggests that the CaMV HC can assist the transmission of FMV.[48] Almost certainly the production of a HC is a feature of all caulimoviruses and, as with the potyviruses, it seems likely that caulimovirus infections consist of populations of variants that include transmission-defective forms. However, there appears at present to be no evidence for caulimoviruses acting as helpers for unrelated transmission-deficient viruses.

C. PEA ENATION MOSAIC VIRUS

Pea enation mosaic virus (PEMV) has 28 nm diameter isometric particles and is transmitted by its principal aphid vector, *Acyrthosiphon pisum,* in a persistent[13] or circulative, nonpropagative[49] manner. PEMV is yet another virus that readily gives rise to isolates that are not aphid-transmissible (NT) except from plants that also contain transmissible (T) isolates. However,

PEMV is unlike the viruses discussed hitherto in that aphids already carrying a T isolate cannot acquire and transmit a NT isolate from a singly infected plant.[50] The particles of T isolates possess two ssRNA species, of mol wt approximately 1.8×10^6 and 1.4×10^6, and two proteins, of 56 and 22 kDa, which are encoded by the larger RNA. In NT isolates there is a deletion of a 0.6×10^6 mol wt portion of this RNA which leads in turn to loss of the larger of the two particle proteins.[51,52] Thus, this virus too appears to encode a special protein needed for aphid-transmissibility, but it is a particle protein rather than a separate HC. As with the potyviruses and caulimoviruses, field isolates of PEMV probably consist of mixed populations of T and NT isolates. PEMV can act as a helper for bean yellow vein banding virus (BYVBV) (see below and Table 1); it seems likely that this interaction involves both PEMV particle proteins, but this has not been investigated.

III. COMPLEXES INVOLVING RELATED VIRUSES WITH DIFFERENT VECTOR SPECIFICITIES

All the viruses that come under this heading are luteoviruses. Viruses in this group have about 25 nm diameter isometric particles and are transmitted in the circulative, nonpropagative manner by aphids. Each member of the group is transmitted by one or very few species of aphid. Thus two of the viruses associated with symptoms of yellow dwarf in barley have been designated after their specific vector: BYDV-MAV, transmitted by *Sitobion* (formerly *Macrosiphum*) *avenae,* and BYDV-RPV, transmitted by *Rhopalosiphum padi.* In fact, these two viruses have no direct serological relationship with each other and RPV is more closely related to beet western yellows virus (BWYV).[53] Rochow[54-56] found that *R. padi* that were allowed to acquire virus from plants mixedly infected with the two viruses, or were injected into the hemocoel with extracts of such plants, transmitted not only RPV but also MAV. In contrast, MAV was not transmitted by *R. padi* allowed to feed sequentially, in either order, on plants infected with each of the viruses alone, or when injected with a mixture of preparations purified from singly infected plants. Treatment of the extracts from mixedly infected plants with RPV antiserum before injection into the aphids prevented the aphids from transmitting both viruses, but treatment with MAV antiserum did not. From this work it was concluded that in mixedly infected plants MAV RNA can become packaged in RPV coat protein to form particles transmissible by *R. padi.* This packaging of the nucleic acid of one virus in shells composed entirely of the coat protein of another has been termed "transcapsidation".[8] Another term, "phenotypic mixing", was coined to denote the coating of the nucleic acid with a mixture of the coat proteins of both viruses. This phenomenon has since been shown[57] to occur in plants mixedly infected with MAV and a serologically related strain, PAV, which is transmitted by both *R. padi* and *S. avenae;* as expected, *R padi* can transmit MAV from such plants.

The importance of virus coat proteins and other virus-coded proteins in determining the vector specificity of plant viruses is well established.[58] For luteoviruses the role of the coat protein was elegantly demonstrated in electron microscope studies of virus-carrying *S. avenae*.[59,60] Virus particles were found in the basal lamina, and in plasmalemma invaginations, of cells of the accessory salivary gland of aphids carrying either MAV or RPV, but particles *within* the salivary gland cells (in coated vesicles and secretory canals) were found only in aphids carrying MAV. Vector specificity therefore seems to depend first on the particles binding to the plasmalemma, and second on their passage from it into coated vesicles within the cytoplasm of the salivary gland cells.

IV. COMPLEXES INVOLVING TRANSMISSION-DEFICIENT VIRUSES

The dependent viruses in the complexes described above are defective or vector-specific strains of transmissible viruses. This section deals with complexes in which the dependent viruses have no transmissible strains but depend entirely on unrelated viruses for transmission by vectors (Table 1).

A. NONPERSISTENT COMPLEXES

The only example of a transmission-deficient virus in the nonpersistent category is potato aucuba mosaic ? *Potexvirus* (PAMV), which has filamentous particles about 580 nm long. The main reason for doubt about the taxonomic status of this virus is that, unlike established potexviruses, it is transmissible in the nonpersistent manner by aphids. However, aphid transmission occurs only in association with a helper virus (PVY) and is mediated by the PVY HC.[16,17,61,62] As with PVY[c], aphids already carrying the PVY HC can acquire and transmit PAMV from singly infected plants. Presumably the PVY HC interacts in a similar way with the PAMV coat protein as with its own coat protein.

B. SEMIPERSISTENT COMPLEXES
1. Parsnip Yellow Fleck Virus and Anthriscus Yellows Virus

Parsnip yellow fleck virus (PYFV) depends on anthriscus yellows virus (AYV) for transmission in the semipersistent manner by the aphids *Cavariella aegopodii* and *C. pastinacae*.[63-65] Aphids that have already acquired AYV on its own can subsequently acquire PYFV from a singly infected plant,[64] or from a preparation of purified virus.[65] PYFV has 30 nm diameter isometric particles containing a single, approximately 10-kb ssRNA and three coat proteins.[66] It somewhat resembles the picornaviruses of vertebrates[67] and has been designated the type member of the newly formed "parsnip yellow fleck virus group". Dandelion yellow mosaic virus (DaYMV),[68] which has similar particle properties to PYFV,[67] is a possible second member of this group. The

Table 1. Viruses Dependent on Unrelated Ones for Transmission by Insect Vectors

Dependent virus	Helper virus	Vector[a]	Dependence mechanism[b]	Ref.
Nonpersistent viruses				
Potato aucuba mosaic ?*Potexvirus* (PAMV)	Potato A *Potyvirus* (PVA) Potato Y *Potyvirus* (PVY)	*Myzus-persicae*	HC	61, 62
Semipersistent viruses				
Parsnip yellow fleck (PYFV) (PYFV group)	Anthriscus yellow (AYV)	*Cavariella aegopodii, C. pastinacae*	HC?	63–65
Rice tungro bacilliform "badnavirus" (RTBV)[c]	Rice tungro spherical (RTSV) (?MCDV group)	*Nephotettix virescens*	HC?	75, 76, 79, 81
Heracleum latent ? *Closterovirus* (HLV)	Heracleum 6 *Closterovirus* (HV6)	*C. theobaldi*	PA	87, 88
Persistent viruses				
Bean yellow veinbanding "umbravirus" (BYVBV)	Bean leaf roll *Luteovirus* (BLRV) Pea enation mosaic (PEMV) (PEMV group)	*Acyrthosiphon pisum*	TC?	96
Carrot mottle "umbravirus" (CMoV)	Carrot red leaf *Luteovirus* (CaRLV)	*C. aegopodii*	TC	103
Groundnut rosette "umbravirus" (GRV)	Groundnut rosette assistor *Luteovirus* (GRAV)	*Aphis craccivora*	TC	107, 108
Lettuce speckles mottle "umbravirus" (LSMV)	Beet western yellows *Luteovirus* (BWYV)	*M. persicae*	TC	94, 95
Sunflower yellow blotch ?"umbravirus" (SYBV)	Suspected virus in *Tridax procumbens*	*Aphis gossypii*	TC?	101, 102
Tobacco mottle "umbravirus" (TMoV)	Tobacco vein distorting ?*Luteovirus* (TVDV)	*M. persicae*	TC?	91
Tobacco yellow vein ?"umbravirus" (TYVV)	Tobacco yellow vein assistor ?*Luteovirus* (TYVAV)	*M. persicae*	TC?	93

[a] All species are aphids (Aphididae) except *Nephotettix virescens*, which is a leafhopper (Cicadellidae).
[b] HC, Helper component; PA, particle association; TC, transcapsidation.
[c] Virus group names in quotation marks have not been officially approved by the ICTV.

scanty published information on the aphid transmission and virus/vector relations of DaYMV[68,69] indicates difficulty in achieving aphid transmission,
and erratic success in attempts to transmit from different source plants. This
suggests that DaYMV, too, might need a helper virus for aphid transmission,
but no experiments have been reported to investigate this.

AYV has 30 nm diameter particles which resemble those of PYFV in
having a single ssRNA of about 10 kb and three (or perhaps four) particle
proteins.[70] However, PYFV is mechanically transmissible whereas its helper
virus, AYV, is not. Correlated with this, PYFV is present throughout the
leaf,[71] whereas AYV is confined to phloem tissue.[72] It follows that aphids
can acquire PYFV, along with AYV, only after lengthy acquisition access
feeds (10 to 15 min or more) that reach the phloem, whereas they can inoculate
it in brief probes of 2 min or so that penetrate only as far as the mesophyll.[64]
However, aphids already carrying AYV can acquire PYFV from mesophyll
cells in 2-min feeds.

Unlike PYFV, AYV induces the formation in infected plant cells of "currant bun" inclusion bodies similar in general appearance to those of caulimoviruses, with virus particles embedded in a densely staining proteinaceous
matrix.[72] The densely staining material resembles that surrounding the VLP
at the retention site in AYV-carrying aphids and perhaps represents a HC
protein. This was suggested[58] because the HC of cauliflower mosaic virus
(CaMV) has been reported[42,44] to copurify with the CaMV inclusion bodies.

AYV is the only known virus with 30 nm diameter isometric particles
that is phloem-limited and is transmitted in the semipersistent manner by
aphids. It is also the only semipersistent aphid-borne virus whose particles
have been detected by electron microscopy at a specific retention site in the
vector aphid.[73] Virus-like particles (VLP) were found at a site in the foregut
of *C. aegopodii* carrying AYV, or AYV plus PYFV, but not in aphids that
had fed on plants infected with PYFV alone, or in aphids that had fed on
healthy plants. The VLP, surrounded immediately by a small amount of
densely staining material, were embedded in a matrix of lightly staining
material ("M-material") which was attached to a 15 to 20 μm long portion
of the lining of the ventral wall of the pharynx where it passes over the
tentorial bar. The M-material, but not the VLP or the densely staining material,
was also found in aphids taken from healthy plants. VLP were found at the
foregut site only after the aphid had fed long enough on infected plants to be
able to transmit AYV, and they were still detectable in aphids that had fed
for 2 h through membranes on virus-free sucrose solutions after having fed
for 24 h on AYV-infected plants. As would be expected of a virus that is
transmitted in the semipersistent manner, both the VLP and the M-material
were cast with the old foregut lining when the aphid molted.

Because AYV differs from PYFV in its distribution in plant tissues and
in its ultrastructural effects in infected cells, it seems unlikely to belong to
the same taxonomic group as PYFV. Rather it resembles, in its particle

properties and ultrastructural effects, two isometric, phloem-limited viruses that are transmitted in the semipersistent manner by leafhoppers: maize chlorotic dwarf virus (MCDV) and rice tungro spherical virus (RTSV). Interestingly, RTSV is the helper virus for rice tungro bacilliform virus (RTBV) (Table 1). AYV may therefore belong, together with RTSV, in the MCDV taxonomic group.

2. Rice Tungro Bacilliform Virus and Rice Tungro Spherical Virus

Tungro is the most important virus disease of rice throughout southeast Asia and is responsible[74] for annual losses estimated at more than $1.5 × 10⁹. The disease is caused by a complex of two viruses, rice tungro bacilliform virus (RTBV) and rice tungro spherical virus (RTSV) (Table 1).[75-77] RTBV has bacilliform particles about 160 to 220 nm long and 30 to 35 nm in diameter. RTSV has isometric particles about 30 nm in diameter. The complex is transmitted in a semipersistent manner by the leafhopper *Nephotettix virescens*.[76,78,79] Both viruses are confined to the phloem tissue of infected rice plants and neither is mechanically transmissible. RTBV particles contain a single dsDNA molecule of about 8.3 kbp and one or two protein species.[80] It is a tentative member of the commelina yellow mottle virus group, for which the name "badnavirus" has been proposed to, though not yet approved by, the International Committee on Taxonomy of Viruses (ICTV). RTSV particles contain a single ssRNA of about 10 kb and two or three protein species.[80] As already mentioned, RTSV seems to fit into the MCDV taxonomic group.

Although RTBV is responsible for most of the symptoms of tungro disease, leafhoppers transmit it only when they have prior or simultaneous access to a source of RTSV.[75,76,79,81] The vector relations of these viruses are very like those of AYV and PYFV and it is interesting that the helper virus, RTSV, and rice waika virus, which is the same or a closely related virus,[75] induce the formation of "currant bun" inclusion bodies,[82,83] which, as mentioned previously for AYV, might resemble those of CaMV in containing a HC. For RTSV there is now some evidence[79,84] that the helper factor is something other than the particles of RTSV (i.e., it is probably a HC protein). First, vector leafhoppers taken from a source of RTSV and RTBV retained transmissible RTSV for 2 days, whereas they retained ability to acquire RTBV for a further 5 days; second, when leafhoppers carrying RTSV were allowed to feed through membranes on anti-RTSV immunoglobulin they lost their ability to transmit RTSV but retained their ability to acquire and transmit RTBV.

3. Heracleum Latent Virus and Heracleum Virus 6

Both HLV and HV6 have been placed together in the somewhat heterogeneous *Closterovirus* taxonomic group. However, HLV, having particles only about 730 nm long,[85] is similar to apple chlorotic leaf spot virus (ACLV), which is now regarded as a tentative member of the group, whereas HV6, having particles about 1400 nm long,[86] is similar to beet yellows virus (BYV) which is a definitive member. HLV is mechanically transmissible but HV6 is not. HV6 is confined to umbelliferous plants whereas HLV infects several species in other families. Both viruses are transmissible by the aphid *Cavariella theobaldi* in the semipersistent manner from wild plants of hogweed *(Heracleum sphondylium)*. However, laboratory experiments[87,88] showed that, although *C. theobaldi* can transmit HV6 from singly infected plants, it transmits HLV only from plants that also contain HV6. Moreover, in contrast to what happens with all the other semipersistent and nonpersistent virus complexes described, aphids already carrying HV6 cannot acquire HLV from a singly infected plant. In this respect the HLV/HV6 complex resembles the persistent virus complexes listed in Table 1 and described in the next section. However, the nature of the interaction between the viruses is completely different.

Electron microscope studies[89,90] showed that in leaf extracts from doubly infected coriander plants the majority of particles could be coated along part of their length with antibodies to HLV and along the remainder of their length with antibodies to HV6. Many of the particles were broken, but of those that appeared to be intact, the portion that became coated with HLV antibodies, was about the same length as a particle of HLV and the portion that became coated with HV6 antibodies was about the same length as a particle of HV6. There was no evidence for the existence of chains of alternating HLV and HV6 particles, i.e., for each virus only one end of the particle seemed to be involved. No such chimeric particles were seen when extracts from plants singly infected with HLV or HV6 were mixed together *in vitro*. Thus, it appears that, in mixedly infected plants, particles of HLV and HV6 become attached together end-to-end in pairs during particle assembly. Although the nature of the linkage is unknown, this association between the particles may well explain the ability of HV6 to assist the aphid transmission of HLV.

C. PERSISTENT COMPLEXES

All the helper viruses in this category in Table 1, except PEMV, are definitive or tentative luteoviruses and are not transmissible mechanically. In contrast, all the dependent viruses are mechanically transmissible. As with the other persistent virus systems described above, the dependent viruses in these persistent virus complexes must be present in the same plant as the helper virus for aphids to be able to transmit them. This distinguishes them from all known nonpersistent and semipersistent virus complexes, with the exception of the HLV/HV6 complex.

Many of the dependent viruses in these persistent complexes are not well studied and their particle morphology is either unknown or uncertain. Indeed, they may not produce true virus particles. Tobacco mottle virus (TMoV) and tobacco vein distorting *Luteovirus* (TVDV), found in Zimbabwe and Malawi, were the first persistent aphid-borne viruses shown to be transmitted as a complex.[91] Tobacco bushy-top virus[92] is possibly a strain of TMoV. The TMoV/TVDV complex and the very similar, possibly related, complex of tobacco yellow vein virus (TYVV) and tobacco yellow vein assistor ?*Luteovirus* (TYVAV),[93] also from Malawi, have been little studied. Lettuce speckles mottle virus (LSMV)[94,95] was troublesome for a time in California, but seems now to have disappeared from the field and is no longer available in laboratory culture.[94a] Its helper virus is beet western yellows *Luteovirus* (BWYV). The importance of bean yellow vein banding virus (BYVBV), which is reported[96] to be dependent on bean leaf roll *Luteovirus* (BLRV) and pea enation mosaic virus (PEMV) in the U.K., is not known, but it may have been responsible for symptoms illustrated in lucerne (alfalfa; *Medicago sativa*) infected with a Netherlands isolate of BLRV[97] and in broad bean (fababean; *Vicia faba*) infected with a U.S.A. isolate of PEMV.[98,99] Sunflower yellow blotch virus (SYBV), reported from sunflower in Kenya,[100] is also responsible for streak necrosis disease of groundnut (peanut)[101,102] and seems common in both these crops in East Africa. Recent studies[101a] suggest the involvement of a helper virus, at present unidentified, in its transmission by *Aphis gossypii*. Carrot mottle virus (CMoV) and its helper, carrot red leaf *Luteovirus* (CaRLV),[103-105] cause carrot motley dwarf disease and are present in most countries where carrots are grown.[106] The disease can have serious effects in the absence of insecticide treatments. Groundnut rosette virus (GRV), assisted by groundnut rosette assistor *Luteovirus* (GRAV),[107,108] causes groundnut rosette disease, which is regarded as the most destructive virus disease of groundnut in Africa, largely because of the unpredictability and severity of epidemics. For example the 1975 epidemic in Nigeria affected over 1 million hectares and was estimated to have caused yield losses of 560000 tonnes.[109] Because of their economic importance, GRV and CMoV are the most thoroughly studied of the dependent viruses that have persistent helpers.

All the dependent viruses in these persistent complexes have similar properties. The similarities are such as to have prompted a proposal[110,111] for a new plant virus taxonomic group ("umbravirus") to contain them. The name was coined from *umbra,* meaning "a shade or shadow, an uninvited guest accompanying an invited one". CMoV is suggested as the type member of the proposed group, with BYVBV, GRV, LSMV, and TMoV as definitive members. The properties so far determined for SYBV[101a] suggest that it too should be a definitive member. TYVV is regarded as a tentative member. This proposed new group has not so far been approved by the ICTV, pending further information about the morphology of the particles (if any) and about the affinities of these viruses to those in existing taxonomic groups. Nevertheless the term "umbravirus", enclosed in quotation marks to indicate its tentative status, will be used in this chapter as a convenient shorthand term.

The most striking property that the definitive "umbraviruses" have in common is that, when they occur alone in plants unaccompanied by the helper virus, no virus-like particles can be found by electron microscopy in leaf extracts. Vesicle-like structures, about 50 nm in diameter, were reported[112,113] at the tonoplast and in the cell vacuoles of *Nicotiana clevelandii* plants infected with CMoV, but it is not certain whether they are virus particles of a kind unusual among plant viruses or merely plant-derived vesicles associated with CMoV infection. Similar structures have been seen in plants infected with BYVBV,[96] LSMV,[95] and GRV.[139] The mechanically transmissible infectivity of CMoV, LSMV, and GRV survived in sap or buffer extracts of leaves for several hours or even up to 1 or 2 days at room temperature[95,112,114] and that of GRV survived for up to 15 days at 4°C.[114] However, the infectivity was abolished by treatment with ether or chloroform. Bentonite-clarified preparations of CMoV[95] consisted mainly of cell membrane in which 50 nm diameter vesicular structures could be found similar to those in the cell vacuoles. Infectivity associated with these preparations was relatively stable, surviving during several days of the purification procedure. Preparations of LSMV[95] and GRV[114] made by similar clarification procedures had very similar properties to those of CMoV. It therefore seems unlikely that the infectivity of these "umbraviruses" exists in plants merely as free nucleic acid. Nevertheless, there is abundant infective ssRNA in plants, as shown by the infectivity of leaf extracts made with phenol.[95,112,114] With CMoV and GRV, no virus-specific bands were detected after polyacrylamide gel electrophoresis (PAGE) of such nucleic acid extracts, but the position of the infective RNA molecules was revealed by slicing the gels and inoculating extracts from the slices to test plants. The infective RNA species of CMoV and GRV were each estimated to have molecular weights of about 1.5×10^6 (4.6 kb).[114,115] With LSMV a virus-specific stained band was observed corresponding to a mol wt of 1.4×10^6 (4.3 kb).[95]

Studies with LSMV[94] and CMoV[116,117] showed that, in mixed infections with their respective helper viruses, some of the particles that are formed consist of the RNA of the dependent virus encapsidated in the coat protein of the helper virus. Particles purified from the mixed infection of CMoV and CaRLV looked in the electron microscope like those of CaRLV but contained manually transmissible infectivity of CMoV; this infectivity could be removed from the preparations, along with the particles of CaRLV, by treatment with CaRLV-specific antiserum.[116] Also, recipient aphids injected with hemolymph from viruliferous donor aphids transmitted both components of the complex to test plants on which they fed, but prior treatment of the hemolymph with antiserum to CaRLV prevented the recipient aphid from transmitting both CaRLV and CMoV. With LSMV and BWYV, aphids that were allowed to feed through membranes on partially purified preparations from doubly infected plants transmitted both viruses to test plants, but prior treatment of the preparations with antiserum to BWYV prevented aphids from transmitting both viruses.[94] Thus, it seems that the mechanism of dependent transmission that operates with these viruses is transcapsidation, as found with RPV and

MAV.[54-56] Indeed, as with those viruses, there is experimental evidence[116] that at least one "umbravirus", CMoV, can exchange its usual helper virus (CaRLV) for another (BWYV or PLRV) and thereby change its specific vector (*Myzus persicae* instead of *Cavariella aegopodii*).

Leaf tissue infected with each of the definitive "umbraviruses" contains abundant dsRNA. Two electrophoretic species, of about 4.6 kbp (dsRNA-1) and 1.3 kbp (dsRNA-2), are common to all of them.[95,101a,114,117] In both CMoV and GRV, dsRNA-1 has extensive sequence homology with dsRNA-2,[118] but no homology has been detected between the dsRNA-1 molecule of CMoV and those of other "umbraviruses".[140] The dsRNA preparations are not infective unless the strands are separated, e.g., by heating, and prevented from reannealing, e.g., by "quenching" rapidly on ice. Only dsRNA-1, which is presumably the ds form of the infective ssRNA, is required to establish an infection producing both dsRNA-1 and dsRNA-2. This suggests that dsRNA-2 represents a subgenomic form of dsRNA-1.

Some "umbraviruses" produce only these two dsRNA species but others form additional species, which may be very abundant, as for example with the 900-bp dsRNA of GRV.[114] This is considered to be a satellite[118] because it can be removed from laboratory cultures of GRV, and seems unnecessary for multiplication of GRV in plants, but cannot multiply on its own. It has no appreciable sequence homology with the genomic dsRNA species of GRV. This satellite RNA is of considerable interest for two reasons. The first is that it is largely or entirely responsible for the symptoms of groundnut rosette disease.[118] Different variants of the satellite are responsible for the two major forms of rosette disease, chlorotic rosette and green rosette;[119] another form, "mosaic" rosette, seems to be caused by mixed infection of groundnut with GRV cultures containing the "chlorotic" satellite variant and a "mottle" variant that itself induces only mild mottle symptoms.[119] The second reason why the GRV satellite RNA is of interest is that its presence in the source plants, as well as that of GRAV, is required for the aphid transmission of GRV.[120] This discovery explains why the satellite RNA is invariably present in field isolates of GRV. It is an interesting debating point whether this molecule, which is not required for normal multiplication of GRV in plants, but seems essential for its survival in nature, should be regarded as a satellite or as a genome part. Further knowledge, both of the satellite and of the genome of GRV, may help our understanding of this question and of the way in which the satellite exerts its effect on aphid transmission of GRV.

V. DISCUSSION

A. INCIDENCE OF DEPENDENT TRANSMISSION

The above account shows that very many plant viruses of diverse types interact with others for transmission by vectors. The phenomenon is evidently much more common than has been generally realized. Indeed, three of the systems listed in Table 1 have been discovered in this laboratory, quite by chance, in the course of other research. One common wild umbellifer in Scotland, *Heracleum sphondylium,* is a host of at least three distinct aphid-transmitted virus complexes: CaRLV and CMoV, PYFV and AYV, and HLV and HV6. Umbellifers are also hosts of other little-studied viruses that may be transmitted in a dependent manner: (1) celery yellow spot virus,[121,122] reportedly transmitted by aphids from some hosts but not others; (2) parsley virus 5,[123] a *Potexvirus* related to potato virus X and transmitted by *Cavariella aegopodii* but apparently only from parsley plants also infected with a CMoV-like virus (and presumably also with CaRLV); and (3) an unnamed "umbra-virus" from parsley[101a] which is not the same as CMoV because its dsRNA has a different electrophoretic band pattern. However, it seems unlikely that umbelliferous plants are any more prone to infection with virus complexes than any other plant family; more probably these reports reflect the interests of the researchers involved.

B. MECHANISMS OF DEPENDENCE

All viruses, like all higher organisms, are collections of genes that depend on each other for their collective survival. If a gene is defective or lacking, the absent function can sometimes be provided by another isolate of the same or even an unrelated virus, a process known as *complementation.* All virus transmission complexes are naturally occurring examples of complementation, and the virus that provides the missing function is the helper virus.

When viral genes are on different parts of a divided genome there is a potential for separation and reassortment of genome parts, with a consequent exchange of phenotypic properties. Experimentally, it is possible to make pseudorecombinant isolates, e.g., of cucumber mosaic *Cucumovirus*[124] or raspberry ringspot *Nepovirus,*[125] in which one or more of the genome parts of one parental isolate receive the coat protein of the other parental isolate, thus acquiring an altered vector specificity. Almost certainly this kind of exchange of genetic material also happens in nature. It is not a very large step from this to the kind of phenomenon found in mixed infections of some luteoviruses, such as RPV and MAV, in which the entire genome of one virus can acquire the coat protein, and thereby the vector specificity, of another distantly related virus. And it is only a small step further to the position found with the "umbraviruses", which have no vector of their own but depend on their genomic RNA molecules becoming packaged in the coat protein of an unrelated helper virus, usually a luteovirus, for transmission by the specific vector of that virus. This phenomenon (transcapsidation[8]) is one of the main mechanisms of dependent transmission.

The other most important mechanism of dependent transmission is found in viruses that require a helper protein (HC). Isolates lacking a functional HC of their own need to borrow one from another virus. A third type of interaction, which seems likely to be more prevalent than is at present reported, is found in the HLV/HV6 complex, in which particles of the dependent and helper viruses become attached together in pairs end to end. The precise mechanism of this interaction is unknown, but it could be a special form of heterologous encapsidation. Finally there is the GRV satellite RNA, which provides an unknown function to assist the GRAV-dependent aphid transmission of GRV. This is the only report of a satellite RNA acting in this way, though it has been shown that a satellite-like RNA is essential for efficient transmission of beet necrotic yellow vein? *Furovirus* by the fungus *Polymyxa betae*.[126]

C. GENERAL GUIDELINES FOR THE RECOGNITION OF TRANSMISSION COMPLEXES

Diagnosis of viruses in transmission complexes does not usually present problems or involve technique different from those used for any other kind of virus. However, recognition that a virus is helper-dependent for vector transmission comes from a conscientious effort to fulfil Koch's third and fourth postulates. The virus, after having been isolated and characterized, should be shown to induce the full disease when returned to the host plant. It should also be reisolated by all the means used originally and shown to have all the characters of the original virus. This should include transmissibility by the vector. Inability to show this would obviously suggest that the virus depends on another agent for vector transmission.

The importance of fulfilling Koch's postulates is especially well shown by recent work[118-120] on the groundnut rosette disease complex. As already mentioned, this complex involves three agents, GRAV, GRV, and the GRV satellite RNA. The GRV satellite RNA is mainly responsible for the disease symptoms but it depends on GRV for its replication and on GRAV for its aphid transmission. GRV in turn depends on GRAV and on the GRV satellite RNA for its transmission by aphids. Thus, all three agents must be returned to groundnut plants to restore all the characters of the disease, including the original rosette disease symptoms and transmissibility by *Aphis craccivora*.

It is important to stress that, in all the virus complexes discussed in this chapter, the function provided by the helper virus is needed by the dependent virus for acquisition, retention, and transmission by the vector, but not for *infection* of the inoculated plant. So far as ability to infect the plant is concerned, the dependent and helper viruses behave independently. For example in a hypothetical experiment in which the frequency of vector transmission of each virus is 50%, 25 out of every 100 plants inoculated would be expected to become infected with both viruses, 25 with each of the viruses alone, with 25 remaining uninfected. Separation of the viruses in this way occurs frequently in experimental work, with the result that vector transmissibility of the dependent virus is readily lost.[79,104,105,108] Chance separations are perhaps less likely to happen in the field where vector populations, and therefore transmission frequencies, may be high. However, instances are known of separation occurring in the field as a result of a crop plant being immune to one of the viruses in the complex. Thus, although PYFV was found to be the commonest virus affecting parsnip crops in the U.K.,[63] this plant is immune to AYV, so that the infection reaches a "dead end" in parsnip crops: all infections observed are primary infections and there is no secondary spread within the crop. The same situation seems to occur in sunflower and groundnut crops infected with SYBV.[101,102,142] Diseased sunflower and groundnut plants are distributed singly or in small clusters of two or three and there is no evidence of secondary spread such as is observed with the rosette disease complex in groundnut, which is of course susceptible to both viruses in the complex, GRV and GRAV. Such a pattern of distribution in the field can therefore be a clue to the presence of a transmission-deficient virus.

If a newly discovered virus is a dependent virus which has become separated from its helper virus, much unproductive effort can be expended in attempting to identify the vector unless it is soon realized that a helper virus is involved. Moreover, even when a helper virus is suspected, its identification can be difficult if the wild host from which the vector transmits the virus complex is unknown, and especially if it is unrelated to the plant in which the dependent virus has been found. In the case of PYFV, a serologically related virus had already been found to be transmissible by the aphid *Cavariella aegopodii* from the wild umbellifer *Anthriscus sylvestris* and so the type of vector was apparent; the realization that failure of *C. aegopodii* to transmit the virus from parsnip was due, not to it being the wrong aphid species, but to lack of a helper virus, took rather longer;[63] moreover, the source plant of parsnip isolates turned out to be, not *A. sylvestris,* but *Heracleum sphondylium.*[127] The discovery that *Tridax procumbens* (Compositae) is a wild host of the virus causing groundnut streak necrosis disease[101,102] and that the vector is *Aphis gossypii* (of which groundnut is not a preferred host) would not have been easily made had it not been for two observations: (1) that the distribution of streak necrosis disease in the field, singly and in small clusters, is similar to that of sunflower yellow blotch disease in the same localities; (2) that the causal agent of each disease has the properties of an "umbravirus". These findings suggested that the two diseases might be caused by the same virus and that this virus might depend on a helper virus not present in groundnut or sunflower for vector transmission. Since SYBV was known to infect *T. procumbens,*[6] and to be transmissible from it by *Aphis gossypii,*[100] it was then but a short step to identifying the source plant and aphid vector of the virus causing streak necrosis disease. These anecdotal remarks are included here because they point to the usefulness of general guidelines for the recognition of helper/dependent virus complexes.

D. VIRUS GROUPS LIKELY TO CONTAIN HELPER VIRUSES

On the basis of existing knowledge, certain kinds of viruses are stronger candidates than others for the role of helper virus. Clearly this applies to the three kinds of viruses that are known to produce HC proteins: potyviruses, caulimoviruses, and the viruses in the AYV/RTSV/MCDV cluster. Although, as already noted, caulimoviruses have not so far been shown to assist the aphid transmission of unrelated transmission deficient viruses in nature, they would seem well qualified for this role. The same would seem true of MCDV. As well as resembling AYV and RTSV in their particle properties and semi-persistent vector relations, MCDV induces the formation in host cells of "currant bun" inclusion bodies[128] and its particles, like those of AYV, have been found at a specific site of retention in the foregut of vector leafhoppers.[129] These similarities suggested[58] that MCDV might produce a HC, and evidence for this has now been reported.[130] Perhaps MCDV too might be found able to assist leafhopper transmission of a dependent virus.

The luteoviruses are another group containing many viruses that act as helper viruses. With these viruses the mechanism of dependent transmission is transcapsidation. There is no apparent reason why the same mechanism should not operate with nonpersistent or semipersistent viruses, but there are no reports of this so far, though there is a recent report of dependent trans-mission of the *Potyvirus* ZYMV resulting from phenotypic mixing.[36] The interparticle association observed in the HLV/HV6 complex is entirely novel. The basis for the association is unknown, but it raises the possibility that "long-particle" closteroviruses in general might be able to act as helpers for "short-particle" ones.

E. VIRUS GROUPS LIKELY TO CONTAIN DEPENDENT VIRUSES

It is also possible to attempt some generalizations about kinds of dependent viruses. The "umbraviruses" seem a fairly well-defined cluster of viruses that depend on persistent aphid-borne viruses, usually luteoviruses, for trans-mission by aphids. Any virus that shares the general properties of "umbra-viruses" (no virus-like particles in extracts of mechanically inoculated plants, infectivity abolished by organic solvents, RNA preparations from infected leaves highly infective, abundant dsRNA in infected leaves with a charac-teristic electrophoretic band pattern) is very likely to be a dependent virus and probably has a luteovirus helper.

The recent discovery of the HLV/HV6 complex suggests that transmis-sion-deficient viruses may also be found among the other closteroviruses in the ACLV subgroup. The mode of transmission of most of these other viruses is unknown, with the exception of grapevine virus A, which is transmitted by mealy bugs.[131-133] HLV has similar particle properties to ACLV[134] and is distantly serologically related to GVA.[135] Moreover, grapevine is host to several other closteroviruses, and closterovirus-like particles coated along only part of their length with GVA antiserum have been reported.[136] HV6 does not share any common hosts with ACLV but BYV, another "long" closterovirus, does. However, BYV did not assist transmission of HLV or ACLV from *Chenopodium quinoa* or sugar beet by *Myzus persicae*.[137] Correlated with this, no chimeric virus particles were found in extracts from plants mixedly infected with BYV and HLV, or with BYV and ACLV.[135] Although these limited experiments gave no new information, further tests with other closteroviruses would seem well worth making.

Obviously, dependent viruses are likely to be found (as defective isolates) within the groups of viruses that produce HC proteins (potyviruses, caulimoviruses, and the viruses in the AYV/RTSV/MCDV cluster) and no further discussion of these seems necessary. However, members of the *Potexvirus* group would also seem to merit consideration as dependent viruses. It is surprising that the PAMV/PVY complex is the only example known of the dependence of a potexvirus on a potyvirus for aphid transmission. Potexviruses are generally thought to spread from plant to plant by mechanical abrasion but it is unclear in many instances how thoroughly this assumption has been tested or whether this is the only method of spread. The possibility that some of them spread in association with a potyvirus helper would be worth critical investigation. Moreover the report[123] of aphid transmission of a potexvirus, parsley virus 5, from plants infected with CMoV (and presumably CaRLV) suggests that other potexviruses might form other kinds of association, perhaps with luteoviruses. In this connection it is of considerable interest that a potexvirus has recently[138] been shown to be associated with strawberry mild yellow edge, an aphid-transmitted disease generally regarded as being caused by strawberry mild yellow edge ?*Luteovirus* (SMYEV). The precise involvement, if any, of these viruses in the etiology of this disease has yet to be elucidated but it is tempting to speculate that the *Potexvirus,* called strawberry mild yellow edge-associated virus, is dependent for aphid transmission on a helper virus, possibly SMYEV.

Finally, what about the "badnaviruses" other than RTBV? Do they need a HC? If so, do they code for such a protein or do they, like RTBV, rely on helper viruses for this function?

F. ENVOI

Many of the comments in this discussion amount to no more than speculation but the purpose of this article has been to draw attention to the extent to which viruses depend upon each other for transmission as well as for other important functions. If the speculation can lead to further fruitful lines of investigation they will have served their purpose.

REFERENCES

1. **Murant, A. F. and Mayo, M. A.,** Satellites of plant viruses, *Annu. Rev. Phytopathol.,* 20, 49, 1982.
2. **Francki, R. I. B.,** Plant virus satellites, *Annu. Rev. Microbiol.,* 39, 151, 1985.

3. **Fritsch, C. and Mayo, M. A.,** Satellites of plant viruses, in *Plant Viruses,* Vol. 1, *Structure and Replication,* Mandahar, C. L., Ed., CRC Press, Boca Raton, FL, 1989, 289.

4. **Kassanis, B.,** Properties and behaviour of a virus depending for its multiplication on another, *J. Gen. Microbiol.,* 27, 477, 1962.

5. **Kassanis, B.,** Portraits of viruses: tobacco necrosis virus and its satellite virus, *Intervirology,* 15, 57, 1981.

6. **Kaper, J. M. and Waterworth, H. E.,** Cucumoviruses, in *Handbook of Plant Virus Infections and Comparative Diagnosis,* Kurstak, E., Ed., Elsevier/North-Holland, Amsterdam, 1981, 257.

7. **Murant, A. F., Mayo, M. A., Harrison, B. D., and Goold, R. A.,** Evidence for two functional RNA species and a ''satellite'' RNA in tomato black ring virus, *J. Gen. Virol.,* 19, 275, 1973.

8. **Rochow, W. F.,** The role of mixed infections in the transmission of plant viruses by aphids, *Annu. Rev. Phytopathol.,* 10, 101, 1972.

9. **Rochow, W. F.,** Dependent virus transmission from mixed infections, in *Aphids as Virus Vectors,* Harris, K. F. and Maramorosch, K., Eds., Academic Press, New York, 1977, 253.

10. **Pirone, T. P.,** Accessory factors in nonpersistent virus transmission, in *Aphids as Virus Vectors,* Harris, K. F. and Maramorosch, K., Eds., Academic Press, New York, 1977, 221.

11. **Falk, B. W. and Duffus, J. E.,** Epidemiology of helper-dependent persistent aphid transmitted virus complexes, in *Plant Diseases and Vectors, Ecology and Epidemiology,* Maramorosch, K. and Harris, K. F., Eds., Academic Press, New York, 1981, 161.

12. **Sylvester, E. S.,** Multiple acquisition of viruses and vector-dependent prokaryotes: consequences on transmission, *Annu. Rev. Entomol.,* 30, 71, 1985.

13. **Watson, M. A. and Roberts, F. M.,** A comparative study of the transmission of *Hyoscyamus* virus 3, potato virus Y and cucumber virus 1 by the vectors *Myzus persicae* (Sulz.), *M. circumflexus* (Buckton) and *Macrosiphum gei* (Koch), *Proc. R. Soc. London Ser. B,* 127, 543, 1939.

14. **Bawden, F. C. and Sheffield, F. M. L.,** The relationships of some viruses causing necrotic diseases of the potato, *Ann. Appl. Biol.,* 31, 33, 1944.

15. **Watson, M. A.,** Evidence for interaction or genetic recombination between potato viruses Y and C in infected plants, *Virology,* 10, 211, 1960.

16. **Kassanis, B. and Govier, D. A.,** New evidence on the mechanism of aphid transmission of potato C and potato aucuba mosaic viruses, *J. Gen. Virol.,* 10, 99, 1971.

17. **Kassanis, B. and Govier, D. A.,** The role of the helper virus in aphid transmission of potato aucuba mosaic virus and potato virus C, *J. Gen. Virol.,* 13, 221, 1971.

18. **Govier, D. A. and Kassanis, B.,** Evidence that a component other than the virus particle is needed for aphid transmission of potato virus Y, *Virology,* 57, 285, 1974.

19. **Govier, D. A. and Kassanis, B.,** A virus-induced component of plant sap needed when aphids acquire potato virus Y from purified preparations, *Virology,* 61, 420, 1974.

20. **Pirone, T. P.,** Efficiency and selectivity of the helper-component-mediated aphid transmission of purified potyviruses, *Phytopathology,* 71, 922, 1981.

21. **Pirone, T. P. and Thornbury, D. W.,** The involvement of a helper component in nonpersistent transmission of plant viruses by aphids, *Microbiol. Sci.,* 1, 191, 1984.

22. **Thornbury, D. W. and Pirone, T. P.,** Helper components of two potyviruses are serologically distinct, *Virology,.* 125, 487, 1983.

23. **Thornbury, D. W., Hellmann, G. M., Rhoads, R. E., and Pirone, T. P.,** Purification and characterization of potyvirus helper component, *Virology,* 144, 260, 1985.

24. **Hobbs, H. A. and McLaughlin, M. R.,** A non-aphid-transmissible isolate of bean yellow mosaic virus-Scott that is transmissible from mixed infections with pea mosaic virus-204-1, *Phytopathology,* 80, 268, 1990.

25. **Maiss, E., Timpe, U., Brisske, A., Jelkmann, W., Casper, R., Himmler, G., Mattanovich, D., and Katinger, H. W. D.,** The complete nucleotide sequence of plum pox virus RNA, *J. Gen. Virol.,* 70, 513, 1989.

26. **Ghabrial, S. A. and Pirone, T. P.,** Physiology of tobacco etch virus-induced wilt of tabasco peppers, *Virology,* 31, 154, 1967.

27. **Simons, J. N.,** Aphid transmission of a non-aphid-transmissible strain of tobacco etch virus, *Phytopathology,* 66, 652, 1976.

28. **Lecoq, H.,** A poorly aphid transmitted variant of zucchini yellow mosaic virus, *Phytopathology,* 76, 1063, 1986.

29. **Antignus, Y., Raccah, B., Gal-On, A., and Cohen, S.,** Biological and serological characterization of zucchini yellow mosaic and watermelon mosaic virus-2 isolates in Israel, *Phytoparasitica,* 17, 289, 1989.

30. **Thornbury, D. W., Patterson, C. A., Dessens, J. T., and Pirone, T. P.,** Comparative sequence of the helper component (HC) region of potato virus Y and a HC-defective strain, potato virus C, *Virology,* 178, 573, 1990.

31. **Pirone, T. P. and Thornbury, D. W.,** Role of virion and helper component in regulating aphid transmission of tobacco etch virus, *Phytopathology,* 73, 872, 1983.

32. **Harrison, B. D. and Robinson, D. J.,** Molecular variation in vector-borne plant viruses: epidemiological significance, *Philos. Trans. R. Soc. London Ser. B,* 321, 447, 1988.

33. **Lain, S., Riechmann, J. L., Mendez, E., and Garcia, J. A.,** Nucleotide sequence of the 3' terminal region of plum pox potyvirus RNA, *Virus Res.,* 10, 325, 1988.

34. **Sako, N. and Ogata, K.,** Different helper factors associated with aphid transmission of some potyviruses, *Virology,* 112, 762, 1981.

35. **Lecoq, H. and Pitrat, M.,** Specificity of the helper-component-mediated aphid transmission of three potyviruses infecting muskmelon, *Phytopathology,* 75, 890, 1985.

36. **Bourdin, D. and Lecoq, H.,** Evidence that heteroencapsidation between two potyviruses is involved in aphid transmission of a non-aphid-transmissible isolate from mixed infections, *Phytopathology,* 81, 1459, 1991.

37. **Sylvester, E. S.,** Beet yellows virus transmission by the green peach aphid, *J. Econ. Entomol.,* 49, 789, 1956.

38. **Markham, P. G., Pinner, M. S., Raccah, B., and Hull, R.,** The acquisition of a caulimovirus by different aphid species: comparison with a potyvirus, *Ann. Appl. Biol.,* 111, 571, 1987.

39. **Lung, M. C. Y. and Pirone, T. P.,** Acquisition factor required for aphid transmission of purified cauliflower mosaic virus, *Virology,* 60, 260, 1974.

40. **Armour, S. L., Melcher, U., Pirone, T. P., Lyttle, D. J., and Essenberg, R. C.,** Helper component for aphid transmission encoded by region II of cauliflower mosaic virus DNA, *Virology,* 129, 25, 1983.

41. **Daubert, S., Shepherd, R. J., and Gardner, R. C.,** Insertional mutagenesis of the cauliflower mosaic virus genome, *Gene,* 25, 201, 1983.

42. **Givord, L., Xiong, C., Giband, M., Koenig, I., Hohn, T., Lebeurier, G., and Hirth, L.,** A second cauliflower mosaic virus gene product influences the structure of the viral inclusion body, *EMBO J.,* 3, 1423, 1984.

43. **Woolston, C. J., Covey, S. N., Penswick, J. R., and Davies, J. W.,** Aphid transmission and a polypeptide are specified by a defined region of the cauliflower mosaic virus genome, *Gene,* 23, 15, 1983.

44. **Rodriguez, D., Lopez-Abella, D., and Diaz-Ruiz, J. R.,** Viroplasms of an aphid-transmissible isolate of cauliflower mosaic virus contain helper component activity, *J. Gen. Virol.,* 68, 2063, 1987.

45. **Harker, C. L., Woolston, C. J., Markham, P. G., and Maule, A. J.,** Cauliflower mosaic virus aphid transmission factor protein is expressed in cells infected with some aphid nontransmissible isolates, *Virology,* 160, 252, 1987.

46. **Woolston, C. J., Czaplewski, L. G., Markham, P. G., Goad, A. S., Hull, R., and Davies, J. W.,** Location and sequence of a region of cauliflower mosaic virus gene 2 responsible for aphid transmissibility, *Virology,* 160, 246, 1987.

47. **Markham, P. G. and Hull, R.,** Cauliflower mosaic virus aphid transmission facilitated by transmission factors from other caulimoviruses, *J. Gen. Virol.,* 66, 921, 1985.

48. **Espinoza, A. M., Markham, P. G., Maule, A. J., and Hull, R.,** *In vitro* biological activity associated with the aphid transmission factor of cauliflower mosaic virus, *J. Gen. Virol.,* 69, 1819, 1988.

49. **Kennedy, J. S., Day, M. F., and Eastop, V. F.,** *A Conspectus of Aphids as Vectors of Plant Viruses,* Commonwealth Institute of Entomology, London, 1962.

50. **Adam, G.,** Die gemeinsame Übertragung eines vektoriellen und eines avektoriellen Stammes des pea enation mosaic virus durch die Blattlaus *Acyrthosiphon pisum, Z. Pflanzenkr. Pflanzenschutz,* 85, 586, 1978.

51. **Hull, R.,** Particle differences related to aphid-transmissibility of a plant virus, *J. Gen. Virol.,* 34, 183, 1977.

52. **Adam, G., Sander, E., and Shepherd, R. J.,** Structural differences between pea enation mosaic virus strains affecting transmissibility by *Acyrthosiphon pisum* (Harris), *Virology,* 92, 1, 1979.

53. **Waterhouse, P. M., Gildow, F. E., and Johnstone, G. R.,** Luteovirus group, *AAB Descriptions of Plant Viruses,* No. 339, Association of Applied Biologists, Wellesbourne, 1988.

54. **Rochow, W. F.,** Apparent loss of vector specificity following double infection by two strains of barley yellow dwarf virus, *Phytopathology,* 55, 62, 1965.

55. **Rochow, W. F.,** Barley yellow dwarf virus: phenotypic mixing and vector specificity, *Science,* 167, 875, 1970.

56. **Rochow, W. F.,** Selective virus transmission by *Rhopalosiphum padi* exposed sequentially to two barley yellow dwarf viruses, *Phytopathology,* 63, 1317, 1973.

57. **Hu, J. S., Rochow, W. F., Palukaitis, P., and Dietert, R. R.,** Phenotypic mixing: mechanism of dependent transmission for two related isolates of barley yellow dwarf virus, *Phytopathology,* 78, 1326, 1988.

58. **Harrison, B. D. and Murant, A. F.,** Involvement of virus-coded proteins in transmission of plant viruses by vectors, in *Vectors in Virus Biology,* Mayo, M. A. and Harrap, K. A., Eds., Academic Press, London, 1984, 1.

59. **Gildow, F. E. and Rochow, W. F.,** Role of accessory salivary glands in aphid transmission of barley yellow dwarf virus, *Virology,* 104, 97, 1980.

60. **Gildow, F. E.,** Coated-vesicle transport of luteoviruses through salivary glands of *Myzus persicae, Phytopathology,* 72, 1289, 1982.

61. **Clinch, P. E. M., Loughnane, J. B., and Murphy, P. A.,** A study of the aucuba or yellow mosaics of the potato, *Sci. Proc. R. Dublin Soc.,* 21, 431, 1936.

62. **Kassanis, B.,** The transmission of potato aucuba mosaic virus by aphids from plants also infected by potato viruses A or Y, *Virology,* 13, 93, 1961.

63. **Murant, A. F. and Goold, R. A.,** Purification, properties and transmission of parsnip yellow fleck, a semi-persistent, aphid-borne virus, *Ann. Appl. Biol.,* 62, 123, 1968.

64. **Elnagar, S. and Murant, A. F.,** Relations of the semi-persistent viruses, parsnip yellow fleck and anthriscus yellows, with their vector, *Cavariella aegopodii, Ann. Appl. Biol.,* 84, 153, 1976.

65. **Elnagar, S. and Murant, A. F.,** The role of the helper virus, anthriscus yellows, in the transmission of parsnip yellow fleck virus by the aphid *Cavariella aegopodii, Ann. Appl. Biol.,* 84, 169, 1976.

66. **Hemida, S. K. and Murant, A. F.,** Particle properties of parsnip yellow fleck virus, *Ann. Appl. Biol.,* 114, 87, 1989.

67. **Murant, A. F.,** Parsnip yellow fleck virus, type member of a proposed new plant virus group, and a possible second member, dandelion yellow mosaic virus, in *The Plant Viruses, Isometric Viruses with Monopartite RNA Genomes,* Koenig, R., Ed., Plenum Press, New York, 1988, 273.

68. **Kassanis, B.,** Studies on dandelion yellow mosaic and other virus diseases of lettuce, *Ann. Appl. Biol.,* 34, 412, 1947.

69. **Bos, L., Huijberts, N., Huttinga, H., and Maat, D. Z.,** Further characterization of dandelion yellow mosaic virus from lettuce and dandelion, *Neth. J. Plant Pathol.,* 89, 207, 1983.

70. **Hemida, S. K., Murant, A. F., and Duncan, G. H.,** Purification and some particle properties of anthriscus yellows virus, a phloem-limited semi-persistent aphid-borne virus, *Ann. Appl. Biol.,* 114, 71, 1989.

71. **Murant, A. F., Roberts, I. M., and Hutcheson, A. M.,** Effects of parsnip yellow fleck virus on plant cells, *J. Gen. Virol.,* 26, 277, 1975.

72. **Murant, A. F. and Roberts, I. M.,** Virus-like particles in phloem tissue of chervil *(Anthriscus cerefolium)* infected with anthriscus yellows virus, *Ann. Appl. Biol.,* 85, 403, 1977.

73. **Murant, A. F., Roberts, I. M., and Elnagar, S.,** Association of virus-like particles with the foregut of the aphid *Cavariella aegopodii* transmitting the semi-persistent viruses anthriscus yellows and parsnip yellow fleck, *J. Gen. Virol.,* 31, 47, 1976.

74. **Herdt, R. W.,** Equity considerations in setting priorities for Third World rice biotechnology research, *Development, Seeds of Change,* 4, 19, 1988.

75. **Hibino, H., Roechan, M., and Sudarisman, S.,** Association of two types of virus particles with penyakit habang (tungro disease) of rice in Indonesia, *Phytopathology,* 68, 1412, 1978.

76. **Hibino, H., Saleh, N., and Roechan, M.,** Transmission of two kinds of rice tungro-associated viruses by insect vectors, *Phytopathology,* 69, 1266, 1979.

77. **Omura, T., Saito, Y., Usugi, T., and Hibino, H.,** Purification and serology of rice tungro spherical and rice tungro bacilliform viruses, *Ann. Phytopathol. Soc. Jpn.,* 49, 73, 1983.

78. **Ling, K. C.,** Nonpersistence of the tungro virus of rice in its leafhopper vector, *Nephotettix impicticeps, Phytopathology,* 56, 1252, 1966.

79. **Hibino, H.,** Relations of rice tungro bacilliform and rice tungro spherical viruses with their vector *Nephotettix virescens, Ann. Phytopathol. Soc. Jpn.,* 49, 545, 1983.

80. **Jones, M. C., Gough, K., Dasgupta, I., Subba Rao, B. L., Cliffe, J., Qu, R., Shen, P., Kaniewska, M., Blakebrough, M., Davies, J. W., Beachy, R. N., and Hull, R.,** Rice tungro disease is caused by an RNA and a DNA virus, *J. Gen. Virol.,* 72, 757, 1991.

81. **Hibino, H.,** Transmission of two rice tungro-associated viruses and rice waika virus from doubly or singly infected source plants by leafhopper vectors, *Plant Dis.,* 67, 774, 1983.

82. **Favali, M. A., Pellegrini, S., and Bassi, M.,** Ultrastructural alterations induced by rice tungro virus in rice leaves, *Virology,* 66, 502, 1975.

83. **Yamashita, S., Doi, Y., and Yora, K.,** Some properties and intracellular appearance of rice waika virus, *Ann. Phytopathol. Soc. Jpn.,* 43, 278, 1977.

84. **Hibino, H. and Cabauatan, P. Q.,** Infectivity neutralization of rice tungro-associated viruses acquired by vector leafhoppers, *Phytopathology,* 77, 473, 1987.

85. **Bem, F. and Murant, A. F.,** Host range, purification and serological properties of heracleum latent virus, *Ann. Appl. Biol.,* 92, 243, 1979.

86. **Bem, F. and Murant, A. F.,** Transmission and differentiation of six viruses infecting hogweed *(Heracleum sphondylium* L.) in Scotland, *Ann. Appl. Biol.,* 92, 237, 1979.

87. **Murant, A. F.,** Dependence of heracleum latent virus on a fellow closterovirus for transmission by the aphid *Cavariella theobaldi, Abstr. 4th Int. Congr. Plant Pathol. Melbourne,* 1983, 121.

88. **Murant, A. F.,** Helper-dependent transmission of heracleum latent virus (HLV) by aphids, *Rep. Scott. Crop Res. Inst. 1982,* 1983, p. 191.

89. **Murant, A. and Duncan, G.,** Nature of the dependence of heracleum latent virus on heracleum virus 6 for transmission by the aphid *Cavariella theobaldi, Abstr. 6th Int. Cong. Virol., Sendai, Japan,* 1984, 328.

90. **Murant, A. F. and Duncan, G. H.,** Heracleum latent virus (HLV) and heracleum virus 6 (HV6), *Rep. Scott. Crop Res. Inst. 1984,* 1985, p. 183.

91. **Smith, K. M.,** The transmission of a plant virus complex by aphides, *Parasitology,* 37, 131, 1946.

92. **Gates, L. F.,** A virus causing axillary bud sprouting of tobacco in Rhodesia and Nyasaland, *Ann. Appl. Biol.,* 50, 169, 1962.

93. **Adams, A. N. and Hull, R.,** Tobacco yellow vein, a virus dependent on assistor viruses for its transmission by aphids, *Ann. Appl. Biol.,* 71, 135, 1972.

94. **Falk, B. W., Duffus, J. E., and Morris, T. J.,** Transmission, host range, and serological properties of the viruses that cause lettuce speckles disease, *Phytopathology,* 69, 612, 1979.

94a. **Duffus, J. E.,** personal communication.

95. **Falk, B. W., Morris, T. J., and Duffus, J. E.,** Unstable infectivity and sedimentable ds-RNA associated with lettuce speckles mottle virus, *Virology,* 96, 239, 1979.

96. **Cockbain, A. J., Jones, P., and Woods, R. D.,** Transmission characteristics and some other properties of bean yellow vein-banding virus, and its association with pea enation mosaic virus, *Ann. Appl. Biol.,* 108, 59, 1986.

97. **Van der Want, J. P. H. and Bos, L.,** Geelnervigheid, een virusziekte van luzerne, *Tijdschr. Plantenziekten,* 65, 73, 1959.

98. **Osborn, H. T.,** Incubation period of pea mosaic in the aphid, *Macrosiphum pisi, Phytopathology,* 25, 160, 1935.

99. **Osborn, H. T.,** Studies on pea virus 1, *Phytopathology,* 28, 923, 1938.

100. **Theuri, J. M., Bock, K. R., and Woods, R. D.,** Distribution, host range and some properties of a virus disease of sunflower in Kenya, *Trop. Pest Manage.,* 33, 202, 1987.

101. **Murant, A. F., Rajeshwari, R., Raschke, J. H., and Roberts, I. M.,** Viruses associated with groundnut rosette disease, *Rep. Scott. Crop Res. Inst. 1987,* 1988, p. 194.

101a. **Murant, A. F.,** unpublished data.

102. **Bock, K. R.,** ICRISAT regional groundnut pathology program: a review of research progress during 1985–87 with special reference to groundnut streak necrosis disease, *Proc. 3rd Regional Groundnut Workshop, 13–18 March 1988, Lilongwe, Malawi,* ICRISAT, Patancheru, India, 1989.

103. **Watson, M., Serjeant, E. P., and Lennon, E. A.,** Carrot motley dwarf and parsnip mottle viruses, *Ann. Appl. Biol.,* 54, 153, 1964.

104. **Elnagar, S. and Murant, A. F.,** Relations of carrot red leaf and carrot mottle viruses with their aphid vector, *Cavariella aegopodii, Ann. Appl. Biol.,* 89, 237, 1978.

105. **Elnagar, S. and Murant, A. F.,** Aphid-injection experiments with carrot mottle virus and its helper virus, carrot red leaf, *Ann. Appl. Biol.,* 89, 245, 1978.

106. **Murant, A. F.,** Occurrence of mottle and redleaf components of carrot motley dwarf disease in British Columbia, *Can. Plant Dis. Surv.,* 55, 103, 1975.

107. **Hull, R. and Adams, A. N.,** Groundnut rosette and its assistor virus, *Ann. Appl. Biol.,* 62, 139, 1968.

108. **Reddy, D. V. R., Murant, A. F., Duncan, G. H., Ansa, O. A., Demski, J. W., and Kuhn, C. W.,** Viruses associated with chlorotic rosette and green rosette diseases of groundnut in Nigeria, *Ann. Appl. Biol.,* 107, 57, 1985.

109. **Reddy, D. V. R.,** Groundnut rosette virus disease: the present situation and research needs, in *Collaborative Research on Groundnut Rosette Virus, Summary Proceedings of the Consultative Group Meeting, 13–14 April, 1985, Cambridge, U.K.,* Reddy, D. V. R., McDonald, D., and Gibbons, R. W., Eds., ICRISAT, Patancheru, India, 1985, 8.

110. **Murant, A. F., Robinson, D. J., Raschke, J. H., and Rajeshwari, R.,** Recent findings on viruses that depend on luteoviruses for transmission by aphids, *Abstr. 5th Int. Cong. Plant Pathol., Kyoto, Japan, 1988,* 1988, 31.

111. **Murant, A. F. and Robinson, D. J.,** Umbraviruses: a proposed new group of viruses that depend on luteoviruses for transmission by aphids, *Abstr. 8th Int. Cong. Virol., Berlin, 1990,* 1990, 499.

112. **Murant, A. F., Goold, R. A., Roberts, I. M., and Cathro, J.,** Carrot mottle — a persistent aphid-borne virus with unusual properties and particles, *J. Gen. Virol.,* 4, 329, 1969.

113. **Murant, A. F., Roberts, I. M., and Goold, R. A.,** Cytopathological changes and extractable infectivity in *Nicotiana clevelandii* leaves infected with carrot mottle virus, *J. Gen. Virol.,* 21, 269, 1973.

114. **Reddy, D. V. R., Murant, A. F., Raschke, J. H., Mayo, M. A., and Ansa, O. A.,** Properties and partial purification of infective material from plants containing groundnut rosette virus, *Ann. Appl. Biol.,* 107, 65, 1985.

115. **Halk, E. L., Robinson, D. J., and Murant, A. F.,** Molecular weight of the infective RNA from leaves infected with carrot mottle virus, *J. Gen. Virol.,* 45, 383, 1979.

116. **Waterhouse, P. M. and Murant, A. F.,** Further evidence on the nature of the dependence of carrot mottle virus on carrot red leaf virus for transmission by aphids, *Ann. Appl. Biol.,* 103, 455, 1983.

117. **Murant, A. F., Waterhouse, P. M., Raschke, J. H., and Robinson, D. J.,** Carrot red leaf and carrot mottle viruses: observations on the composition of the particles in single and mixed infections, *J. Gen. Virol.,* 66, 1575, 1985.

118. **Murant, A. F., Rajeshwari, R., Robinson, D. J., and Raschke, J. H.,** A satellite RNA of groundnut rosette virus that is largely responsible for symptoms of groundnut rosette disease, *J. Gen. Virol.,* 69, 1479, 1988.

119. **Murant, A. F. and Kumar, I. K.,** Different variants of the satellite RNA of groundnut rosette virus are responsible for the chlorotic and green forms of groundnut rosette disease, *Ann. Appl. Biol.,* 117, 85, 1990.

120. **Murant, A. F.,** Dependence of groundnut rosette virus on its satellite RNA as well as on groundnut rosette assistor luteovirus for transmission by *Aphis craccivora, J. Gen. Virol.,* 71, 2163, 1990.

121. **Freitag, J. H. and Severin, H. H. P.,** Transmission of celery-yellow-spot virus by the honeysuckle aphid, *Rhopalosiphum conii* (Dvd.), *Hilgardia,* 16, 375, 1945.

122. **Van Dijk, P. and Bos, L.,** Survey and biological differentiation of viruses of wild and cultivated Umbelliferae in the Netherlands, *Neth. J. Plant Pathol.,* 95 (Suppl. 2), 1, 1989.

123. **Frowd, J. A. and Tomlinson, J. A.,** The isolation and identification of parsley viruses occurring in Britain, *Ann. Appl. Biol.,* 72, 177, 1972.

124. **Mossop, D. W. and Francki, R. I. B.,** Association of RNA3 with aphid transmission of cucumber mosaic virus, *Virology,* 81, 177, 1977.

125. **Harrison, B. D., Murant, A. F., and Mayo, M. A.,** Two properties of raspberry ringspot virus determined by its smaller RNA, *J. Gen. Virol.,* 17, 137, 1972.,

126. **Tamada, T. and Abe, H.,** Evidence that beet necrotic yellow vein virus RNA-4 is essential for efficient transmission by the fungus *Polymyxa betae, J. Gen. Virol.,* 70, 3391, 1989.

127. **Hemida, S. K. and Murant, A. F.,** Host ranges and serological properties of eight isolates of parsnip yellow fleck virus belonging to the two major serotypes, *Ann. Appl. Biol.,* 114, 101, 1989.

128. **Bradfute, O. E., Gingery, R. E., Gordon, D. T., and Nault, L. R.,** Tissue ultrastructure, sedimentation and leafhopper transmission of a virus associated with a maize dwarfing disease, *J. Cell Biol.,* 55, 25a, 1972.

129. **Childress, S. A. and Harris, K. F.,** Localization of virus-like particles in the foreguts of viruliferous *Graminella nigrifrons* leafhoppers carrying the semi-persistent maize chlorotic dwarf virus, *J. Gen. Virol.,* 70, 247, 1989.

130. **Hunt, R. E., Nault, L. R., and Gingery, R. E.,** Evidence for infectivity of maize chlorotic dwarf virus and for a helper component in its leafhopper transmission, *Phytopathology,* 78, 499, 1988.

131. **Rosciglione, B., Castellano, M. A., Martelli, G. P., Savino, V., and Cannizzaro, G.,** Mealybug transmission of grapevine virus A, *Vitis,* 22, 331, 1983.

132. **Rosciglione, B. and Castellano, M. A.,** Further evidence that mealybugs can transmit grapevine virus A (GVA) to herbaceous hosts, *Phytopathol. Mediterr.,* 24, 186, 1985.

133. **Engelbrecht, D. J. and Kasdorf, G. G. F.,** Association of a closterovirus with grapevines indexing positive for grapevine leafroll disease and evidence for its natural spread in grapevine, *Phytopathol. Mediterr.,* 24, 101, 1985.

134. **Bem, F. and Murant, A. F.,** Comparison of particle properties of heracleum latent and apple chlorotic leaf spot viruses, *J. Gen. Virol.,* 44, 817, 1979.

135. **Murant, A. F., Duncan, G. H., and Roberts, I. M.,** Heracleum latent virus (HLV) and heracleum virus 6 (HV6), *Rep. Scott. Crop Res. Inst. 1984,* 1985, p. 182.

136. **Milne, R. G., Conti, M., Lesemann, D. E., Stellmach, G., Tanne, E., and Cohen, J.,** Closterovirus-like particles of two types associated with diseased grapevines, *Phytopathol. Z.,* 110, 360, 1984.

137. **Murant, A. F.,** Heracleum latent virus (HLV) and heracleum virus 6 (HV6), *Rep. Scott. Crop Res. Inst. 1983,* 1984, p. 189.

138. **Jelkmann, W., Martin, R. R., Lesemann, D.-E., Vetten, H. J., and Skelton, F.,** A new potexvirus associated with strawberry mild yellow edge disease, *J. Gen. Virol.,* 71, 1251, 1990.

139. **Lesemann, D.-E.,** personal communication.

140. **Murant, A. F. and Robinson, D. J.,** unpublished data.

141. **Bock, K. R. and Murant, A. F.,** unpublished data.

INDEX

A

Abiotic transmission in soil, 76–77
Acridine orange, 308
Additives in mechanical inoculation, 60–61
African cassava mosaic virus (ACMV), 26
Agarose gel electrophoresis, 302–303
Agglutination tests, 181–182
Agropyron mosaic virus (AgMV), 26
Air brush in mechanical inoculation, 62
Alfalfa latent virus (ALV), 26
Alfalfa mosaic virus (AMV)
  vs. *Cucumovirus*, 21
  hosts and nonhosts, 22, 26
  inclusions in diagnosis, 104, 109
  purification, 133
Alkali, 307
Alkaline phosphatase
  immunoglobulins, 175
  nucleic acid hybridization, 264–266
American hop latent virus (AHLV), 26
American plum line pattern virus (APLPV),
  26
Ammonium molybdate, 228
Ammonium sulfate precipitation, 172
Andean potato mottle virus (APMV), 26
Anthriscus sylvestris, 348
Anthriscus yellows virus and parsnip yellow
  fleck virus, 338–341, 349
Antibodies
  decoration methods, 233–236
  in diagnosis, 197–198
  ELISA and other solid-phase assays,
    184–193
  gold labeling, 236–243
  immunoblotting, 194–195
  immunodiffusion tests, 182–184
  immunosorbent electron microscopy,
    195–196, 229–233, 231
  labeling, 175–176
  precipitation, 180–182
Antibodies, monoclonal, see Monoclonal
  antibodies
Antibody dilution buffers, 231
Antibody fragment purification, 172–175
Antibody reagents
  antisera production, 165–167
  immunoglobulin labeling, 175–177

immunoglobulins and antibody fragment
  purification, 172–175
monoclonal antibody production, 167–172
monoclonal vs. polyclonal antibodies,
  177–179
Antigens
  ELISA, 189–190, 193
  viral proteins, 160–165
Antiserum
  production, 165–167
  protein A and protein G, 232–233
Aphid-transmitted viruses
  anthriscus yellows, 340
  barley yellow dwarf, 337–338
  cauliflower mosaic, 336
  mechanical inoculation, 63–64
  parnsip yellow fleck, 340, 348
  pea enation mosaic virus, 336–337
  potato Y, 335
  seed transmission, 67
Aphis gossypii, 348
Apple chlorotic leaf spot virus (ACLV), 26,
  349–350
Apple mosaic virus (ApMV), 26
Apple stem grooving virus (ASGV), 134
Approach graft, 52–53
Arabis mosaic virus (ArMV), 26
Arracacha A virus (AVA), 26
Arracacha B virus (AVB), 26
Artichoke Italian latent virus (AILV), 26
Artichoke vein banding virus (AVBV), 26
Artichoke yellow ringspot virus (AYRSV),
  27
Asparagus virus 2 (AV2), 27
Assays, see specific tests
Association of Applied Virologists, 11
Avidin, 265
Azure A
  Comovirus, 119
  in diagnosis, 108–110
  Geminivirus, 112
  Potexvirus, 118
  preparation, 128

B

Back inoculation, 17, 25
Bacterial diseases, 11–12